Hodder Gibson

Scottish Examination Materials

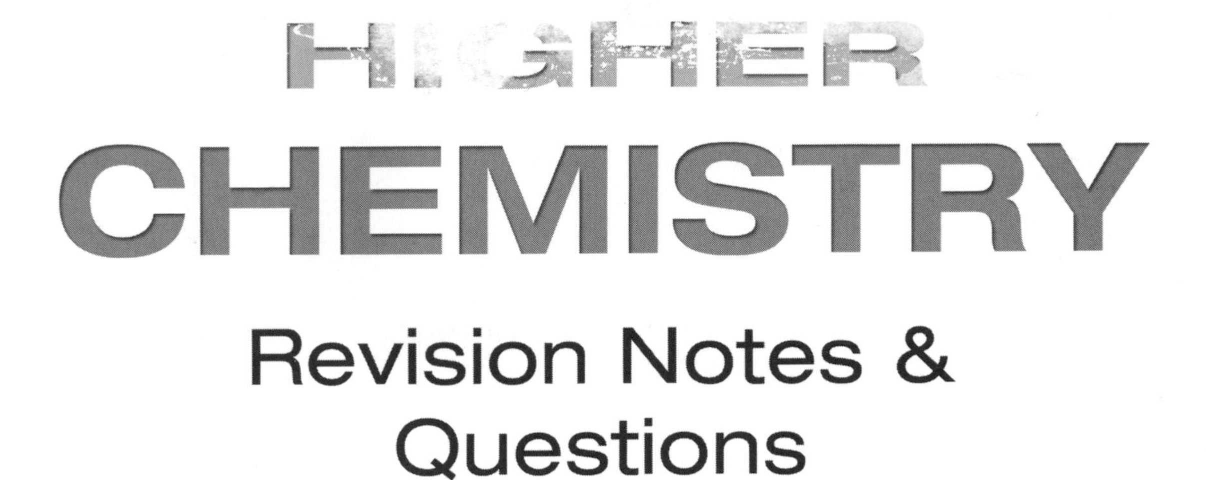

Revision Notes & Questions

Eric Allan and John Harris

The Publishers would like to thank the following for permission to reproduce copyright material:

Acknowledgements
Extracts from Question Papers are reprinted by permission of the Scottish Qualifications Authority.

Orders: please contact Bookpoint Ltd, 130 Milton Park, Abingdon, Oxon OX14 4SB. Telephone: (44) 01235 827720. Fax: (44) 01235 400454. Lines are open from 9.00 – 6.00, Monday to Saturday, with a 24-hour message answering services. Visit our website at www.hoddereducation.co.uk. Hodder Gibson can be contacted direct on: Tel: 0141 848 1609; Fax: 0141 889 6315; email: hoddergibson@hodder.co.uk

First edition published in 2000 by
Hodder Gibson, a member of the Hodder Headline Group
2a Christie Street
Paisley PA1 1NB

Impression number 10 9 8 7 6 5 4 3 2 1
Year 2010 2009 2008 2007 2006 2005

Cover photo from Beken of Cowes Collection
Typeset in Garamond BE Regular 10.5/12 by Fakenham Photosetting Ltd, Fakenham, Norfolk
Printed and bound in Great Britain by Martins the Printers, Berwick upon Tweed.

A catalogue record for this title is available from the British Library

ISBN-10: 0 340-90562-X
ISBN-13: 978-0-340-90562-3

Contents

Unit 1 Energy Matters

Unit 2 The World of Carbon

Unit 3 Chemical Reactions

Preface

Revision Notes and Questions for Higher Chemistry is intended to be used by candidates who already use our *New Higher Chemistry* textbook (0340 72479X), as well as those who use other resources. We have adhered to the same chapter headings as in the textbook and we have used the same layout and method for worked examples, although the content is different in the majority of cases.

With regard to the questions, our principal aim has been to provide candidates with a variety of experience of both types of questions now encountered in the final exam, namely multiple choice and extended answer questions. Many questions are appropriate examples from past exam papers or are slight modifications of them. Past paper questions, included in the end-of-chapter sections, are denoted by a star (*) throughout. We make no apology that some questions are harder than strictly necessary for Higher. These are intended to challenge the best candidates.

In this edition, answers have been included at the back and grid questions have been removed or replaced by multiple choice questions on similar topics. Some changes have been made to end-of-chapter questions, most notably in Chapter 17, Section A. The questions in Chapter 12 have been reorganised into Section A (Fats and Oils) and Section B (Proteins). We have introduced summary charts for Structure and Bonding and for Basic Organic Chemistry. Data values used in this book are from the SQA Higher and Advanced Higher Data Booklet (1999).

Another significant change in this edition is to replace the End of Course questions by a Practice Prelim paper based on Units 1 and 2 as well as a test on Unit 3. Nearly all the questions in these items have been taken from NQ 2000 to 2004 exam papers with a few modifications to avoid repetition. In Section A of the Unit 1 and 2 Practice Prelim, questions 1–8 are on Standard Grade/Intermediate 2 topics, questions 9–24 are on Unit 1 and questions 25–40 are on Unit 2. Schools running prelims based on Units 1 and 3 can replace the Unit 2 questions in the Practice Prelim by the Unit 3 test. We are grateful to SQA for permission to reprint questions from previous examination papers.

We wish to thank our former colleagues at George Heriot's School, Edinburgh for constructive help and advice in producing both editions of this book. Several schools were contacted for comments about proposed changes in this second edition and we are grateful to those teachers who responded.

ERA & JHH
March 2005

Unit 1

ENERGY MATTERS

1 Reaction Rates

Following the course of a reaction

- Reactions can be followed by measuring changes in concentration, mass or volume of reactants or products.

A suitable reaction to study is that between marble chips (calcium carbonate) and hydrochloric acid using the apparatus shown in Figure 1.

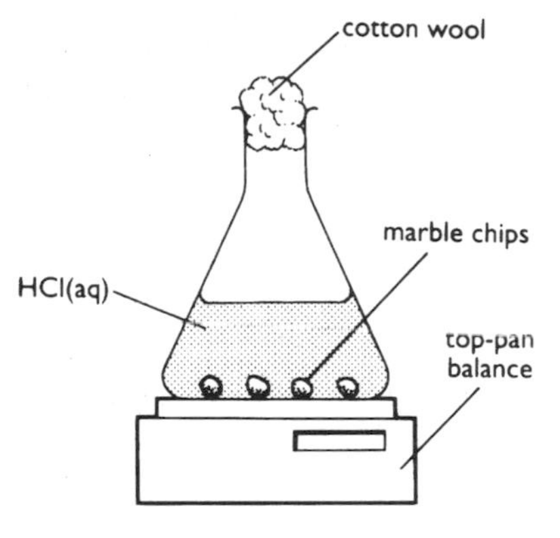

Figure 1

As the reaction proceeds carbon dioxide gas is released and the mass of flask and contents decreases. A cotton wool 'plug' prevents loss of acid spray during effervescence whilst allowing the gas to escape. Equation:

$$CaCO_3(s) + 2HCl(aq) \rightarrow CaCl_2(aq) + CO_2(g) + H_2O(l)$$

Specimen results from an experiment in which 15 g of marble chips, an excess, were added to 50 cm^3 of 4 mol l^{-1} hydrochloric acid are given. The decrease in mass is the mass of carbon dioxide released and this quantity can be plotted against time (Figure 2). From the loss in mass it is also possible, using the equation, to find the concentration of the acid at various times. These calculated results are plotted against time (Figure 3).

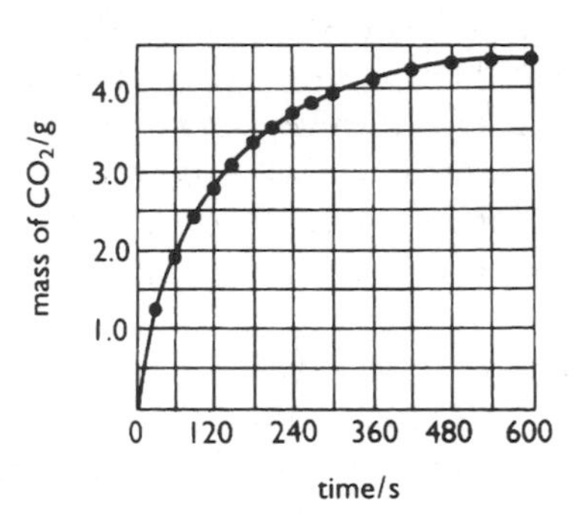

Figure 2 Mass of CO_2 against time

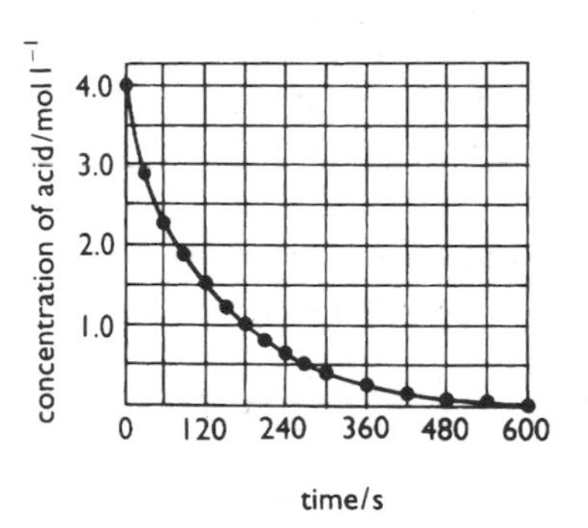

Figure 3 Concentration of acid against time

- The **rate of reaction** is the change in concentration of reactants or products in unit time.

In Figures 2 and 3, the slope of the graph is steepest at the beginning of the reaction and levels off as time passes, i.e. the rate of reaction is greatest initially and decreases with time.

- The rate of a reaction is **proportional** to the reciprocal of the time taken.
- The **average rate** over a certain period of time can be calculated in this experiment from the loss in mass or decrease in acid concentration which occurs in a certain time interval.

Worked Example 1.1

a) Use the data given in Figure 2 to calculate the average rate of reaction during the period from 240 seconds to 300 seconds, in terms of the mass of carbon dioxide produced.

Mass of CO_2 released between 240 s and 300 s = 4.0 – 3.7 = 0.3 g

$$\text{Average rate} = \frac{\text{mass of } CO_2}{\text{time interval}}$$

$$= \frac{0.3}{60}$$

$$= 0.005 \text{ g s}^{-1}$$

b) Use the data given in Figure 3 to calculate the average rate of reaction during the period from 120 seconds to 180 seconds, in terms of the decrease in the concentration of hydrochloric acid.

Decrease in concentration of HCl(aq) between 120 s and 180 s

$$= 1.5 - 1.0$$

$$= 0.5 \text{ mol l}^{-1}$$

$$\text{Average rate} = \frac{\text{decrease in acid concentration}}{\text{time interval}}$$

$$= \frac{0.5}{60}$$

$$= 0.0083 \text{ mol l}^{-1} \text{ s}^{-1}$$

Factors affecting the rate of a reaction

- Reactions, in which one of the reactants is a solid, can be speeded up or slowed down by altering the **particle size** of the solid.

In the reaction discussed in the previous section, using smaller marble chips increases the surface area and speeds up the reaction, as shown by a steeper slope in Figure 4.

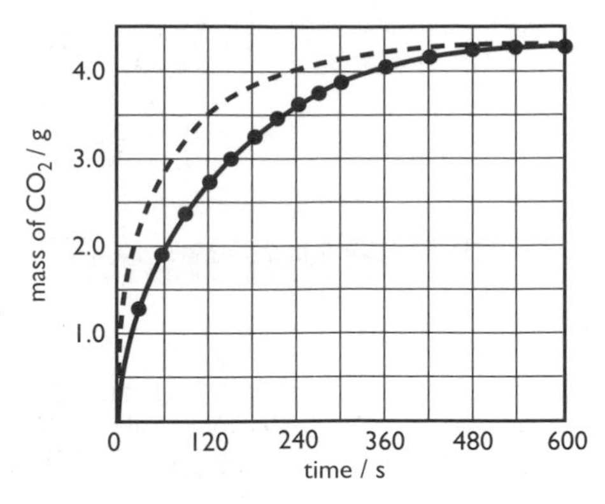

Figure 4

- **Smaller particle size**, or **larger surface area**, of solid reactants increases reaction rate.
- Higher **concentration of reactants** in solution increases reaction rate.

The reaction between marble and hydrochloric acid, for example, can be speeded up by increasing the concentration of the acid.

- Higher **temperature** increases reaction rate.

Prescribed Practical Activity

To investigate the relationship between rate of reaction and concentration of a reactant, a **'clock reaction'** may be used. This is a reaction in which a time lapse occurs before a sudden end-point is reached.

The reaction between hydrogen peroxide and acidified potassium iodide solution is used to show how rate depends on concentration of iodide ions. The equation for the reaction is:

$$H_2O_2(aq) + 2H^+(aq) + 2I^-(aq) \rightarrow 2H_2O(l) + I_2(aq)$$

Starch solution and sodium thiosulphate solution, $Na_2S_2O_3(aq)$, are also in the reaction mixture. Iodine produced in the reaction is immediately changed back into iodide ions by reacting with thiosulphate ions:

$$I_2(aq) + 2S_2O_3^{2-}(aq) \rightarrow 2I^-(aq) + S_4O_6^{2-}(aq)$$

While this is happening the reaction mixture is colourless.

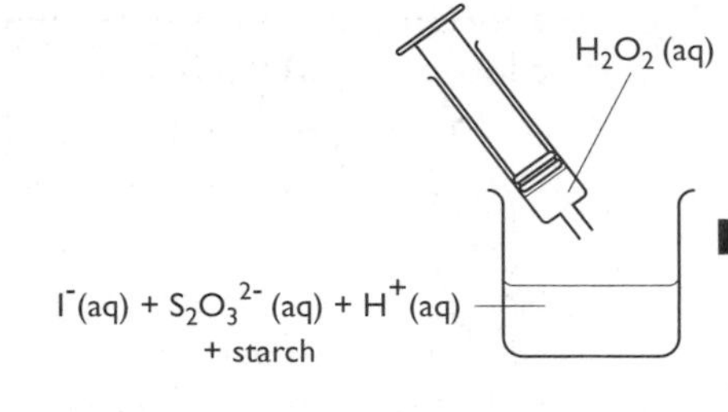

Figure 5

When all of the thiosulphate ions have reacted a blue–black colour suddenly appears as iodine is detected by starch.

As Figure 5 shows potassium iodide solution, starch, sodium thiosulphate solution and dilute sulphuric acid are mixed. Hydrogen peroxide solution is added and the time taken for the mixture to turn blue–black is measured. The experiment is repeated using smaller volumes of the iodide solution but adding water so that the total volume of the reacting mixture is always the same. The concentrations and volumes of all other solutions are kept constant.

The number of moles of thiosulphate ions is the same in each experiment so that when the blue–black colour appears the same extent of reaction has occurred. Since rate is inversely proportional to time, the reciprocal of time (1/t) is taken to be a measure of the rate of the reaction. A graph of rate against volume of KI(aq) is shown in Figure 6.

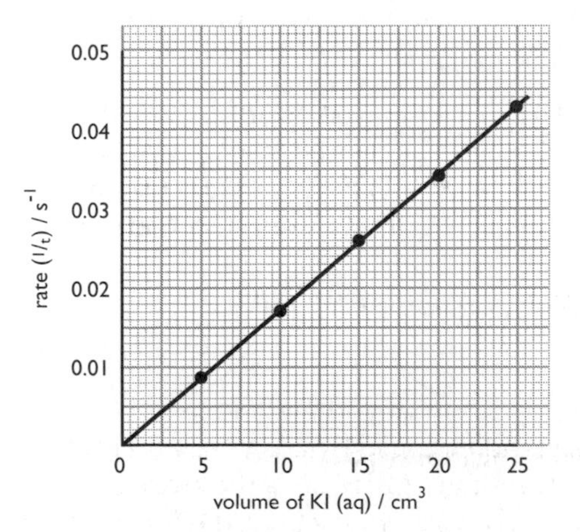

Figure 6

The graph of rate against volume of potassium iodide solution shows a straight line. Since the total volume is always the same, the volume of KI(aq) is a measure of iodide ion concentration.

The straight-line graph means that the rate of this reaction is directly proportional to the concentration of iodide ions.

Prescribed Practical Activity

The effect of temperature on rate can be studied using the following reaction. Acidified potassium permanganate solution, containing purple permanganate ions, MnO_4^-, is decolourised by an aqueous solution of oxalic acid, $(COOH)_2$. This reaction is very slow at room temperature but is almost instantaneous above 80°C. The equation is:

$$5(COOH)_2(aq) + 6H^+(aq) + 2MnO_4^-(aq) \rightarrow 2Mn^{2+}(aq) + 10CO_2(g) + 8H_2O(l)$$

This experiment is carried out at temperatures ranging from about 40°C to about 70°C.

Volumes and concentrations of all the reactants are kept constant. As shown in Figure 7, the reaction starts when the oxalic acid is added to the permanganate solution, which is acidified with dilute sulphuric acid. The time taken for the solution to become colourless is measured.

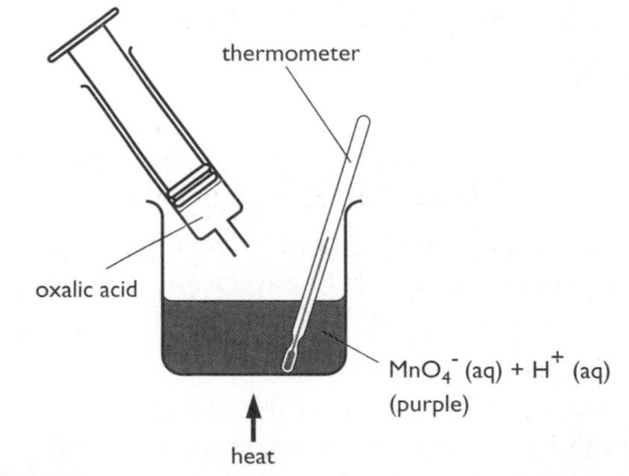

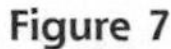

Figure 7

As the number of moles of permanganate ions is the same in each experiment, the same amount of reaction has occurred when the end-point has been reached. The reciprocal of the time taken to reach the end-point (1/t) is taken as a measure of the rate of reaction.

A graph of rate against temperature is shown in Figure 8. The rate of reaction increases with rising temperature, but since the graph is a curve the rate is not directly proportional to the temperature.

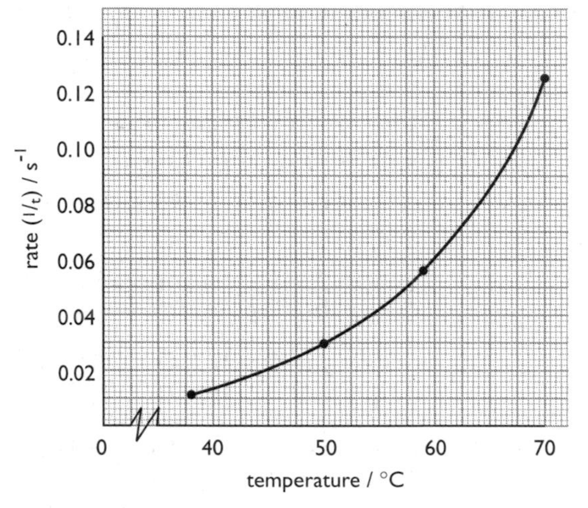

Figure 8

Collision theory

- ❍ All substances are made up of very small particles, which are called atoms, ions or molecules. These particles are continually moving, the speed and extent of the motion is related to whether the substance is a gas, a liquid, a solid or in solution. This description is often referred to as the '**kinetic model of matter**'.
- ❍ For a chemical reaction to occur, the particles of reactants must collide.
- ❍ Any factor which increases the number of collisions per second between the particles of the reactants is likely to increase the rate of reaction.
- ❍ More collisions occur if the particle size of a solid reactant is decreased, since its overall surface area is consequently increased.
- ❍ If the concentration of a reactant is increased, more collisions between particles will take place.
- ❍ Raising the temperature at which the reaction occurs not only increases the number of collisions between particles. Temperature is a measure of the average kinetic energy of the particles in a substance. At a higher temperature the particles have greater kinetic energy and collide with greater force.
- ❍ Reactions occur when reactant particles collide but not all collisions result in a successful reaction.

Activation energy and energy distribution

- ❍ For a reaction to occur the colliding particles must have a minimum amount of kinetic energy, the **activation energy**. This varies from one reaction to another. If the activation energy is high, only a few particles will have enough energy for collisions between them to be successful and hence the reaction will be slow, but a reaction with a low activation energy will be fast.
- ❍ At a given temperature individual molecules of a gas have widely different kinetic energies.

The distribution of kinetic energy values is illustrated in Figure 9. The kinetic energy of individual molecules will continually change due to collisions with other molecules but at constant temperature the overall distribution of energies remains the same.

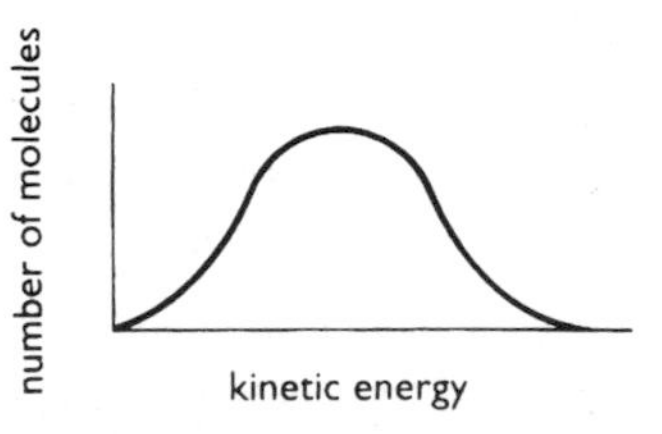

Figure 9 Distribution of energy

Figure 10 also includes the activation energy, E_A. The shaded area represents the number of molecules which have energy greater than the activation energy, i.e. it represents the proportion of molecules which have sufficient energy to react. If the activation energy is greater then the shaded area would be smaller thus representing a smaller proportion of the total number of molecules.

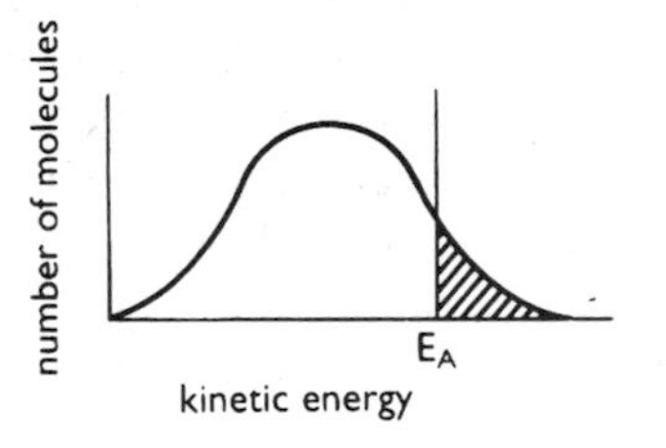

Figure 10 Distribution of energy including activation energy

- The energy distribution changes when the temperature changes.

The effect of a small rise in temperature, from T_1 to T_2, is shown in Figure 11. The average energy is increased but the most significant feature is the considerable increase in the area that is shaded. This is the real reason why a small change in temperature has such a marked effect on the rate of a reaction.

- An increase in temperature causes a significant increase in the number of molecules which have energy greater than the activation energy.

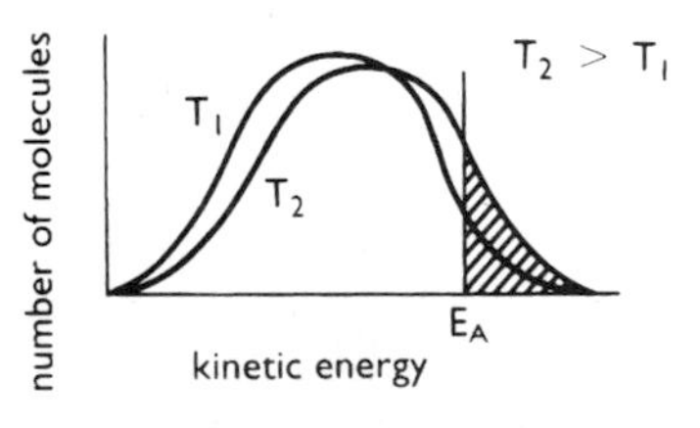

Figure 11 Distribution of energy at different temperatures

Photochemical reactions

- In some chemical reactions light energy is used to increase the number of molecules which have energy greater than the activation energy.
- In photosynthesis light energy is absorbed by chlorophyll to convert carbon dioxide and water into glucose and oxygen.
- In black and white photography when a film is exposed to light silver ions are reduced to silver atoms.
- Some reactions are 'set off' or initiated by light. A mixture of hydrogen and chlorine gases explodes when exposed to a light source of high enough energy.

Excess reactant

In previous work balanced equations have enabled the calculation of the mass of a product from the mass of a reactant, or vice versa. In reactions involving two reactants this calculation can only be done if there is more than enough of the other reactant, i.e. it is present in **excess**. An example of a question of this type is, '*4.46 g of lead(II) oxide were reacted with excess dilute nitric acid. Calculate the mass of lead(II) nitrate produced.*'

In the experiment described on page 1 calcium carbonate, in the form of marble chips, was reacted with hydrochloric acid. Worked example 1.2 refers to this reaction.

Worked Example 1.2

15 g of calcium carbonate were reacted with 50 cm^3 of 4 mol l^{-1} hydrochloric acid.

a) Show by calculation which reactant was present in excess.

b) Calculate the mass of carbon dioxide produced.

$$CaCO_3 + 2HCl \rightarrow CaCl_2 + CO_2 + H_2O$$

$CaCO_3$	$2HCl$		CO_2
1 mol	**2 mol**		**1 mol**
(100 g)			**(44 g)**

Answers:

a) Number of moles of $CaCO_3$,

$$n = \frac{m}{gfm} = \frac{15}{100} = 0.15 \text{ mol}$$

where m = mass of substance (in g) and gfm = gram formula mass.

Number of moles of HCl,

$$n = C \times V = 4 \times \frac{50}{1000} = 0.2 \text{ mol}$$

where C = concentration and V= volume.

According to the equation, 1 mol of $CaCO_3$ neutralises 2 mol of HCl.

Hence, 0.1 mol of $CaCO_3$ neutralises 0.2 mol of HCl.

Since there is more than 0.1 mol of $CaCO_3$ present, *this* reactant is in excess.

b) To calculate the mass of carbon dioxide produced we use the quantity of the reactant which is completely reacted (i.e. the acid) and not the one which is present in excess.

According to the equation, 2 mol of HCl produce 1 mol of CO_2.

Hence, 0.2 mol of HCl produce 0.1 mol of CO_2.

$= 0.1 \times 44 = 4.4$ g of CO_2.

Catalysts

- A catalyst is a substance which alters the rate of a reaction without being used up in the reaction
  e.g. manganese(IV) oxide catalyses the decomposition of hydrogen peroxide solution.

$$2H_2O_2(aq) \rightarrow 2H_2O(l) + O_2(g)$$

- Catalysts play an important part in many industrial processes. Table 1 summarises some of these processes.
- The catalysts listed in Table 1, along with manganese(IV) oxide in the decomposition of hydrogen peroxide, are said to be **heterogeneous**, since they are in a *different physical state* from the reactants.
- A catalyst which is in the *same physical state* as the reactants is said to be **homogeneous**.

An example of this is illustrated in Figure 12. The reaction between aqueous solutions of potassium sodium tartrate and hydrogen peroxide is slow, even when the mixture is heated. An aqueous solution containing cobalt(II) ions catalyses it. The immediate colour change to green and the return of the pink colour at the end of the reaction shows that a catalyst may undergo a temporary chemical change during its catalytic activity.

Catalyst	Process	Reaction	Importance
Vanadium(V) oxide	Contact	$2SO_2 + O_2 \rightleftharpoons 2SO_3$	Manufacture of sulphuric acid
Iron	Haber	$N_2 + 3H_2 \rightleftharpoons 2NH_3$	Manufacture of ammonia
Platinum	Catalytic oxidation of ammonia	$4NH_3 + 5O_2 \rightleftharpoons 4NO + 6H_2O$	Manufacture of nitric acid
Nickel	Hydrogenation	Unsaturated oils + $H_2 \rightarrow$ saturated fats	Manufacture of margarine
Aluminium silicate	Catalytic cracking	Breaking down long-chain hydrocarbon molecules	Manufacture of fuels and monomers for the plastics industry

Table 1

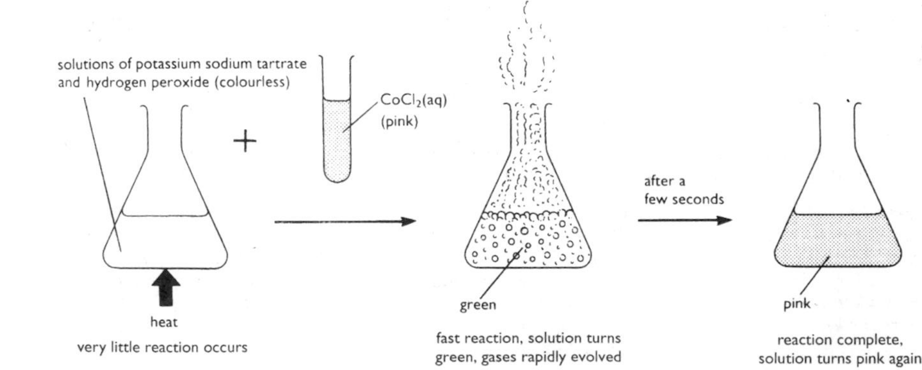

Figure 12

How heterogeneous catalysts work

- It is an advantage if a heterogeneous catalyst has a large surface area. Catalysis occurs on the surface at certain points called **active sites**. At these sites molecules of at least one of the reactants are **adsorbed**.
- How a heterogeneous catalyst works can be represented in three stages (Figure 13).
- A catalyst can become **poisoned** if certain molecules are preferentially adsorbed or even permanently attached to the surface of the catalyst. This reduces the number of active sites available and makes the catalyst ineffective.

Carbon monoxide is a catalyst poison in the Haber Process. In a catalysed industrial process, impurities in reactants can cause additional costs if the catalyst has to be regenerated or renewed.

During the catalytic cracking of long-chain hydrocarbons, carbon is deposited on the surface of the catalyst reducing its efficiency. The catalyst can be regenerated by burning off the carbon in a plentiful supply of air.

Cars with petrol engines have catalytic converters fitted as part of their exhaust systems. The converter contains a ceramic material covered with metals such as platinum and rhodium which catalyse the conversion of carbon monoxide to carbon dioxide and nitrogen oxides to nitrogen. Catalytic converters should only be used in cars which run on unleaded petrol, otherwise the lead compounds will poison the catalyst.

Enzymes

- Many biochemical reactions in the living cells of plants and animals are catalysed by enzymes. Examples include amylase, which catalyses the hydrolysis of starch, and catalase, present in blood, which catalyses the decomposition of hydrogen peroxide.
- The molecular shape of an enzyme usually plays a vital role in its function as a catalyst. It operates most effectively at a certain optimum temperature and within a narrow pH range.
- Enzymes are usually highly **specific**. Maltose and sucrose are disaccharides but are hydrolysed by different enzymes.
- Industrial applications of enzymes include:
  - yeast to provide enzymes for the fermentation of glucose to ethanol
  - invertase to hydrolyse sucrose to make soft-centred chocolates
  - rennin in cheese production
  - amylase in removing starch from fabrics, a process called desizing.

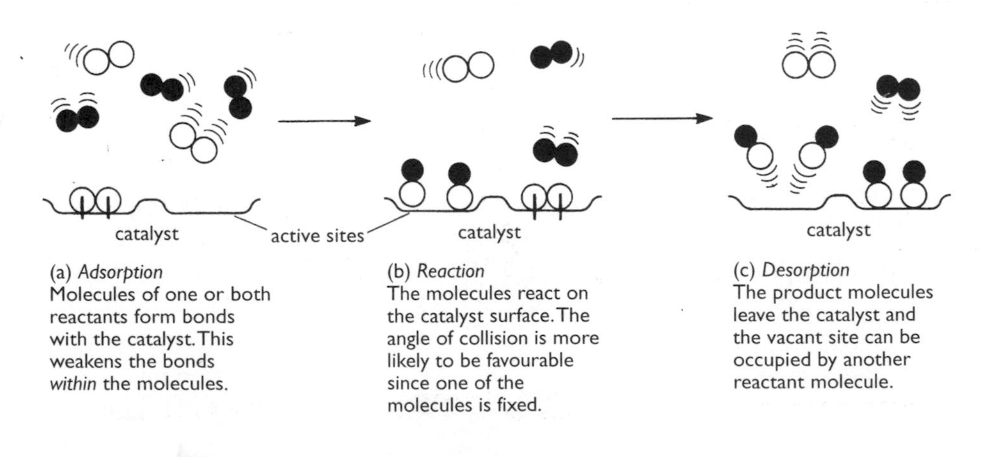

Figure 13 Heterogeneous catalysis

Questions

Questions 1 and **2** relate to the following graph.

Graph **X** was obtained when 1 g of calcium carbonate powder reacted with excess dilute hydrochloric acid at 20°C.

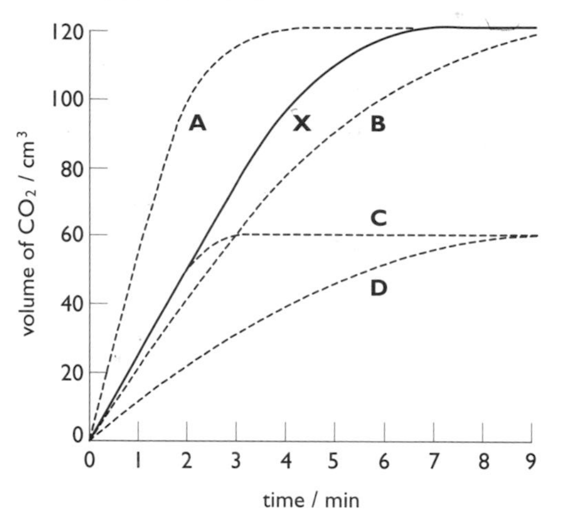

1* Which curve would best represent the reaction of 0.5 g lump calcium carbonate with excess of the same dilute hydrochloric acid at 20°C?

2 Which curve would best represent the reaction of 1 g of calcium carbonate powder with excess of the same dilute hydrochloric acid at 15°C?

3

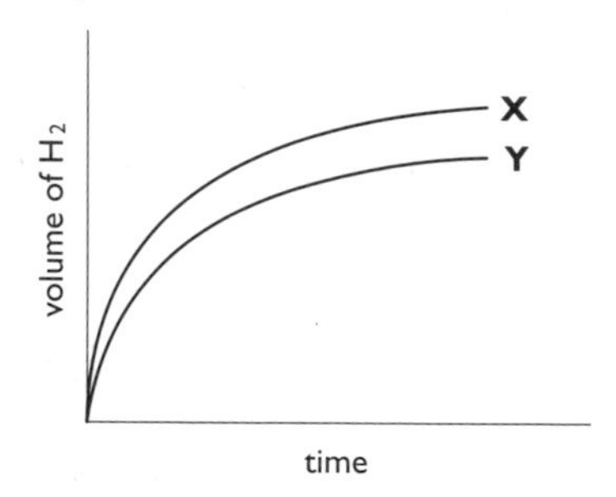

When zinc is reacted with excess dilute sulphuric acid, hydrogen is released. The curves shown above were obtained under different conditions.

The change from **X** to **Y** could be achieved by

A increasing the concentration of the acid
B decreasing the mass of zinc
C decreasing the particle size of the zinc
D adding a catalyst.

4* The graph shows the volume of hydrogen given off against time when an excess of magnesium ribbon is added to 100 cm³ of hydrochloric acid (concentration 1 mol l^{-1}) at 20°C.

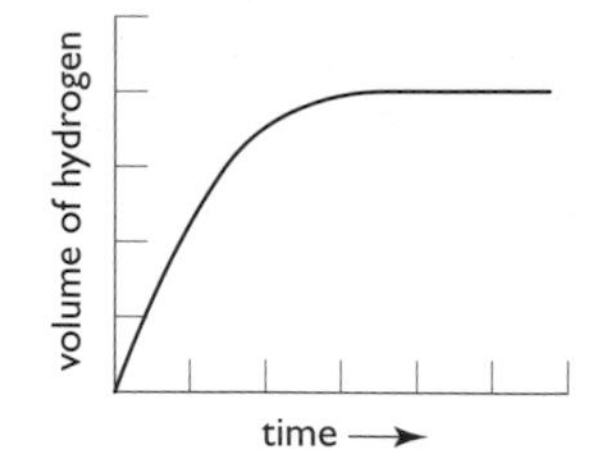

Which graph would show the volume of hydrogen given off when an excess of magnesium ribbon is added to 50 cm³ of hydrochloric acid of the same concentration at 30°C?

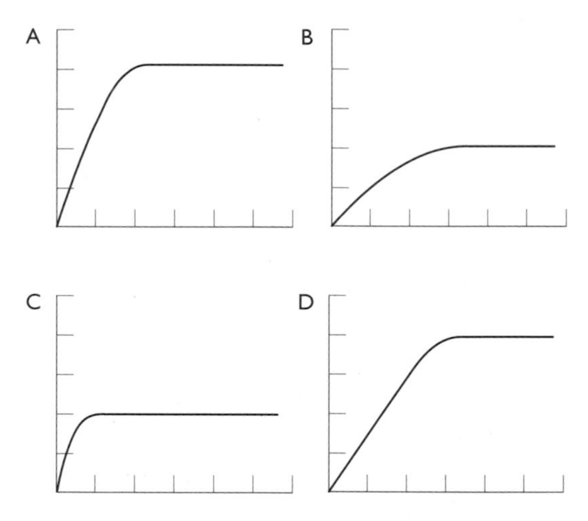

5* For any chemical, the temperature is a measure of

A the average kinetic energy of the particles which react
B the average kinetic energy of all the particles
C the activation energy
D the minimum kinetic energy required before reaction occurs.

6* Two identical samples of zinc were added to an excess of two solutions of sulphuric acid, concentrations 2 mol l^{-1} and 1 mol l^{-1} respectively. Which of the following would have been the same for the two samples?

A The total mass lost
B The total time for the reaction
C The initial reaction rate
D The average rate of evolution of gas

Questions 7 and **8** refer to the following graph on which the relative volumes of hydrogen produced in a given time are plotted for three different reactions.

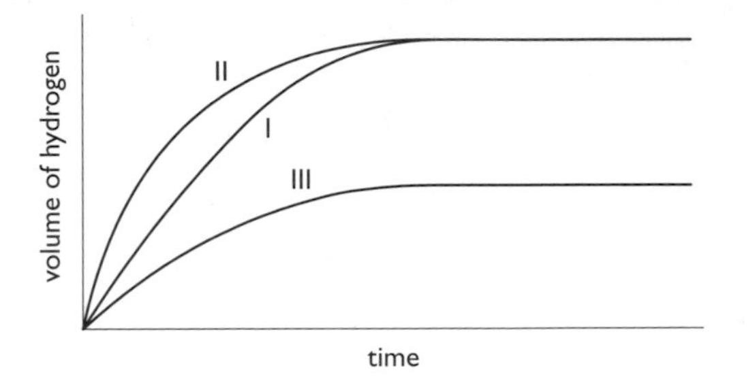

Curve I is for the reaction of excess zinc powder with 100 cm^3 of 0.1 mol l^{-1} hydrochloric acid.

7 Curve II could represent the reaction of 100 cm^3 of 0.1 mol l^{-1} hydrochloric acid with excess powdered samples of

A copper
B iron
C lead
D magnesium.

8 Curve III could represent the reaction of excess zinc powder with

A 50 cm^3 of 0.1 mol l^{-1} hydrochloric acid
B 25 cm^3 of 0.1 mol l^{-1} sulphuric acid
C 100 cm^3 of 0.05 mol l^{-1} hydrochloric acid
D 100 cm^3 of 0.05 mol l^{-1} sulphuric acid

9 A pupil added 6.54 g of zinc to 100 cm^3 of 0.5 mol l^{-1} copper(II) sulphate solution. Identify the correct statement.

A All of the zinc reacts.
B 0.05 mol of copper is displaced.
C 0.1 mol of zinc(II) sulphate is formed.
D Copper(II) sulphate is in excess.

10 The reaction between acidified potassium permanganate solution and oxalic acid solution was studied in one of the prescribed practicals.

a) What colour change occurs in this reaction?

b) The reaction was carried out at various temperatures between 40°C and 70°C. It was found that the time taken for this colour change to occur approximately halved for every 10°C rise in temperature.

i) How would you ensure that a fair comparison was being made in this experiment?

ii) Draw a graph of rate against temperature to show the results of this experiment.

c) Explain why a small rise in temperature has such a marked effect on the rate of a reaction. Refer to activation energy and include a diagram showing the kinetic energy distribution of reacting particles in your answer.

d) This reaction can be catalysed by $MnSO_4(aq)$. Is the catalyst heterogeneous or homogeneous? Explain your choice.

11 A pupil added 10 g of $CaCO_3$ to 100 cm^3 of 1.0 mol l^{-1} HCl. In this experiment

A all of the $CaCO_3$ dissolved
B 4.4 g of CO_2 were produced
C excess acid was used
D 5 g of $CaCO_3$ did not react.

12 For each of the following reactions indicate whether the catalyst is heterogeneous or homogeneous.

a) $CH_3CHO(g) \rightarrow CH_4(g) + CO(g)$
Catalyst: $I_2(g)$

b) $KClO_3(s) \rightarrow KCl(s) + \frac{3}{2}O_2(g)$
Catalyst: $MnO_2(s)$

c) $H_2(g) + I_2(g) \rightarrow 2HI(g)$
Catalyst: Au(s)

13 The following graph shows how the concentration of a dilute acid varies with time during reaction with powdered copper carbonate.

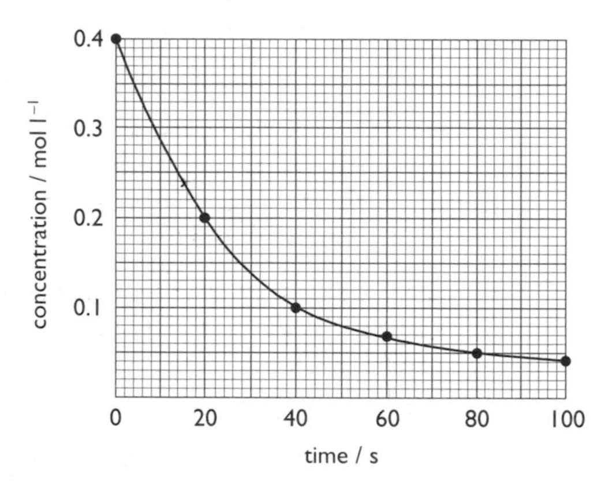

a) Calculate the average reaction rate during
i) the first 15 s
ii) the second 15 s
iii) the first minute.

b) The graph levels off after 140 s at a concentration of 0.03 mol l^{-1}.
Which reactant is present in excess?

14 a) In an experiment on reaction rate, 2.43 g of magnesium ribbon was added to 100 cm^3 of 2 mol l^{-1} sulphuric acid.
i) Write the balanced equation for this reaction.
ii) Show by calculation which reactant was in excess.

b) The following graph was obtained from the experimental results.

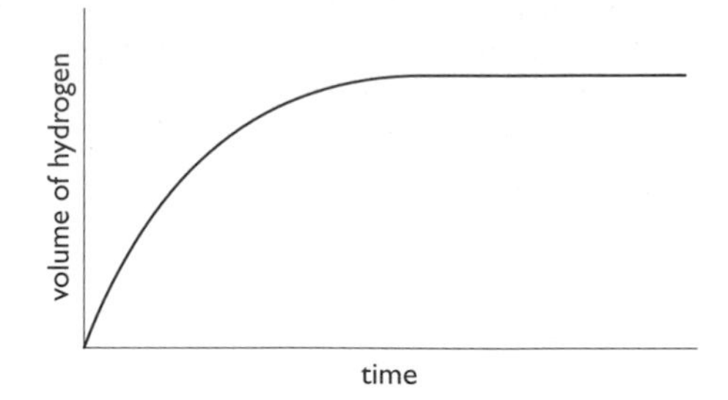

Copy the graph and add lines to show the graphs obtained when the experiment is repeated using
i) 0.81 g of magnesium powder
ii) 100 cm^3 of 2 mol l^{-1} hydrochloric acid.

Questions 15 and 16 refer to the graph and statements below.

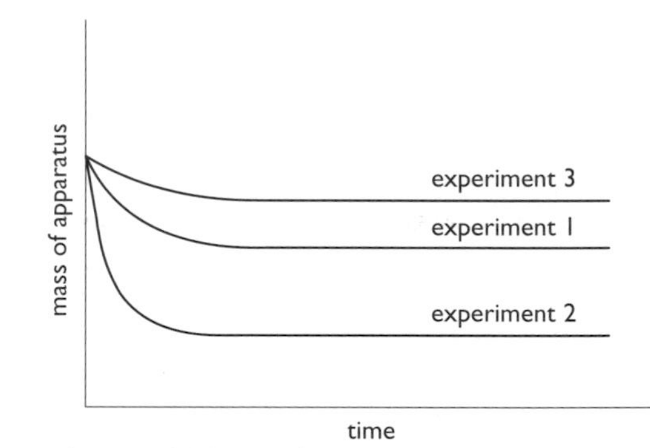

The graph shows the results of three experiments involving the reaction of excess powdered magnesium carbonate with dilute acid.

100 cm^3 of 0.5 mol l^{-1} sulphuric acid were added in experiment 1.

A 100 cm^3 of 0.5 mol l^{-1} hydrochloric acid
B 50 cm^3 of 2.0 mol l^{-1} sulphuric acid
C 100 cm^3 of 1.0 mol l^{-1} hydrochloric acid
D 50 cm^3 of 2.0 mol l^{-1} hydrochloric acid

15 Which solution was used in experiment 2?

16 Which solution was used in experiment 3?

17* Three experiments were carried out in a study of the rate of reaction between magnesium (in excess) and dilute hydrochloric acid. A balance was used to record the mass of the reaction flask and its contents. The results of experiment 1, using 0.4 mol l^{-1} acid, are shown in the graph.

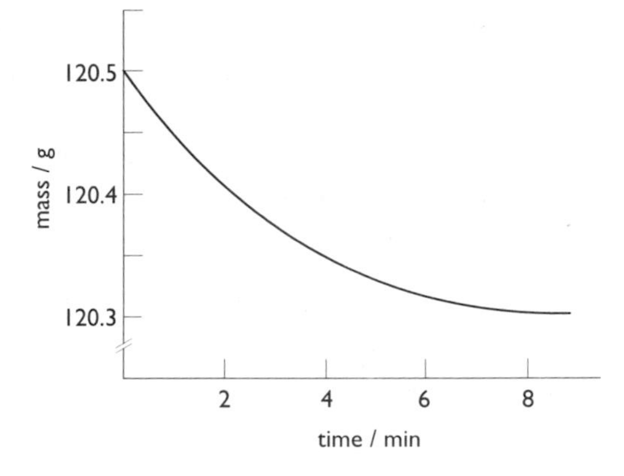

a) Why did the balance record a decrease in mass during the reaction?

b) The **only** difference between experiment 2 and experiment 1 was the use of a catalyst. On a copy of the graph, sketch a curve that could be expected for experiment 2.

c) The **only** difference between experiment 3 and experiment 1 was the use of 0.2 mol l^{-1} acid. On your graph, sketch a curve that could be expected for experiment 3.

18* Marble chips, calcium carbonate, reacted with excess dilute hydrochloric acid. The rate of reaction was followed by recording the mass of the container and the reaction mixture over a period of time. The results of an experiment are shown in the following graph.

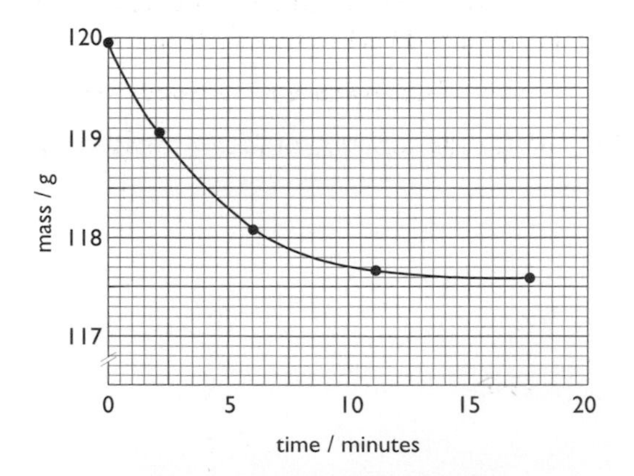

a) Write a balanced equation for the reaction.

b) Calculate the average rate of reaction over the first five minutes.

c) Why does the average rate of reaction decrease as the reaction proceeds?

d) The half-life of the reaction is the time taken for half of the calcium carbonate to be used up. Calculate the half-life for this reaction.

e) Sketch a curve showing how the volume of gas produced changes over the same period of time.

19 A group of S5 pupils were asked to carry out experiments using the following chemicals:

A 0.5 mol l^{-1} silver(I) nitrate solution
B 1.0 mol l^{-1} hydrochloric acid
C Copper powder.

a) One pupil mixed 20 cm^3 of solution **A** with 15 cm^3 of solution **B** and filtered off the precipitate of silver chloride. The equation for the reaction is:

$$AgNO_3(aq) + HCl(aq) \rightarrow AgCl(s) + HNO_3(aq)$$

i) Show by calculation which reactant is in excess.

ii) Calculate the mass of precipitate which was produced.

b) Another pupil added 0.25 g of substance **C** to 20 cm^3 of solution **A**. The equation for this reaction is:

$$Cu(s) + 2AgNO_3(aq) \rightarrow 2Ag(s) + Cu(NO_3)_2(aq)$$

i) Show by calculation which reactant is in excess.

ii) Calculate the mass of silver displaced in this experiment.

20* Air bags in cars are intended to prevent injuries in a car crash. They contain sodium azide, NaN_3, which produces nitrogen if an impact is detected.
The reaction which generates nitrogen is:

$$2NaN_3(s) \rightarrow 3N_2(g) + 2Na(s)$$

a) Other chemicals are present in air bags. These chemicals take part in further reactions. Suggest why these reactions are necessary.

b) Calculate the mass of sodium azide required to produce 84 g of nitrogen.

c) In order to provide protection, the gas must be generated very rapidly. The graph shows how the volume of nitrogen produced changes over a period of time.

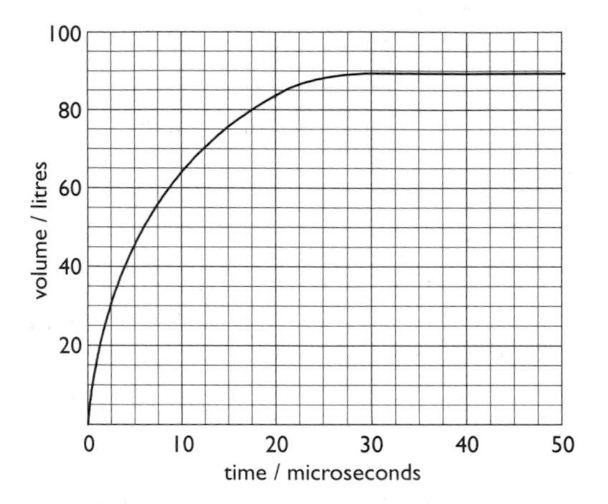

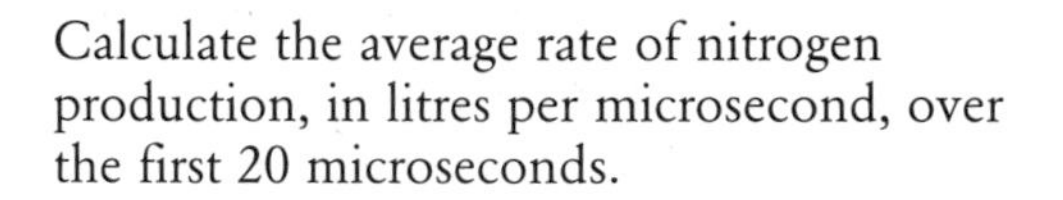

Calculate the average rate of nitrogen production, in litres per microsecond, over the first 20 microseconds.

2 Enthalpy

Potential Energy

Exothermic and endothermic reactions

- Most common reactions involve a release of energy to the surroundings, usually in the form of heat, and are said to be **exothermic**.

Examples include: combustion of fuels, neutralisation of acids by alkalis and reactive metals, displacement of metals.

- Energy may also be released in the form of light or sound.
- Reactions in which heat is absorbed from the surroundings are said to be **endothermic**.

Examples include: dissolving certain salts in water (e.g. ammonium nitrate), neutralising ethanoic acid with sodium hydrogencarbonate.

- During an exothermic reaction energy possessed by the reactants, i.e. potential energy, is released to the surroundings. The products have less potential energy than the reactants.

A potential energy diagram (Figure 1) shows the energy pathway as the reaction proceeds from reactants to products.

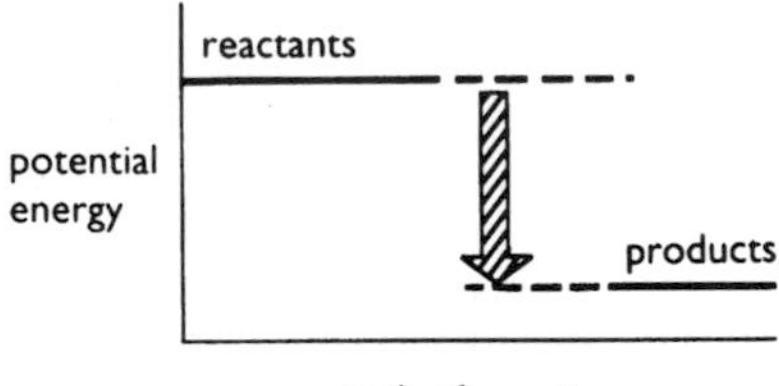

Figure 1 Exothermic reaction

- In an endothermic reaction the reactants absorb energy from the surroundings so that the products possess more energy than the reactants. See Figure 2.

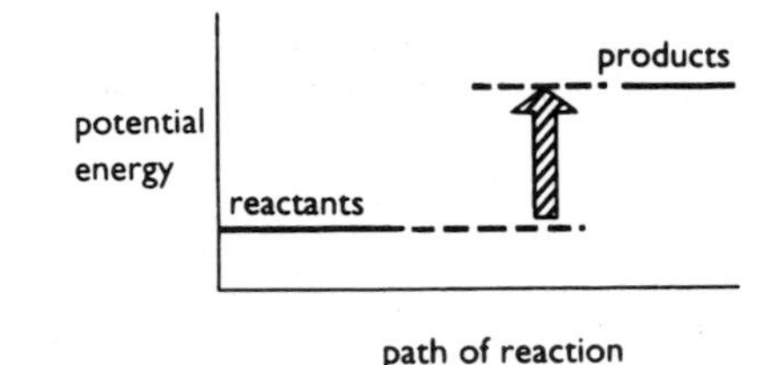

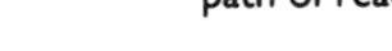

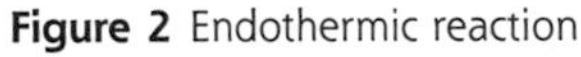

Figure 2 Endothermic reaction

Enthalpy change

- The difference in potential energy between reactants and products is called the **enthalpy change**, symbol: ΔH.
- Enthalpy changes are usually quoted in kJ mol^{-1}.
- Since the reactants lose energy in an exothermic reaction, ΔH is negative (Figure 3).

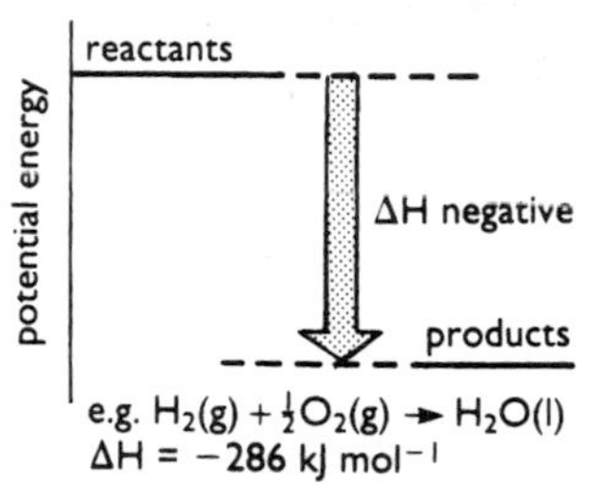

Figure 3 ΔH for an exothermic reaction

- In an endothermic reaction the products possess more energy than the reactants. An endothermic change has a positive ΔH value (Figure 4).

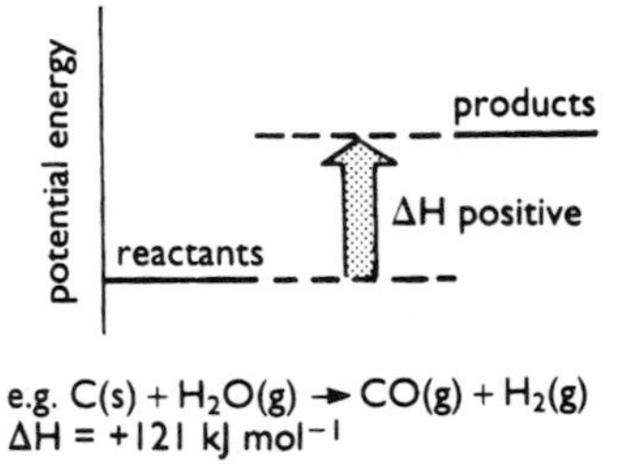

Figure 4 ΔH for an endothermic reaction

- The minus sign must appear in front of the numerical value for the enthalpy change if the reaction is exothermic. The absence of a sign from the ΔH value indicates that the reaction is endothermic.

Activation energy and activated complex

- In Chapter 1, **activation energy** was defined as the minimum kinetic energy required by colliding molecules for a reaction to occur.
- In the potential energy diagrams (Figures 5 and 6) the activation energy appears as an 'energy barrier' which has to be overcome as the reaction proceeds from reactants to products.
- The rate of a reaction will depend on the height of this barrier. The higher the barrier, the slower the reaction. The rate of reaction does *not* depend on the enthalpy change.

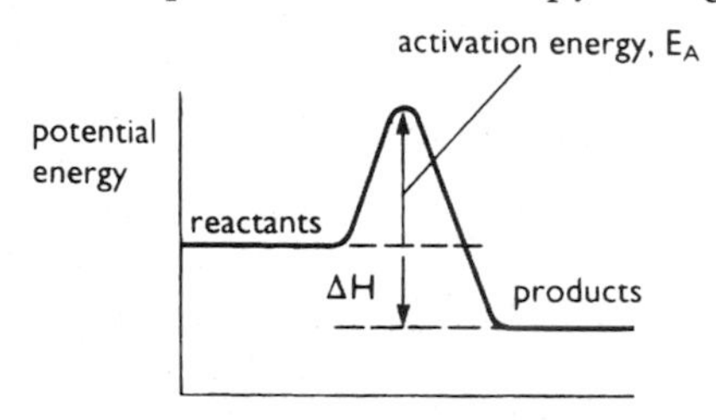

Figure 5 The activation energy for an exothermic reaction

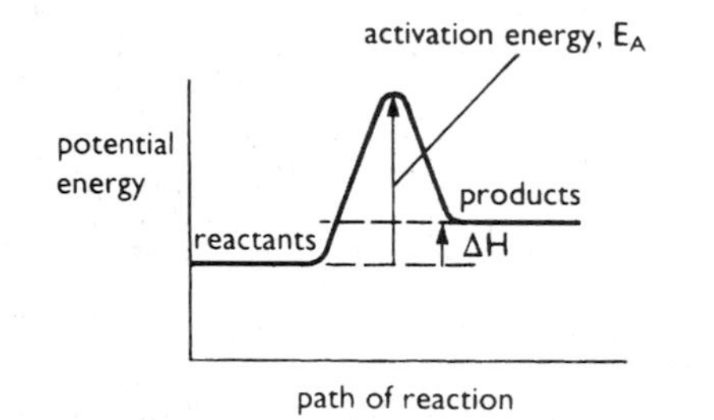

Figure 6 The activation energy for an endothermic reaction

- As the reaction proceeds from reactants to products an intermediate stage is reached at the top of the activation energy barrier at which a highly unstable species called an **activated complex** is formed. See Figure 7, which also shows that the activation energy can be redefined as the energy needed by colliding particles to form the activated complex.

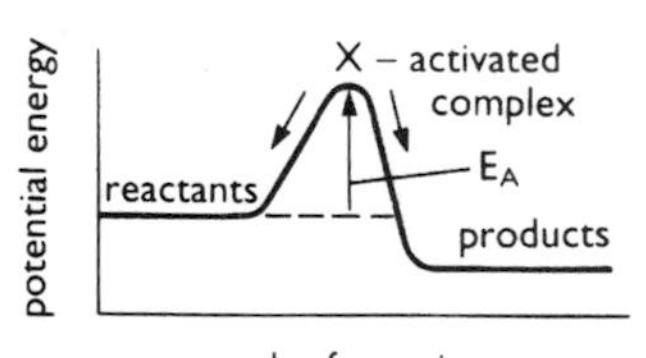

Figure 7

- Potential energy diagrams when drawn to scale can be used to calculate the enthalpy change and/or the activation energy of a reaction.

Catalysts

- In general, catalysts provide alternative reaction pathways involving less energy, i.e. **a catalyst lowers the activation energy of a reaction**. See Figure 8.

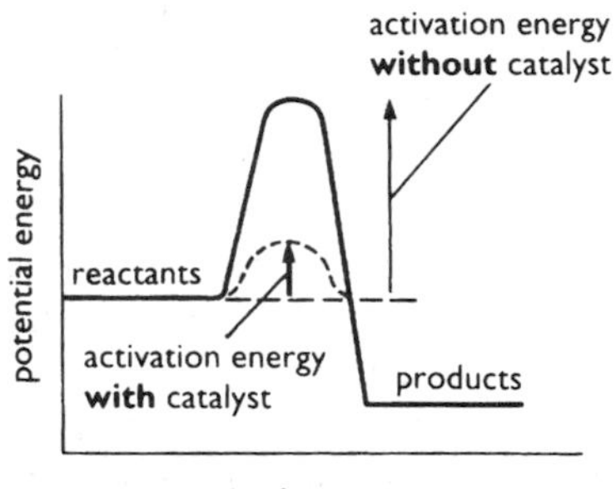

Figure 8 The lowering of the activation energy by a catalyst

Note the contrasting ways in which the use of a catalyst and the use of heat affect the rate of a reaction. Heating speeds up a reaction by increasing the number of molecules which have energy greater than the activation energy. A catalyst speeds up a reaction by lowering the activation energy. Catalysts save energy in many industrial processes.

Experimental measurement of enthalpy changes

Enthalpy of combustion

- The **enthalpy of combustion** of a substance is the enthalpy change when one mole of the substance is burned completely in oxygen.

The equation for the complete combustion of ethane is given below along with its enthalpy of combustion.

$$C_2H_6(g) + \tfrac{7}{2}O_2(g) \rightarrow 2CO_2(g) + 3H_2O(l) \qquad \Delta H = -1560 \text{ kJ mol}^{-1}$$

Note

1. the negative sign in the ΔH value since combustion is exothermic,
2. the equation is written showing one mole of the substance that is burning.

The equation shows that 3.5 moles of oxygen are needed per mole of ethane. If the equation is doubled, the ΔH value is doubled and is shown in kJ only.

Prescribed Practical Activity

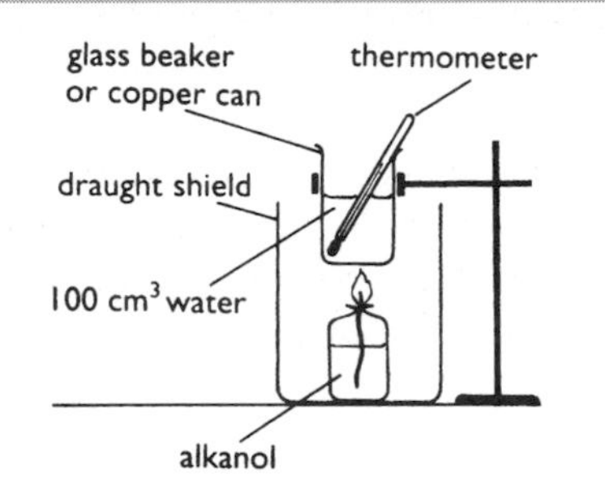

Figure 9

The enthalpy of combustion of a simple alkanol can be determined by experiment, using an apparatus like that shown in Figure 9. The burner containing the alkanol is weighed before and after burning. The alkanol is allowed to burn until the temperature of the water in the beaker has been raised by, say, 10°C before extinguishing the flame.

The heat energy (E_h) gained by the water in the beaker can be calculated from the formula:

$$\mathbf{E_h = cm\,\Delta T}$$

where **c** is the specific heat capacity of water, 4.18 kJ kg^{-1} $°C^{-1}$

m is the mass of water heated, in kg

ΔT is the change in temperature of the water.

This is the heat released by the burning alkanol and from this the enthalpy of combustion of the alkanol can be calculated. The method of calculation is shown in Worked Example 2.1 using specimen data for the burning of ethanol. The result obtained using the above apparatus will usually be considerably less than the accepted figure given in the SQA Data Book, mainly because of heat losses.

Worked Example 2.1

Enthalpy of combustion of ethanol, C_2H_5OH

Data:

Mass of burner + ethanol before burning = 53.85 g

Mass of burner + ethanol after burning = 53.49 g

Mass of water heated, m = 100 g = 0.1 kg

Temperature rise of water, ΔT= 10°C

Calculation:

Heat energy released, $E_h = cm\Delta T$
$= 4.18 \times 0.1 \times 10$
$= 4.18$ kJ

Gram formula mass of ethanol, C_2H_5OH = 46 g

Mass of ethanol burned = 0.36 g

Number of moles of ethanol burned, n $= \frac{0.36}{46}$

Heat energy released per mole,

$$\frac{E_h}{n} = \frac{4.18}{n} = \frac{4.18 \times 46}{0.36}$$

= 534 kJ

Enthalpy of combustion of ethanol, ΔH = – 534 kJ mol^{-1}

Note that, since the reaction is exothermic, a negative sign is inserted in the final result.

Enthalpy of solution

- The **enthalpy of solution** of a substance is the enthalpy change when one mole of the substance dissolves in water.

The enthalpy of solution can be determined experimentally as illustrated in Figure 10. The temperature of the water before adding a weighed amount of solute is measured and so is the temperature of the final solution. The method of calculating the enthalpy change is shown in Worked Example 2.2.

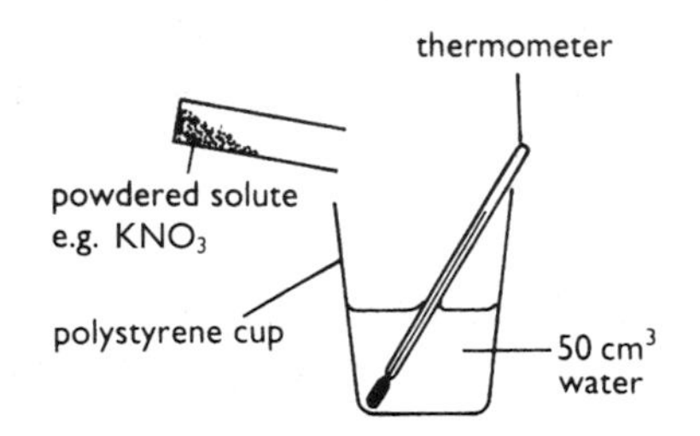

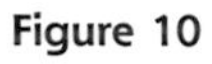

Figure 10

Enthalpy of neutralisation

- The **enthalpy of neutralisation** of an acid is the enthalpy change when the acid is neutralised to form one mole of water.
- When an acid such as hydrochloric acid is neutralised by an alkali such as sodium hydroxide, a salt (in this case sodium chloride) and water are formed.

$$\mathbf{HCl(aq) + NaOH(aq) \rightarrow NaCl(aq) + H_2O(l)}$$

- When any acid is neutralised by any alkali the reaction can be expressed by the following equation (with spectator ions omitted).

$$\mathbf{H^+(aq) + OH^-(aq) \rightarrow H_2O(l)}$$

The enthalpy of neutralisation of an acid by an alkali can be found by experiment as shown in Figure 11. The temperature of each reactant is measured before mixing so that the average initial temperature can be calculated. The solutions are then mixed and the highest temperature of the neutral solution is noted. The method of calculating the enthalpy change is shown in Worked Example 2.3.

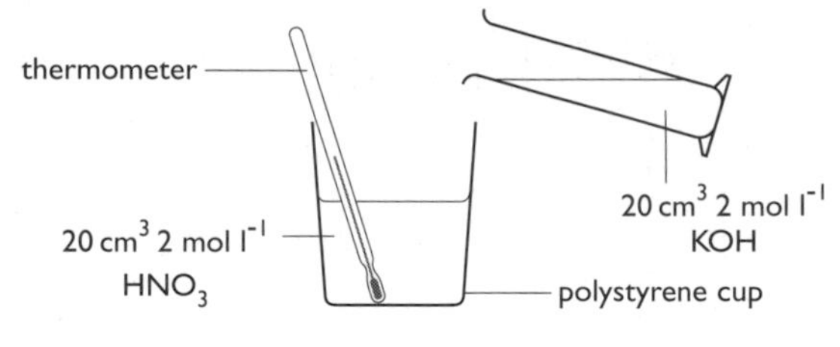

Figure 11

Worked Example 2.2

Enthalpy of solution of KNO_3

Data:

Mass of solute (potassium nitrate) $= 1.00$ g

Mass of water used, m $= 50$ g $= 0.05$ kg

Temperature of water initially $= 20.4°C$

Temperature of solution $= 18.7°C$

Calculation:

Temperature fall, $\Delta T = 1.7°C$

Heat energy absorbed,

$$E_h = cm\Delta T = 4.18 \times 0.05 \times 1.7 \text{ kJ}$$

Gram formula mass of potassium nitrate $= 101.1$ g

Number of moles of solute used, n

$$= \frac{1.00}{101.1} = 0.0099$$

Heat energy absorbed per mole,

$$\frac{E_h}{n} = \frac{4.18 \times 0.05 \times 1.7}{0.0099}$$

$$= 35.9 \text{ kJ}$$

Enthalpy of solution of potassium nitrate, $\Delta H = 35.9 \text{ kJ mol}^{-1}$

As the reaction is endothermic, a plus sign is not needed before the numerical value.

Worked Example 2.3

Enthalpy of neutralisation of $HNO_3(aq)$ by KOH(aq)

Data:

Solutions used: 20 cm^3 of 2 mol l^{-1} HNO_3
20 cm^3 of 2 mol l^{-1} KOH

Temperature of acid before mixing
= 19.5°C

Temperature of alkali before mixing
= 18.5°C

Temperature of solution after mixing
= 32.0°C

Calculation:

Average initial temperature $= \frac{19.5 + 18.5}{2}$
= 19.0°C

Temperature rise, ΔT = 32.0 – 19.0
= 13.0°C

Total volume of solution
= 40 cm^3

Mass of solution heated
= 40 g = 0.04 kg

Heat energy released,
$E_h = cm\Delta T = 4.18 \times 0.04 \times 13.0$ kJ

Number of moles of H^+ ions
= number of moles of water produced.

$n = C \times V$ $= 2 \times 0.02$
= 0.04

Heat energy released per mole

$$\frac{E_h}{n} = \frac{4.18 \times 0.04 \times 13.0}{0.04}$$

= 54.34 kJ mol^{-1}

Enthalpy of neutralisation,
$\Delta H = -54.34$ kJ mol^{-1}

In the calculation above 0.04 moles of HNO_3 are neutralised by 0.04 moles of KOH producing 0.04 moles of water. Hence in the calculation, $cm\Delta T$ is divided by n = 0.04 so that the ΔH value obtained refers to one mole of acid being neutralised to form one mole of water.

Note that in Worked Examples 2.2 and 2.3 two approximations are being made:

1) the density of a dilute aqueous solution is the same as that of water, i.e. 1 g cm^{-3} at room temperature, and

2) the specific heat capacity of a dilute aqueous solution is the same as that of water, i.e. 4.18 kJ kg^{-1} K^{-1}.

If sulphuric acid, a dibasic acid, is used the enthalpy of neutralisation is about the same as for hydrochloric acid, since one mole of sulphuric acid produces two moles of water when completely neutralised by sodium hydroxide as shown by the following equation.

$$H_2SO_4(aq) + 2NaOH(aq) \rightarrow Na_2SO_4(aq) + 2H_2O(l)$$

Questions

1*

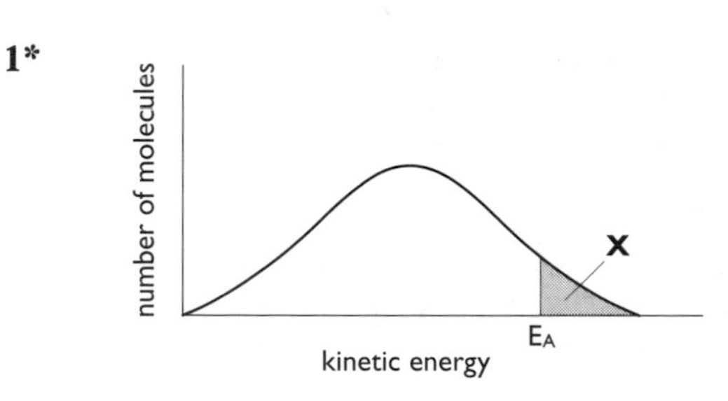

In area **X**

A molecules always form an activated complex
B no molecules have the energy to form an activated complex
C collisions between molecules are always successful in forming products
D all molecules have the energy to form an activated complex.

Questions 2 and **3** refer to the following potential energy diagram for a chemical reaction.

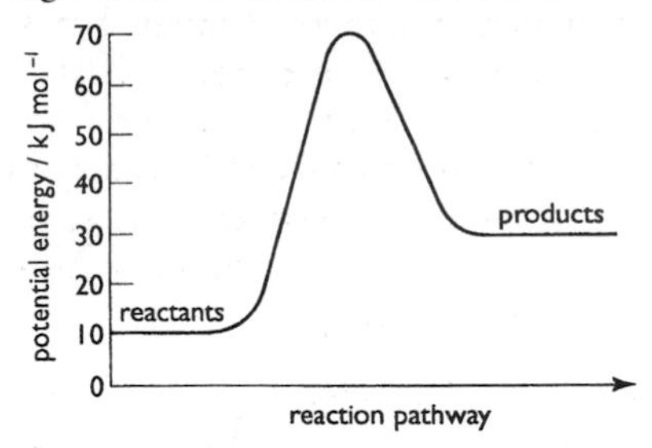

2 The activation energy, in kJ mol^{-1}, for this reaction is

A 30 B 40 C 60 D 70.

3 The enthalpy change, in kJ mol^{-1}, for this reaction is

A −20 B + 60 C − 60 D + 20.

4

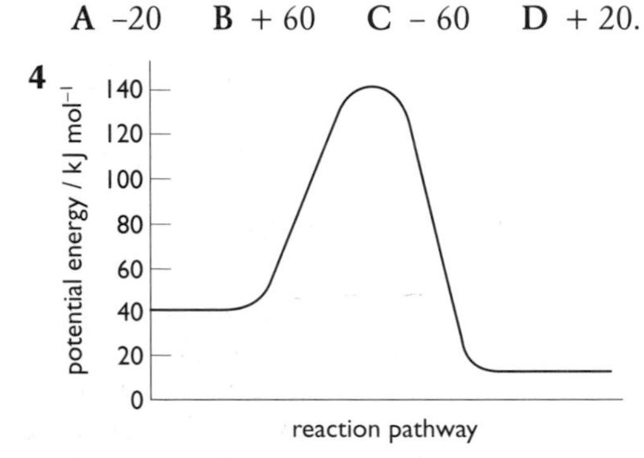

Which set of data applies to this reaction?

	Enthalpy change	E_a / kJ mol^{-1}
A	Exothermic	100
B	Endothermic	100
C	Exothermic	140
D	Endothermic	140

5* Which of the following is **not** a correct statement about the effect of a catalyst?

The catalyst

A provides an alternative route to the products
B lowers the energy which molecules need for successful collisions
C provides energy so that more molecules have successful collisions
D forms bonds with reacting molecules.

6 Which of the following statements describes the effect of a catalyst in a reaction?

A ΔH decreases and E_a increases
B ΔH does not change and E_a decreases
C ΔH decreases and E_a does not change
D ΔH increases and E_a decreases

7 The potential energy diagram below refers to a reversible reaction.

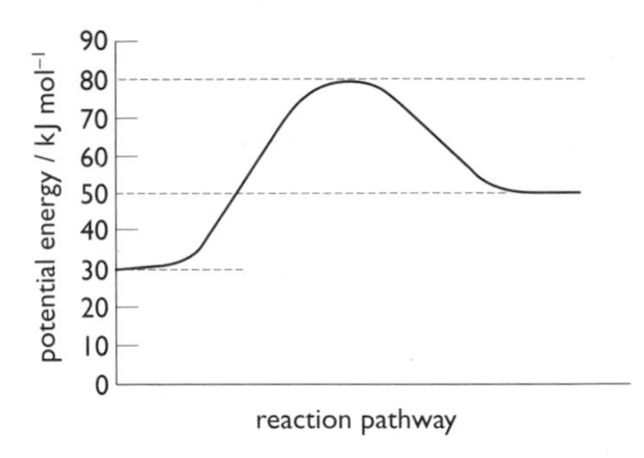

The enthalpy change, in kJ mol^{-1}, for the **reverse** reaction is

A −20 B −30 C + 30 D + 20.

8 The boiling point of pentane is 36°C. Which equation illustrates the enthalpy of combustion of pentane?

A $C_5H_{12}(l) + 8O_2(g) \rightarrow 5CO_2(g) + 6H_2O(l)$

B $C_5H_{12}(l) + \frac{11}{2}O_2(g) \rightarrow 5CO(g) + 6H_2O(l)$

C $C_5H_{12}(g) + 8O_2(g) \rightarrow 5CO_2(g) + 6H_2O(g)$

D $C_5H_{12}(l) + 5O_2(g) \rightarrow 5CO_2(g) + 6H_2(g)$

9 A pupil added 0.06 mol of sodium nitrate to 100 cm^3 of water at 20°C.
The enthalpy of solution of sodium nitrate is + 20.5 kJ mol^{-1}.
After dissolving the solute, the temperature of the solution will be

A 17°C B 20°C C 23°C D 26°C.

10 When 3.6 g of glucose, $C_6H_{12}O_6$, was burned, 56 kJ of energy was released.
From this data, what is the enthalpy of combustion of glucose, in kJ mol^{-1}?

A −15.6 B + 15.6 C −2800 D +2800

11* Consider the energy diagram:

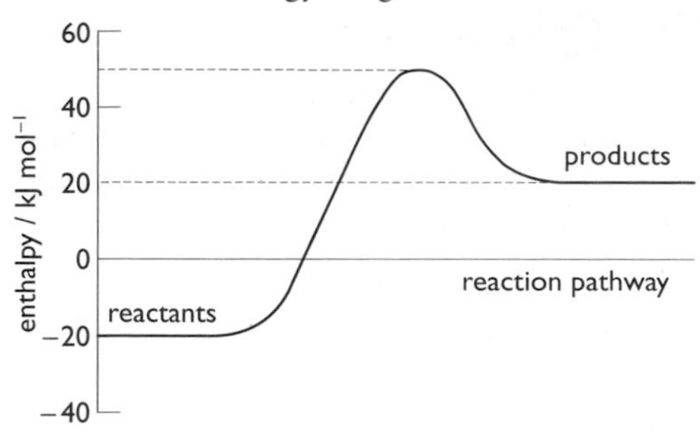

a) From the diagram, find values for
- **i)** the enthalpy change and the activation energy of the forward reaction;
- **ii)** the enthalpy change and the activation energy of the reverse reaction.

b) Which, if any, of the above values would be altered by the use of a catalyst?

12 $2N_2O(g) \rightarrow 2N_2(g) + O_2(g)$
$\Delta H = -\ 80\ kJ\ mol^{-1}$

The equation for the decomposition of dinitrogen monoxide (nitrous oxide) is shown above. The activation energy for this reaction when catalysed by platinum is 140 kJ mol^{-1}.

a) Is the catalyst heterogeneous or homogeneous?

b) Draw a potential energy diagram for the catalysed decomposition of $N_2O(g)$, with both energy values clearly indicated.

c) Elements in their natural state, such as the products of the above reaction, are assigned a value of 0 kJ mol^{-1} on a potential energy diagram.
What would be the energy value of the activated complex for the decomposition of $N_2O(g)$ when catalysed by platinum?

d) If this reaction is carried out without a catalyst, what effect (if any) would this have on the
- **i)** enthalpy change
- **ii)** activation energy?

13 In 1780, Lavoisier and Laplace published details of their '*ice calorimeter*' which they had used to measure the heat produced when carbon is burned. The method involved burning carbon in a crucible surrounded by ice in a larger container. The heat of reaction melted some of the ice, forming water that flowed out of the bottom of the container into a pre-weighed beaker. The beaker was reweighed to find the mass of water obtained.

They found that for every gram of carbon burned, 98 g of water were collected.

a) Calculate the mass of water that would be collected when one mole of carbon was burned in the ice calorimeter.

b) One kilogram of ice at 0 °C requires 334 kJ to convert it into water. Calculate the enthalpy of combustion of carbon using the ice calorimeter.

14 Sodium hydrogencarbonate is soluble in water and it also neutralises acids, including ethanoic acid, CH_3COOH.
The equation for this reaction is:

$$NaHCO_3(s) + CH_3COOH(aq) \rightarrow CH_3COO^-(aq) + Na^+(aq) + CO_2(g) + H_2O(l)$$

a) Calculate the enthalpy of solution of $NaHCO_3$ from the following data.

4.2 g of solute were added to 50 cm^3 of water in a polystyrene cup.
The temperature of the water (before mixing) was 20.0 °C and the temperature of the solution (after mixing) was 16.2 °C.

b) The enthalpy of neutralisation of ethanoic acid by solid $NaHCO_3$ is 26.3 kJ mol l^{-1}. If 2.1g of solid $NaHCO_3$ were added to 25.0 cm^3 of 1.0 mol l^{-1} ethanoic acid at 19.8°C, what would be the final temperature of the mixture?

15 An experiment was carried out using a methanol burner and an apparatus like that shown in Figure 9 page 15.

a) Write the equation for the complete combustion of methanol, $CH_3OH(l)$, and write down its enthalpy of combustion from your Data Book.

b) If the apparatus used is 50% efficient, calculate the mass of methanol burned when 80 cm^3 of water (in the copper can) rises in temperature from 20.5°C to 31.0°C.

3 Patterns in the Periodic Table

The greatest step in the progress towards a Periodic Law was taken by Mendeleev in 1869. His main points were:

- The elements fall into a repeating pattern of similar properties if arranged in order of increasing **atomic mass**.
- The list was arranged into vertical and horizontal sequences, called groups and periods. The groups contained elements which were chemically similar.
- Blanks were left to prevent dissimilar elements coming together. Mendeleev predicted the properties for the missing elements and, after their later discovery, the predictions were found to be accurate.
- In the modern Periodic Table, the elements are arranged by increasing **atomic number**, each new horizontal row, or period, commences when a new layer of electrons in the atom starts to fill. Each main vertical column contains elements with the same number of electrons in the outermost layer.

Trends in physical properties

- Melting and boiling points and densities of the elements vary across a period and down a group.

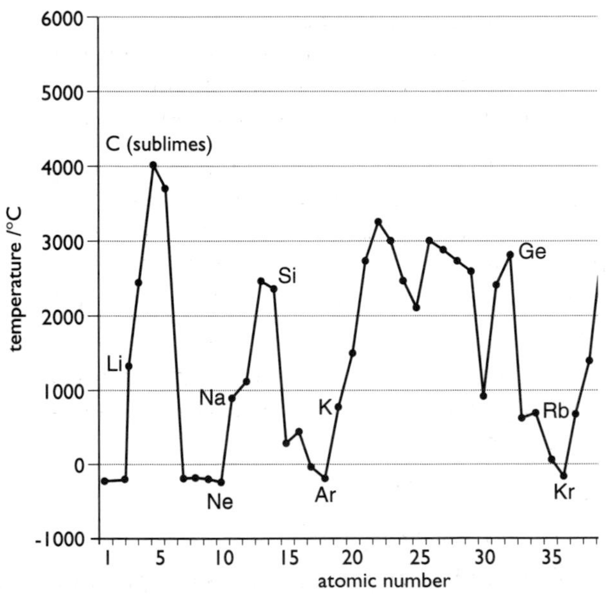

Figure 1 Variation of boiling point with atomic number

In Figures 1, 2 and 3, a repeating pattern of high and low values can be seen.

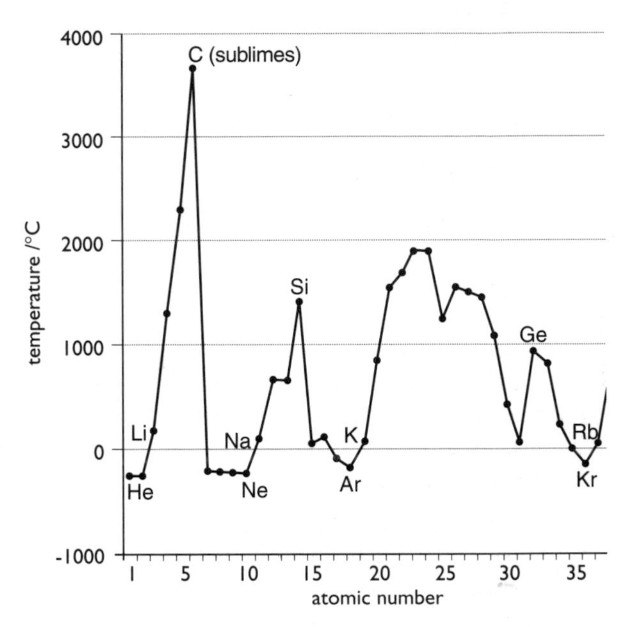

Figure 2 Variation of melting point with atomic number

Atomic Size

Covalent radius is a useful indication of atomic size.

Li	Be	B	C	N	O	F	Ne
134	129	90	77	75	73	71	–
Na	Mg	Al	Si	P	S	Cl	Ar
154	145	130	117	110	102	99	–
K	Ca	Ga	Ge	As	Se	Br	Kr
196	174	120	122	121	117	114	–
Rb	Sr	In	Sn	Sb	Te	I	Xe
216	191	150	140	143	135	133	–

increase in value in each column ↓

decrease in value in each row →

Table 1 Covalent atomic radii (10^{-12} m)

- In a **horizontal row** (period) of the Periodic Table, atomic size **decreases from left to right** because the atoms being considered all have

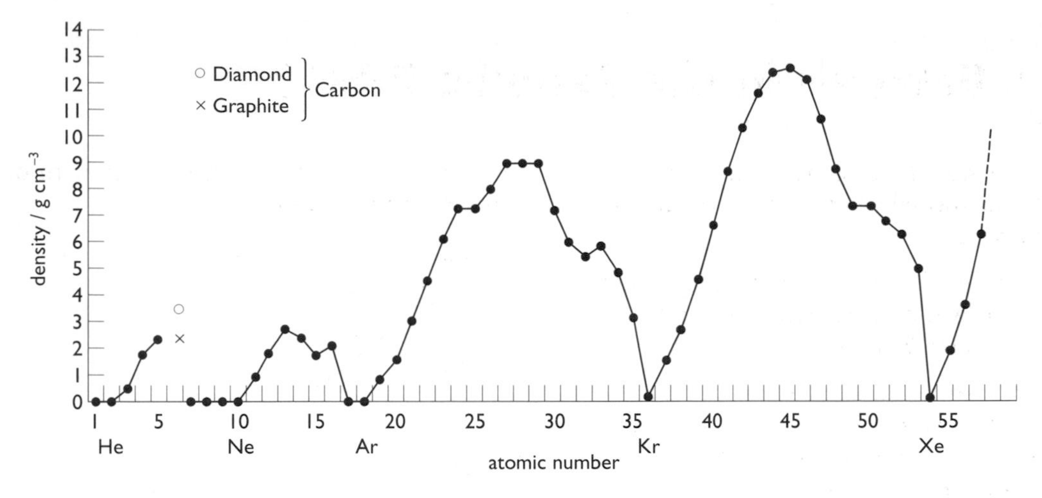

Figure 3 Variation of density (g cm^{-3}) with atomic number

the same number of occupied energy levels whilst there is an increase in nuclear charge from one element to the next. This exerts an increasing attraction on the electrons resulting in the atom decreasing in size.

❍ In any **vertical column** (group) all the elements have the same number of outer electrons, but one more energy level is occupied by electrons in each succeeding element, hence atomic size **increases down a group**.

First ionisation energy (or enthalpy)

❍ The energy change involved in creating one mole of singly-charged positive ions from one mole of atoms in the gaseous state is called the first ionisation energy.

e.g **K(g) → K⁺(g) + e⁻**

ΔH = (+) 425 kJ mol⁻¹

❍ Down groups of elements, there is a decrease of first ionisation energy. The electron is being removed from the outermost layer of electrons which is increasingly distant from the nuclear attraction and so less energy is required to remove the electron.

❍ In each period there is an overall increase of first ionisation energy from left to right. The electron being removed is in the same layer for any element in the same period. The nuclear charge is increasing along each period therefore the outermost electrons are more strongly held and so the energy required to remove them increases along each period.

Li	Be	B	C	N	O	F	Ne
526	905	807	1090	1410	1320	1690	2090
Na	Mg	Al	Si	P	S	Cl	Ar
502	744	584	792	1020	1010	1260	1530
K	Ca	Ga	Ge	As	Se	Br	Kr
425	596	577	762	947	941	1150	1350
Rb	Sr	In	Sn	Sb	Te	I	Xe
409	556	556	709	834	870	1020	1170

decrease down group ↓

overall increase along period →

Table 2 First ionisation energies (kJ mol^{-1})

❍ The screening effect of electrons in inner orbitals reduces the attraction of the nucleus for outermost electrons, reducing the ionisation energy from the value expected.

e.g. Gallium's additional 11 protons should increase its ionisation energy compared with calcium's but the additional 11 electrons are in internal orbitals and screen the outer electrons from the nucleus, in fact lowering the ionisation energy.

❍ The second ionisation energy is the enthalpy change associated with the loss of a second electron.

e.g **K⁺(g) → K²⁺(g) + e**

ΔH = (+) 3060 kJ mol⁻¹

Third and successive ionisation energies can be defined in a similar way.

Electronegativities

- In a covalent bond the relative powers of the atoms in the bond to attract bonding electrons to themselves are different and are defined as their **electronegativities**.

H							
2.2							
Li	Be	B	C	N	O	F	
1.0	1.5	2.0	2.5	3.0	3.5	4.0	
Na	Mg	Al	Si	P	S	Cl	decrease
0.9	1.2	1.5	1.9	2.2	2.5	3.0	down group
K	Ca	Ga	Ge	As	Se	Br	
0.8	1.0	1.6	1.8	2.2	2.4	2.8	
Rb	Sr	In	Sn	Sb	Te	I	
0.8	1.0	1.7	1.8	2.1	2.2	2.6	
Cs	Ba						
0.8	0.9						
increase across period →							

Table 3 Electronegativity values

- The electronegativity increases from left to right along a period since nuclear charge increases in the same direction.
- The electronegativity decreases down a group of the periodic table since the atomic size increases down the group.
- The difference in electronegativity values for the atoms joined gives an indication of the relative degrees of polarity in covalent bonds.

Questions

1 Which equation represents the first ionisation enthalpy of fluorine?

A $F(g) + e^- \rightarrow F^-(g)$

B $F(g) \rightarrow F^+(g) + e^-$

C $F^-(g) \rightarrow F(g) + e^-$

D $F^+(g) + e^- \rightarrow F(g)$

2 Which equation represents the second ionisation enthalpy of copper?

A $Cu^+(g) \rightarrow Cu^{2+}(g) + e^-$

B $Cu(g) \rightarrow Cu^{2+}(g) + 2e^-$

C $Cu^+(s) \rightarrow Cu^{2+}(s) + e^-$

D $Cu(s) \rightarrow Cu^{2+}(g) + 2e^-$

3* Potassium has a larger covalent radius than sodium because potassium has

A a larger nuclear charge
B a larger nucleus
C more occupied energy levels
D a smaller ionisation enthalpy.

4 $Al(g) \rightarrow Al^{3+}(g) + 3e^-$

The energy, in kJ mol^{-1}, required to bring about the above change is

A 1752 **B** 2414 **C** 2760 **D** 5174.

5 Between 98°C and 883°C sodium exists as a liquid. Which of the following elements exists as a liquid over the greatest temperature range?

A Magnesium **B** Aluminium
C Silicon **D** Phosphorus

6 The bar graph shows how the first ionisation energies of 14 consecutive elements in the Periodic Table vary with atomic number. The element at which the bar graph starts is not specified.

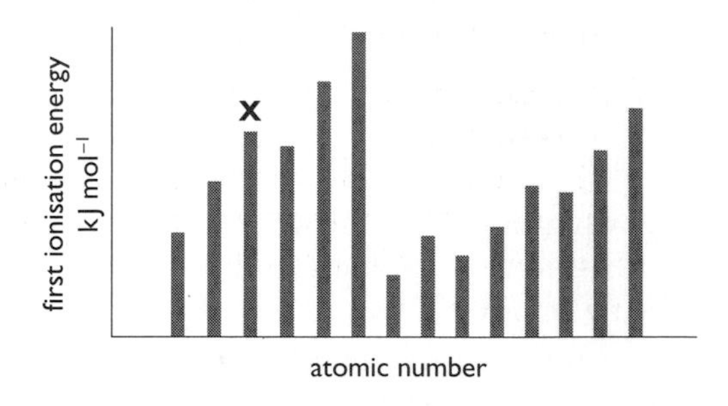

In which group of the Periodic Table is element **X**?

A 3 **B** 4 **C** 5 **D** 6

7* The difference between the covalent radius of sodium and silicon is mainly due to the difference in the

A number of electrons
B number of protons
C number of neutrons
D mass of each atom.

8 Going down Group 6 of the Periodic Table the

A density decreases
B covalent radius increases
C nuclear charge decreases
D electronegativity increases.

9 Which statement correctly describes a trend in the Periodic Table?

A The metallic bond strengths decrease down Group 1.
B The first ionisation energies decrease from sodium to argon.
C The covalent radii increase from lithium to fluorine.
D The strength of the van der Waal's forces decrease down Group 0.

10 Which trend would occur as the relative atomic mass of the halogens increases?

A The covalent radius decreases.
B The density decreases.
C The boiling point decreases.
D The van der Waal's forces become stronger.

11 a) Write the equation, including state symbols, which represents the first ionisation energy of potassium.

b) Explain why the first ionisation energy of potassium is less than
i) the first ionisation energy of calcium
ii) the first ionisation energy of sodium
iii) the **second** ionisation energy of potassium.

12* The diagram shows the first ionisation energies of successive elements (A–T), plotted against their atomic numbers.

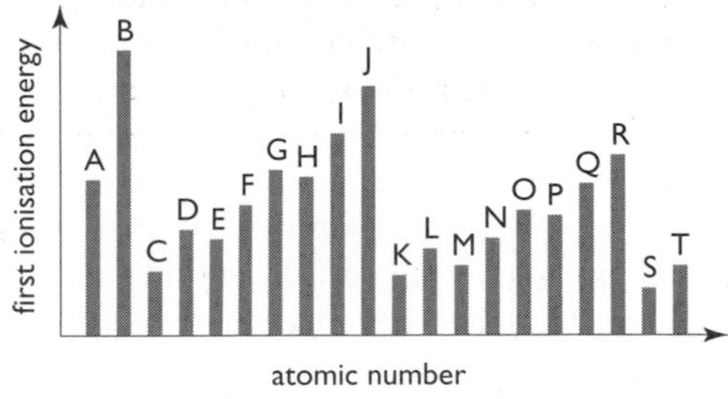

a) Which Group of elements is represented by the letters B, J and R?
b) Why is the first ionisation energy of element L greater than that of element K?
c) Why is the **second** ionisation energy of element L considerably less than that of element K?

13* The table gives the ionisation energies of some of the alkali metals and some of the halogens.

Element	Ionisation Energies/kJ mol^{-1}		
	First	Second	Third
lithium	526	7310	11800
fluorine	1690	3380	6060
sodium	502	4560	6920
chlorine	1260	2310	3840
potassium	425	3060	4440
bromine	1150	2100	3480

a) Why is the first ionisation energy of each alkali metal much less than that of the halogen **in the same period**?
b) Why is the second ionisation energy of each alkali metal much greater than that of the halogen **in the same period**?
c) Calculate the energy per mole required to bring about the change

$$K^{+}(g) \rightarrow K^{3+}(g)$$

14 The covalent radius of technetium, atomic number 43, is not known.
a) How would you expect the covalent radius of technetium to compare with that of
i) manganese
ii) cadmium (atomic number 48)?
b) Explain your answer to **a) i)**.

15 Explain why the covalent radius
a) increases from magnesium to barium
b) decreases from magnesium to sulphur.

4 Bonding, Structure and Properties of Elements

Bonding is a term describing the mechanism by which atoms join together. **Structure** describes the way in which the atoms, or particles derived from them, are arranged. The resulting characteristics, whether physical or chemical, of the substances are their **properties.**

Types of bonding in elements

Metallic bonding

The mechanism of metallic bonding is described under the heading 'Groups 1, 2 and 3' below.

Covalent bonding

- Covalent bonding is **electrostatic**, the atoms being held together by the attraction between their positive nuclei and negatively charged shared pairs of electrons. Each shared pair of electrons constitutes a covalent bond.
- In an element all the atoms are alike in terms of protons and electrons so the bonding electrons are shared equally.
- When more than one covalent bond is formed by an atom, the bonds are orientated in specific directions to each other.

Van der Waals' bonding

- Van der Waals' bonding is very weak bonding between molecules. Its mechanism is explained under the heading 'Noble Gases'.

Bonding in the first 20 elements

The Noble Gases – monatomic

- These elements, with the exception of helium, have an outer layer of eight electrons which is an especially stable arrangement.
- The Noble Gases are monatomic, i.e. their molecules consist of only one atom.
- The uneven distribution of the constantly moving electrons around the nuclei of the atoms causes the formation of **temporary dipoles** on the atoms (Figure 1). The atoms then attract each other.

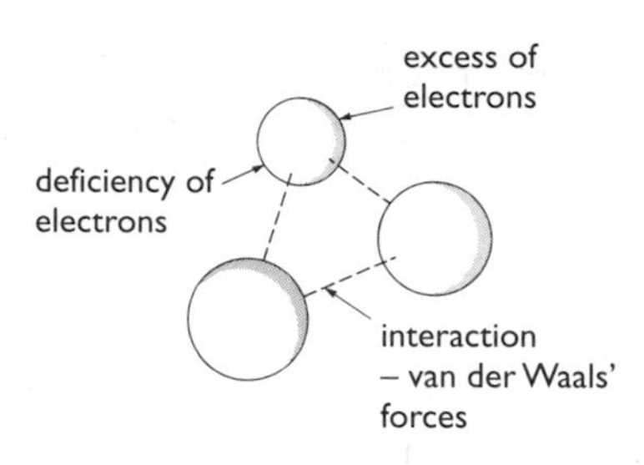

Figure 1

- These forces are van der Waals' forces and are the weakest type of bond but are strong enough to allow the Noble Gases to liquify and solidify if they are cooled.
- Helium, with only two electrons per atom, has the weakest van der Waals' forces between its atoms and is the element with the lowest melting point and boiling point. The other Noble Gases have increasing numbers of electrons so the van der Waals' forces, and hence melting and boiling points, increase.

Groups 7, 6 and 5 – discrete covalent molecules

- In these groups, covalent bond formation results in eight outer electrons. These elements form small, discrete molecules.

Group 7 elements

- Halogen atoms have one unpaired outer electron and can form one covalent bond, forming diatomic molecules F_2, Cl_2, Br_2 and I_2.
- These molecules interact only weakly by van der Waals' forces so that all the elements are volatile, and fluorine and chlorine are gaseous.

van der Waals' forces (weak)

Cl–Cl ··· Cl–Cl

covalent bonds (strong)

Figure 2 Bonding in halogens

Group 6 elements

Oxygen

- Each oxygen atom uses its two unpaired electrons to form two covalent bonds with one other oxygen atom.
- The molecules interact by van der Waals' bonding but since the interaction is weak, O_2 is gaseous.

Sulphur

- The atoms bond to two other atoms. Closed eight membered puckered rings are found in the crystalline forms, and zig-zag chains are found in plastic sulphur.
- Van der Waals' forces between the molecules are strong enough to make sulphur solid at room temperature.

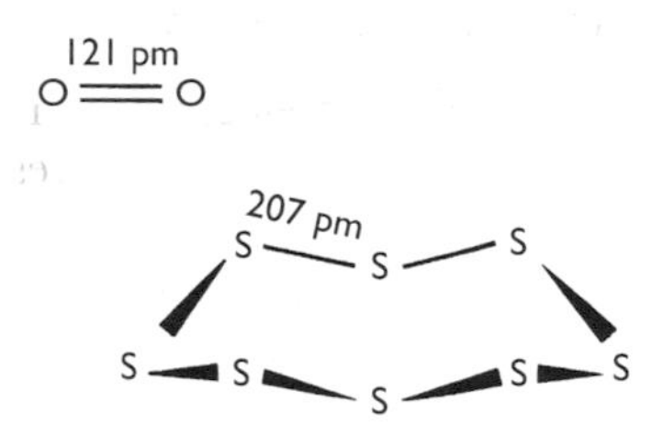

Figure 3 Bonding in oxygen and sulphur

Group 5 elements

Nitrogen

- Nitrogen atoms form diatomic molecules with a triple bond, and weak van der Waals' interaction.

Phosphorus

- Phosphorus makes use of single bonds to three other atoms to form tetrahedral P_4 molecules.

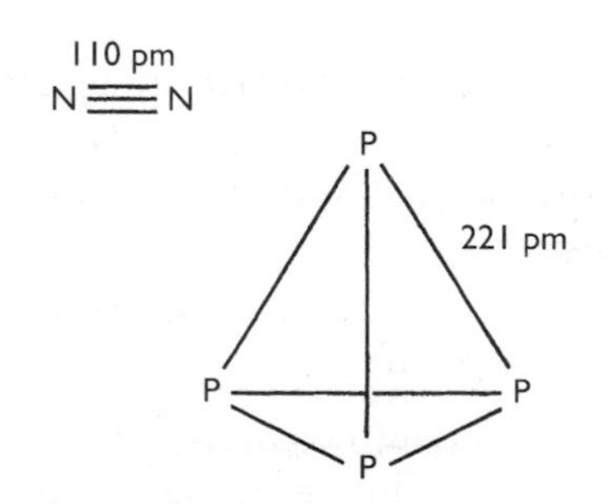

Figure 4 Bonding in nitrogen and phosphorus

- In the elements of Groups 7, 6 and 5, the **intramolecular** forces, i.e. the bonds **within** the molecule are **covalent**.
- The **intermolecular** forces, those **between** the molecules are the very weak van der Waals' forces.

Group 4 elements – usually covalent networks

Diamond

- The standard structure is an infinite three dimensional network or lattice as in diamond and silicon, where each atom bonds covalently to four other atoms. The resultant structure is exceptionally hard and rigid. There are no discrete molecules.
- There are no free electrons to allow conduction but in diamonds, for example, 'tunnels' between the atoms allow light to pass through, thus making them transparent.
- These properties make diamond suitable as an abrasive and a gemstone.

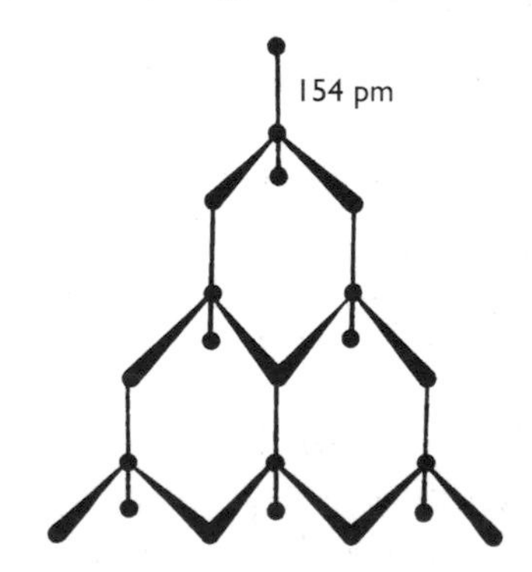

Figure 5 Diamond

Graphite

- This other well-known variety of carbon has three covalent bonds from each atom in one plane, forming layers of hexagonal rings. The fourth unpaired electron from each carbon atom is delocalised. The result is strong bonding within the layers but only weak interaction between the layers.
- Since the delocalised electrons are held quite weakly, they can flow across the layers. Graphite therefore conducts, like a metal. The layers separate easily so graphite is flaky. The layers are offset with respect to each other, light cannot pass through, so graphite is opaque.

- These properties result in the use of graphite in pencils and as a lubricant in electric motors.

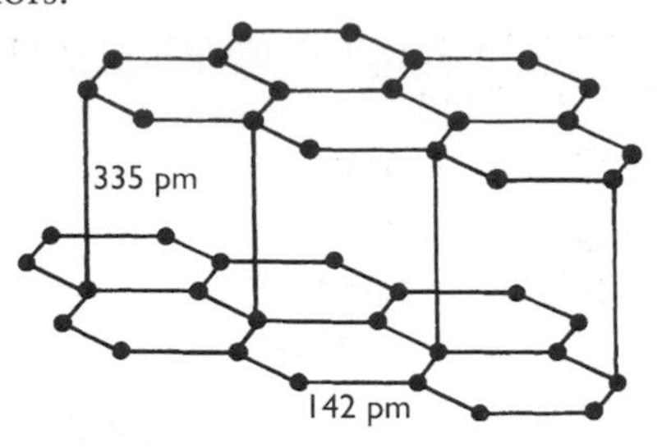

Figure 6 Graphite

Fullerenes

- Discovered in 1985, fullerenes are discrete covalently bonded molecules. The smallest is spherical, C_{60}. Other molecules have elongated shapes e.g. C_{70}, and there are also much longer 'nanotubes'. All contain 5 and 6 membered carbon atom rings.
- The properties of these molecules are under intensive investigation to discover uses.

Figure 7 Fullerene

Groups 1, 2 and 3 – usually metallic bonding

- Elements in these groups have insufficient electrons to allow the achievement of an octet of electrons in their outer layer by covalent bonding.
- Their outer electrons are delocalised, and act as a binding medium for the resultant positive ions. This is **metallic bonding.**
- The bonding is less directional than covalent bonding, and the elements are therefore malleable and ductile.
- The outer electrons move easily and hence the elements are electrical conductors.
- Elements of Groups 1, 2 and 3 are typical metals, except boron which forms a structure made up of B_{12} groups, which are interbonded with other groups. The result is an element almost as hard as diamond.

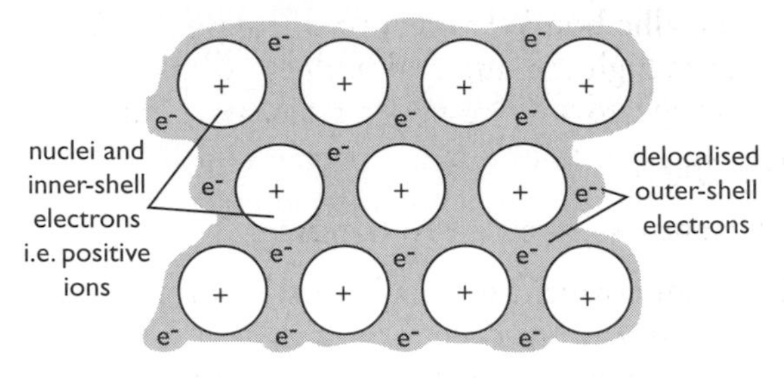

Figure 8 Metallic bonding

Summary of structures of the first 20 elements

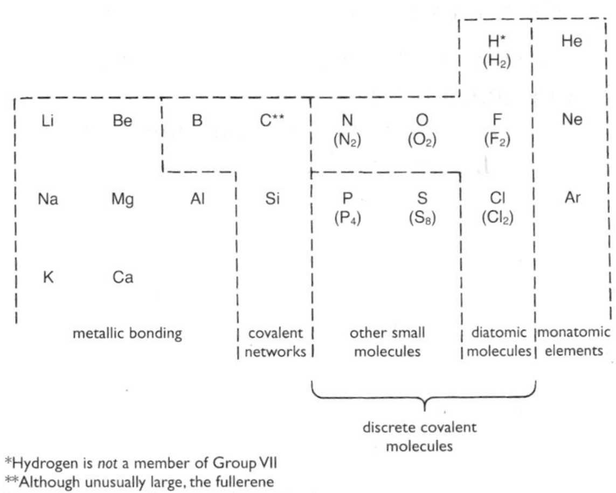

Figure 9

Specific physical properties of elements related to bonding

Melting and boiling points

- Where the elements consist of discrete molecules (the monatomic and diatomic gases and P_4 and S_8) the melting and boiling points are low because only the weak intermolecular van der Waals' forces have to be overcome in melting and boiling the element. The strong, covalent intramolecular forces are unaffected.
- In the covalent network solids, such as carbon and silicon, strong covalent bonds must be broken when melting or boiling

takes place. Melting and boiling points are therefore much higher.

- For Group 1, 2 and 3 elements, strong metallic bonds have to be overcome so they have high melting and boiling points compared with covalent molecular elements.

Hardness

- Hardness is related to bonding.
- Giant covalent molecules have all their atoms interlinked by directional bonds so substances such as diamond are very hard, but the bonds will break on impact so that the substances are brittle.
- Small covalent molecules like S_8 and P_4 are only attracted to each other by van der Waals' forces so the substances are soft.
- Metallic bonds are strong but they are not directional like covalent bonds. Metals can be distorted by impact or pressure i.e. they are malleable and ductile.

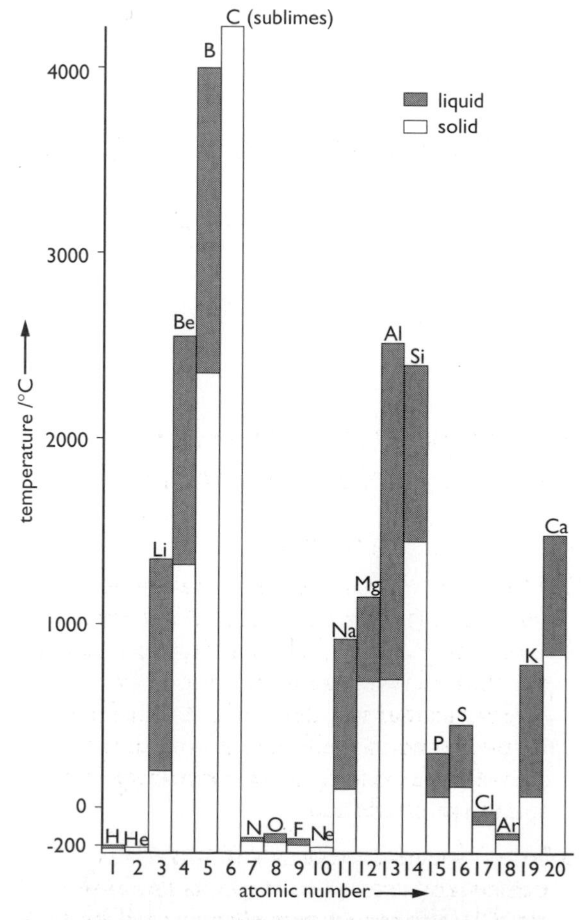

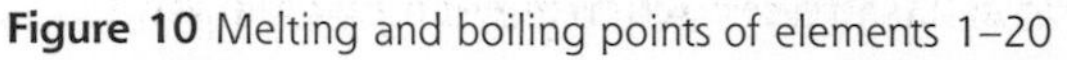

Figure 10 Melting and boiling points of elements 1–20

Questions

1 Which equation represents the first ionisation energy of bromine?

A $Br(l) + e^- \rightarrow Br^-(g)$
B $Br(g) \rightarrow Br^+(g) + e^-$
C $\frac{1}{2} Br_2(l) \rightarrow Br^+(g) + e^-$
D $\frac{1}{2} Br_2(g) + e^- \rightarrow Br^-(g)$

Questions 2 and **3** relate to the following table of data about certain elements.

Element	Melting point/°C	Boiling point/°C	Conduction when solid?
A	44	280	no
B	1083	2567	yes
C	1410	2355	no
D	114	184	no

2 Which of these elements is most likely to have a covalent network structure?

3 Which of these elements is most likely to have delocalised electrons?

4 Which element is a solid at room temperature and consists of discrete molecules?

A Carbon **B** Silicon **C** Sulphur **D** Boron

5* The diagram shows the melting points of successive elements across a period in the Periodic Table.

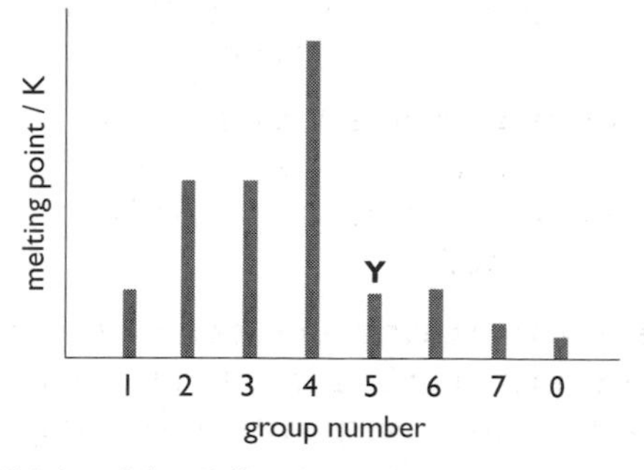

Which of the following is a correct reason for the low melting point of element **Y**?

A It has weak ionic bonds.
B It has weak covalent bonds.
C It has weakly-held outer electrons.
D It has weak forces between molecules.

Use these equations to answer **Questions 6** and **7**.

A $I2(s) \rightarrow I2(g)$

B $I(g) \rightarrow I+(g) + e-$

C $I-(g) \rightarrow I(g) + e-$

D $I2(g) \rightarrow 2I(g)$

6 Which equation represents the first ionisation energy of iodine?

7 Which equation represents a change in which van der Waal's forces are overcome?

Use these structural types to answer **Questions 8** and **9**.

A Discrete covalent molecular solid.

B Covalent network.

C Monatomic gas.

D Positive ions surrounded by delocalised electrons.

8 At room temperature, which structure is possessed by platinum?

9 At room temperature, which structure is possessed by phosphorus?

10 This question refers to elements in the third period of the Periodic Table, i.e. elements with atomic numbers 11 to 18 inclusive.

a) Which of these elements is
 i) diatomic
 ii) monatomic?

b) Why does sulphur have a much lower melting point than silicon?

c) Suggest a reason for aluminium having a higher melting point than sodium.

11 For each of the following changes, name the type of bonding being broken.

a) $C(s) \rightarrow C(g)$
b) $Ne(l) \rightarrow Ne(g)$
c) $Na(l) \rightarrow Na(g)$
d) $F_2(g) \rightarrow 2F(g)$
e) $P_4(l) \rightarrow P_4(g)$

12* The first twenty elements of the Periodic Table are as follows.

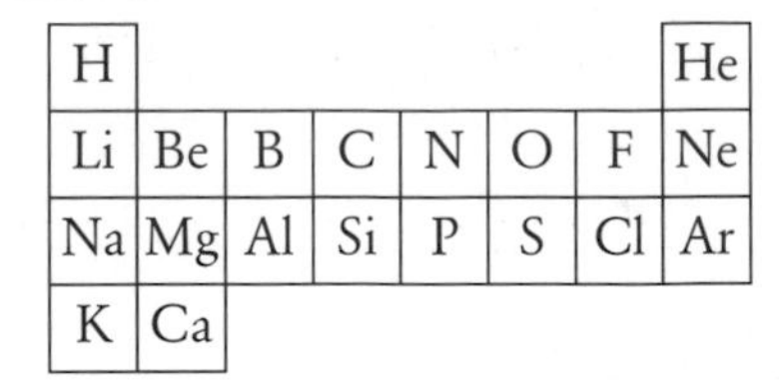

H							He
Li	Be	B	C	N	O	F	Ne
Na	Mg	Al	Si	P	S	Cl	Ar
K	Ca						

a) From the first twenty elements of the Periodic Table, give **one** example of a covalent molecular solid at room temperature.

b) **i)** Name the type of structure in silicon.
 ii) Why do elements with this type of structure have high melting points?

c) Sodium and magnesium both contain metallic bonding. Why is the bonding in magnesium stronger than that in sodium?

13 The following table shows the radii, measured in picometres (pm), of ions of certain elements.

Na^+	Mg^{2+}	Al^{3+}	Si^{4-}	P^{3-}	S^{2-}	Cl^-	K^+	Ca^{2+}
95	65	50	271	212	184	181	133	99

a) Explain why an aluminium ion is much smaller than an aluminium atom.

b) Why is there a large increase in ionic radius between aluminium and silicon?

c) What do all the ions from silicon to calcium have in common?

d) Why do the ionic radii decrease from silicon to calcium?

5 Bonding, Structure and Properties of Compounds

Types of bonding in compounds

Ionic bonding

Ionic bonding is an electrostatic attraction between the positive ions of one element and the negative ions of another element.

Polar covalent bonding

Polar covalent bonding results from electron sharing but if the atoms being joined are not of the same element, electrons are not always shared equally, so producing a polar bond.

Intermolecular bonds

Van der Waals' forces mentioned in the last chapter are one type of intermolecular bond. Other types of these bonds include **permanent dipole-permanent dipole** interactions and **hydrogen bonds**.

Ionic bonding and structure

- Different elements have different attractions for bonding electrons i.e. different electronegativities.
- The greater their difference in electronegativity, the more likely are two elements to form ionic bonds. The element with the greater electronegativity is more likely to gain electrons to form a negative ion and the element with the smaller to lose electrons to form a positive ion.
- Elements far apart in the Periodic Table are more likely to form ionic bonds i.e. ionic compounds result from metals combining with non metals (e.g. sodium fluoride and magnesium oxide). Caesium fluoride has the greatest degree of ionic bonding.
- Electrostatic attraction holds the oppositely charged ions together in appropriate numbers so that the total charge is zero. For example in sodium chloride, there are equal numbers of Na^+ and Cl^- ions, in calcium fluoride there are twice as many F^- ions as Ca^{2+}.
- Ionic compounds do **not** form molecules. Instead the positive and negative ions aggregate into various three dimensional structures called **lattices**.

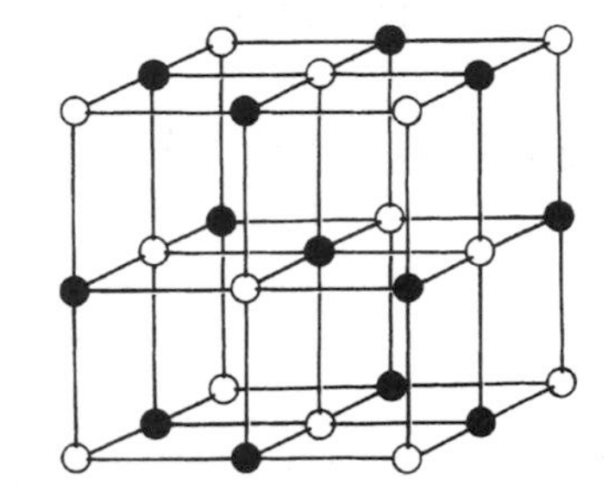

Figure 1 Sodium chloride lattice

Covalent compounds and structure

- Most covalent compounds are formed by combinations of non-metallic elements.
- Most are **molecular** compounds such as methane and carbon dioxide.

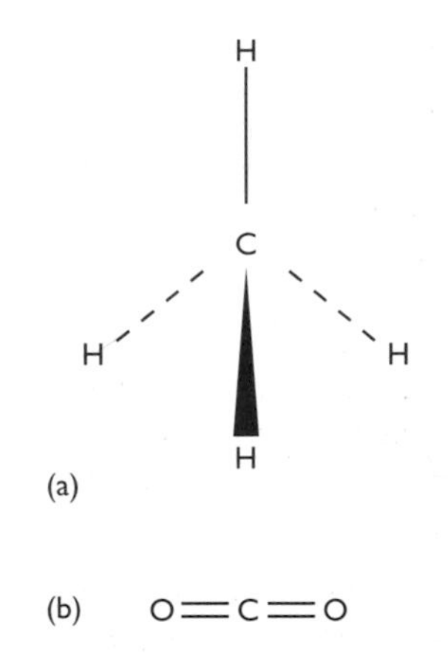

Figure 2 Bonding in methane (a) and carbon dioxide (b)

- Molecular compounds such as CH_4, CO_2 and SiH_4 (silane) are gaseous at room temperature since there are only weak **intermolecular** forces. Melting and boiling of these compounds require only the overcoming of these weak intermolecular forces, whilst the covalent **intramolecular** forces remain intact.

- A number of covalent compounds occur as **network** structures, e.g. silicon carbide and silicon dioxide.
- Silicon carbide, SiC, has a structure similar to that of diamond and is also an abrasive.
- Silicon dioxide, SiO_2, is formed by bonding silicon atoms to four oxygen atoms to give SiO_4 tetrahedra which are linked by sharing of each of their oxygen atoms between two silicon atoms.

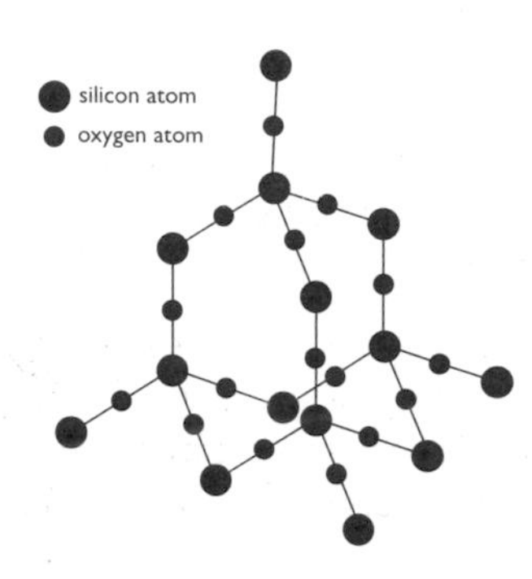

Figure 3 Structure of silicon dioxide

- Network solids have very high melting points since melting requires the breaking of strong covalent bonds. They are very hard, although brittle, because of these same strong directional bonds.

Polar covalent bonding

- Most compounds which are covalent are formed by elements with different electronegativities, although not so different as those forming ionic bonds.
- In these compounds, the bonding electrons are not shared equally. The atom with the greater share of electrons has a slight negative charge by comparison with the other atom.

For example chlorine, oxygen and nitrogen are all more electronegative than hydrogen.

Hydrogen chloride, water and ammonia can be represented as in Figures 4 (a), (b) and (c). The symbols δ^+ and δ^- mean 'slightly positive' and 'slightly negative'.

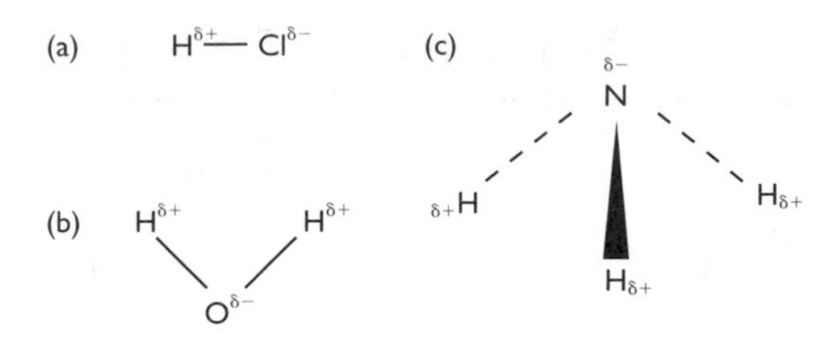

Figure 4 Bonding in hydrogen chloride (a), water (b) and ammonia (c)

- Covalent bonds with unequal electron sharing are called **polar covalent bonds.**
- Some molecules containing such polar bonds have an overall polarity because the bonds are not arranged symmetrically, e.g. HCl, H_2O and NH_3. Such molecules are said to have a **permanent dipole** and are described as **polar**.
- Other molecules have a symmetrical arrangement of polar bonds and the polarity cancels out on the molecules as a whole, as in carbon dioxide and tetrachloromethane (Figure 5).

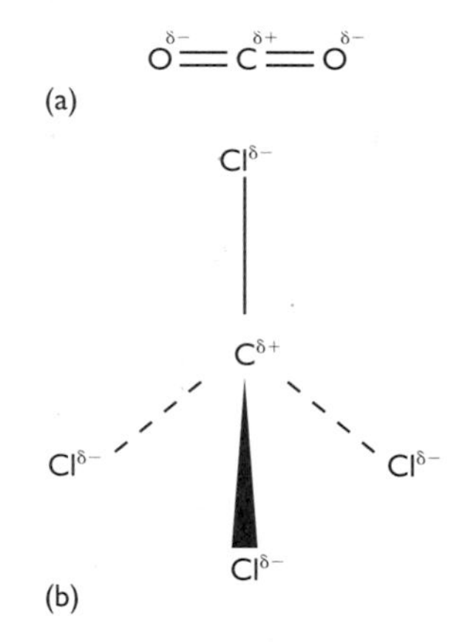

Figure 5 Bonding in carbon dioxide (a) and tetrachloromethane (b)

- Heptane has almost **non-polar** bonds and the molecule is non-polar as a whole. Ethanol has one very polar O–H bond giving the molecule an overall polarity. Chloroform, unlike CCl_4 is polar overall because of the unsymmetrical arrangement of the C–Cl bonds (Figure 6).

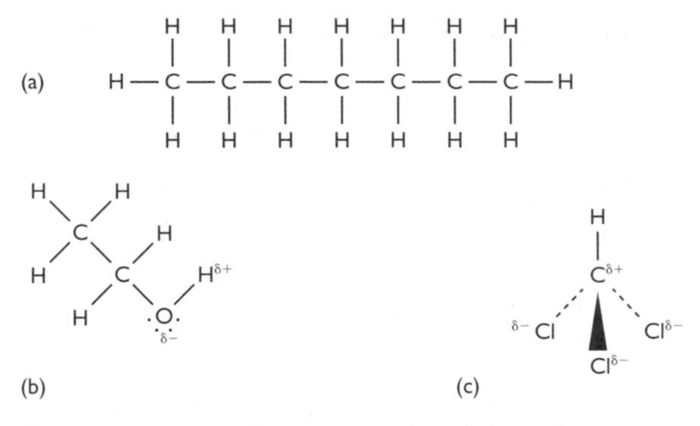

Figure 6 Bonding in heptane (a), ethanol (b) and trichloromethane (chloroform) (c)

Melting and boiling points

- Polar molecules have melting and boiling points higher than those of non-polar molecules of a similar molecular mass. The intermolecular forces are increased by the mutual attraction of the permanent dipoles on neighbouring molecules. Propanone and butane (Figure 7) are good examples of this behaviour.

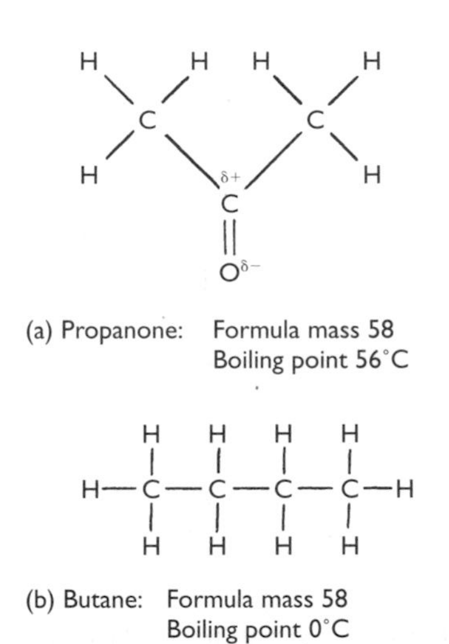

Figure 7

- **All** covalent molecular compounds interact by van der Waals' bonding.
- Permanent dipole-permanent dipole attractions are stronger than van der Waals' bonds for molecules of equivalent size.

Solvent action

- Because it is polar, water can dissolve other polar and ionic substances.
- Polar and ionic substances are more likely to be soluble in polar solvents and insoluble in non-polar solvents e.g. salt dissolves in water but not in heptane.
- Non-polar substances are more likely to be soluble in non-polar solvents and insoluble in polar solvents e.g. wax dissolves in heptane but not in water.

Anomalous physical properties of some hydrides

Boiling points

- Group IV hydrides have, as expected, an increase in melting and boiling points with molecular mass.
- In the Groups V, VI and VII, the values of melting and boiling points for NH_3, H_2O and HF (and HCl to some extent) are higher than would be expected for their molecular mass.

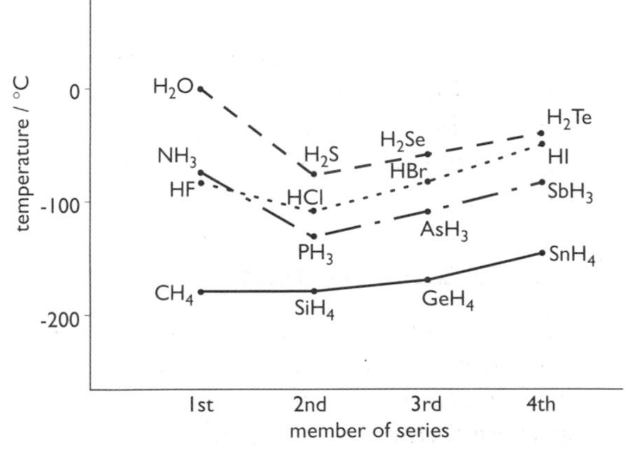

Figure 8 Melting points of hydrides of groups IV, V, VI and VII

- These anomalous properties indicate stronger bonding between the molecules than the expected van der Waals' bonding and simple permanent dipole-permanent dipole attraction.
- The compounds showing these anomalous properties all contain bonds which are very polar i.e. O–H, N–H, F–H (and Cl–H) as shown by the electronegativity differences in Table 1 on the next page and can therefore interact in the fashion shown in Figure 10.

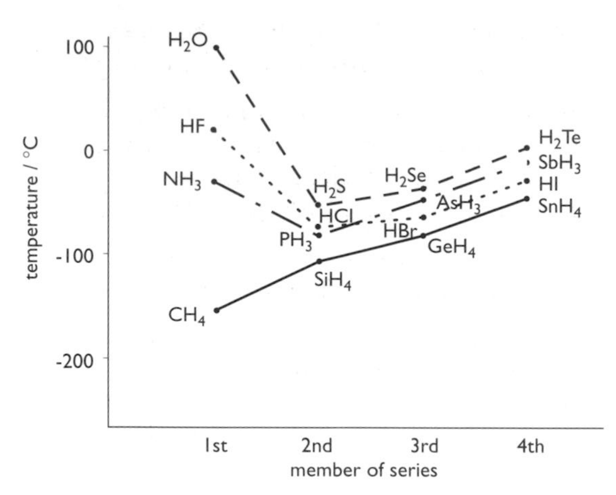

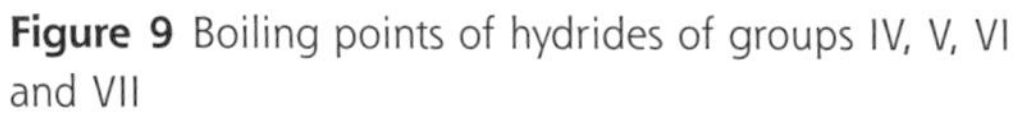
Figure 9 Boiling points of hydrides of groups IV, V, VI and VII

H–C	H–N	H–O	H–F
0.3	0.8	1.3	1.8
	H–P	H–S	H–Cl
	0.0	0.3	0.8
		H–Se	H–Br
		0.2	0.6
		H–Te	H–I
		0.0	0.4

Table 1 Electronegativity differences

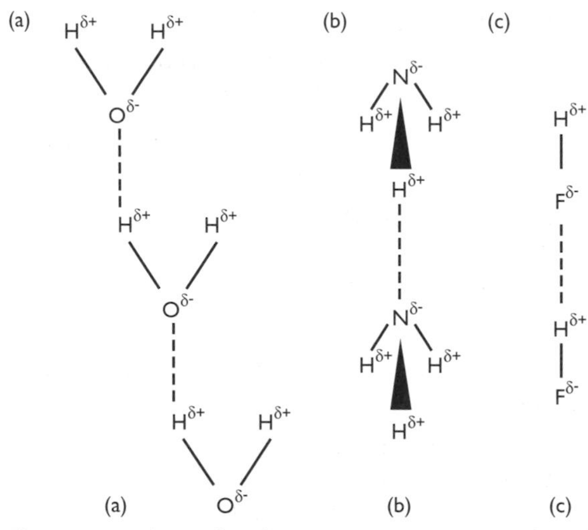

Figure 10 Hydrogen bonding in water (a), ammonia (b) and hydrogen fluoride (c)

- This interaction is called **hydrogen bonding** since it occurs only for compounds containing a strongly electronegative element linked to hydrogen. The pull of electrons away from the hydrogen results in a positive charge located on a small atom, and hence a high positive charge density capable of interacting with the negative end of other molecules.
- Hydrogen bonding is stronger than van der Waals' bonding, and stronger than ordinary permanent dipole-permanent dipole attractions but weaker than covalent bonds.

Viscosity

Viscosity normally increases with molecular mass but molecules with, for example, OH groups show higher viscosity than expected, i.e. hydrogen bonding occurs between molecules and increases viscosity.

Miscibility

- Miscible liquids mix thoroughly without any visible boundary between them, e.g. ethanol and water, but water and hexane are immiscible, with the hexane forming a visible upper layer.
- Hydrogen bonding owing to –OH groups in water and ethanol aids miscibility, but other polar liquids like propanone, although without –OH groups, are frequently miscible with water.

Density of water

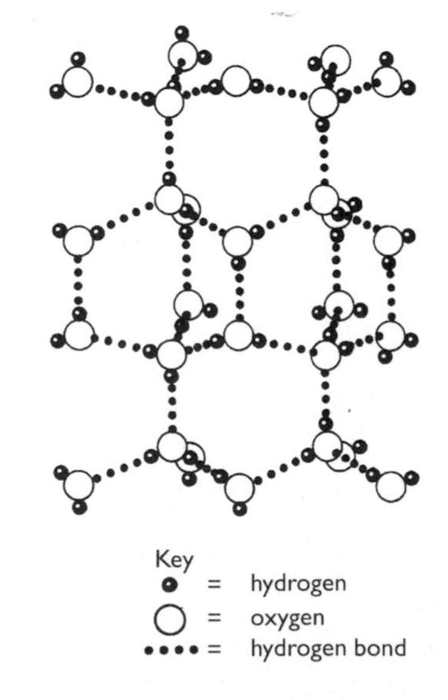

Figure 11 Structure of ice

- As with all liquids, water contracts on cooling, but when it reaches 4°C it begins to expand again, and at its freezing point is less dense than the water which is about to freeze. The reason for this is the ordering of molecules, caused by hydrogen bonding, into an open lattice.
- As a result ice floats on water, seas freeze from the top downwards, allowing fish to survive in unfrozen water beneath and pipes burst when water freezes inside them.

A summary of Bonding and Structure in elements and compounds is provided on pages 139–141.

Questions

1 Which of the following fluorides is likely to have the **most** ionic character?

A NaF B MgF_2 C CsF D BaF_2

2* What type of bond is broken when ice is melted?

A Ionic B Polar covalent
C Hydrogen D Non-polar covalent

3 Which type of structure describes a substance which conducts electricity when solid and melts at 1500°C?

A Covalent network B Monatomic
C Ionic lattice D Metallic

4 Which of the following is an example of intramolecular bonding?

A Polar covalent B Hydrogen
C Van der Waals' D Ionic

5 Which of the following solids is most likely to dissolve in tetrachloromethane, CCl_4?

A Calcium chloride
B Copper(II) chloride
C Potassium chloride
D Phosphorus(V) chloride

6* Silicon carbide can be used as

A a lubricant
B a tip for cutting/grinding tools
C a substitute for pencil 'lead'
D an electrical conductor.

7* Which of the following shows the types of bonding in **decreasing** order of strength?

A Covalent : hydrogen : van der Waals'
B Covalent : van der Waals' : hydrogen
C Hydrogen : covalent : van der Waals'
D Van der Waals' : hydrogen : covalent

8* Carbon dioxide is a gas at room temperature while silicon dioxide is a solid because

A van der Waals' forces are much weaker than covalent bonds
B carbon dioxide contains double covalent bonds and silicon dioxide contains single covalent bonds
C carbon–oxygen bonds are less polar than silicon–oxygen bonds
D the relative formula mass of carbon dioxide is less than that of silicon dioxide.

9 Which of the following compounds has non-polar molecules?

A HBr B CO_2 C H_2O D CH_3Cl

Questions **10** and **11** relate to the following equations.

A $H_2(g) \rightarrow 2H(g)$
B $H_2(l) \rightarrow H_2(g)$
C $H_2O(g) \rightarrow 2H(g) + O(g)$
D $H_2O(l) \rightarrow H_2O(g)$

10 Which equation represents a reaction in which hydrogen bonds are broken?

11 Which equation represents a reaction in which non-polar covalent bonds are broken?

12 Which type of intramolecular bond is present in a sample of phosphine, PH_3?

A van der Waals'.
B covalent.
C polar covalent.
D ionic.

13 Which statement applies to ammonia but not to methane?

A It conducts electricity.
B It has covalent bonding.
C It has hydrogen bonding.
D Van der Waal's forces are present.

14 Iodine monochloride, ICl, is a brown liquid produced when iodine reacts with chlorine.

a) Write a balanced equation, including state symbols, for this reaction.

b) Copy and complete the following table in which iodine monochloride is compared with bromine, which is also a brown liquid.

	Iodine monochloride	**Bromine**
Molecular mass		159.8
Boiling point/°C	97	
Type of bonding within molecules		Covalent (non-polar)
Type of bonding between molecules		

c) Explain why iodine monochloride has a much higher boiling point than bromine.

15 The structural formula of an amino acid called cysteine is shown below.

```
      H   H   O
      |   |   ||
  H—N—C—C—O—H
          |
      H—C—H
          |
          S—H
```

Refer to the SQA Data Book to help you decide which of the following bonds present in cysteine is

a) non-polar **b)** the most polar.

H – N N – C C – S S – H

16 The following flow diagram summarises one way of purifying silicon.

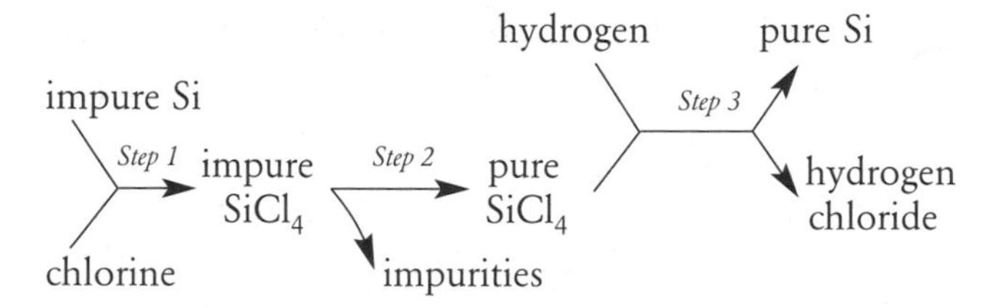

a) Write balanced equations for the reactions occurring at

i) *step 1* **ii)** *step 3*.

b) Which of the chemicals shown in the flow chart has the most polar bonds?

c) $SiCl_4$ is a liquid at room temperature while silicon is a solid with a very high melting point. Why are these substances so different?

d) Draw a diagram to show the shape of a molecule of $SiCl_4$.

e) Hydrogen chloride is very soluble in water.

i) What are the products of electrolysis of this solution?

ii) Why might electrolysis of this solution affect the economics of the method of purifying silicon described above?

17 The following table contains data relating to the hydrides of Group 5 elements.

Compound	Ammonia NH_3	Phosphine PH_3	Arsine AsH_3	Stibnine SbH_3
Molecular mass	17	34	77.9	124.8
Boiling point/°C	-33	-88	-55	-17

a) Why does arsine have a higher boiling point than phosphine?

b) On a graph of temperature against molecular mass, plot the boiling points of phosphine, arsine and stibnine.

c) Predict the boiling point of ammonia, **if** it followed the trend shown on the graph.

d) Why is the boiling point of ammonia much higher than this prediction?

18 Although propane and ethanol have similar molecular masses, the alkane is a gas at room temperature while the alcohol is a liquid. This difference in boiling points is due to the different strengths of the intermolecular forces in the two compounds.

Explain why propane is a gas at room temperature while ethanol is a liquid. In your answer you should name the intermolecular forces involved in each compound and explain how they arise.

6 The Mole

The Avogadro Constant

- A **mole** of a substance is its **gram formula mass**, gfm.
- One mole of any substance contains Avogadro's constant, **L**, i.e. 6.02×10^{23}, formula units. The term 'formula unit' relates to the type of particle present in the substance.

Metals and monatomic species e.g Noble Gases.

- Here the formula unit is an **atom**.

Thus, 20.2 g of neon and 23.0 g of sodium each contain **L** (6.02×10^{23}) atoms.

Covalent substances

- Here the formula unit is a **molecule**.
- The total number of atoms in a mole of molecules can be found by multiplying **L** by the number of atoms in each molecule.

For example, 32 g oxygen, O_2, has L molecules. The number of atoms present is 2L.

44 g propane, C_3H_8, has L molecules. The number of atoms present is 11L (3L of C & 8L of H).

Ionic compounds

- Here the formula unit consists of the ratio of **ions** expressed by the ionic formula of the compound. The total number of ions depends on the number of each kind of ion in the formula.

For example, 74.6 g K^+Cl^- contains L formula units comprising L K^+ & L Cl^-, i.e. 2L ions.

400 g $(Fe^{3+})_2(SO_4^{2-})_3$ contains L formula units comprising 2L Fe^{3+} & 3L SO_4^{2-} i.e. 5 L ions.

- Equimolar quantities of different substances contain equal numbers of formula units. Each of the following quantities represents 0.2 moles of the substance and as a result contains 0.2L formula units.

4.6 g of sodium, Na, contain 0.2L **atoms**.

5.6 g of nitrogen gas, N_2, contain 0.2L **molecules**.

22.2 g of calcium chloride, $CaCl_2$, contain 0.2L **formula units**.

Worked Example 6.1

1 g H_2 0.2 mol C_2H_6 6 g H_2O

Which of the quantities given above contains

a) the most molecules

b) the most hydrogen atoms?

1 g H_2 = 0.5 mol, so it contains 0.5L molecules or L H atoms.

0.2 mol C_2H_6 contains 0.2L molecules, 1.2L H atoms (6×0.2).

6 g H_2O = 0.33 mol, so it contains 0.33L molecules, 0.66L H atoms.

Answers: **a)** 1 g H_2 **b)** 0.2 mol C_2H_6

Worked Example 6.2

a) What mass of $Ca_3(PO_4)_2$ contains 6.02×10^{23} positive ions?

b) What volume of 0.5 mol l^{-1} $Cu(NO_3)_2$ contains 3.01×10^{23} negative ions?

Answers:

a) 1 mole of $Ca_3(PO_4)_2$ = 310 g and contains 3L Ca^{2+} ions, so 103.3g of the compound contains 6.02×10^{23} positive ions.

b) 1 mole of $Cu(NO_3)_2$ contains 2L nitrate ions, so 0.25 mol $Cu(NO_3)_2$ contains $\frac{L}{2}$, 3.01×10^{23} nitrate ions.

$V = \frac{n}{c} = \frac{0.25}{0.5} = 0.5$

Hence 0.5 litres of this solution contains 3.01×10^{23} nitrate ions.

Molar volume

- With gases it is usually more appropriate to measure volume rather than mass. The molar volume of any gas at 0°C and one atmosphere pressure can be calculated by dividing its molar mass by its density at that temperature and pressure.

For example, fluorine, F_2, gfm = 38 g and density at 0°C and 1 atmosphere pressure = 1.7 g litre^{-1}.

Hence, molar volume = $\frac{38}{1.7}$ = 22.4 litres.

- The densities of gases increase in proportion to their gram formula masses. Consequently the molar volumes of different gases are approximately the same for all gases at the same temperature and pressure.
- Since the volume of a gas changes if the temperature and/or the pressure changes, it is important to specify the temperature and pressure at which a volume is being measured. At room temperature and pressure, i.e. 20°C and 1 atmosphere pressure, the molar volume of any gas is approximately 24 litres mol^{-1}.

Worked Example 6.3

Calculate the molar volume of oxygen from the following data obtained at 20°C and 1 atmosphere pressure.

Mass of empty flask	= 205.42 g
Mass of flask + oxygen	= 206.70 g
Volume of flask	= 960 cm^3

Calculation:

Mass of oxygen = 1.28 g
(206.70 – 205.42)
1.28 g of oxygen occupies a volume of 960 cm^3.
Gram formula mass of O_2 = 32 g

32 g of O_2 occupies a volume of

$$\frac{960 \times 32 \text{ cm}^3}{1.28} = 24\,000 \text{ cm}^3$$

Hence, the molar volume of O_2 is 24.0 litres at 20°C & 1 atmosphere pressure.

- The relationship between volume of a gas, number of moles and molar volume can be expressed as follows.

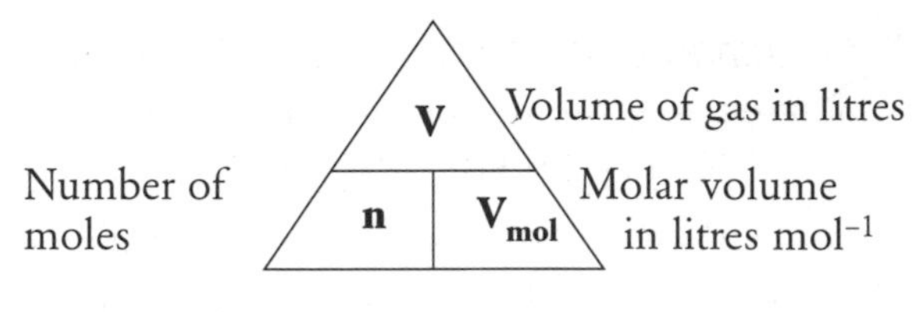

Hence, volume of gas, $\mathbf{V = n \times V_{mol}}$ and number of moles, $\mathbf{n = \frac{V}{V_{mol}}}$

Worked Example 6.4

The molar volume at 0°C and 1 atmosphere pressure is 22.4 litres mol^{-1}.

Calculate:

a) the volume of 0.125 mol of hydrogen

b) the number of moles of oxygen in 560 cm^3 under these conditions.

Answers:

a) Volume of hydrogen,
$V = n \times V_{mol}$
$= 0.125 \times 22.4$
$= 2.8$ litres

b) 560 cm^3 = 0.560 litres.

Number of moles of oxygen,

$$n = \frac{V}{V_{mol}} = \frac{0.560}{22.4} = 0.025$$

Reacting volumes

There are two main types of calculation involving molar volumes.

1 Calculations involving mass and volume

Worked Example 6.5

Calculate the volume of carbon dioxide released when 0.4 g of calcium carbonate is dissolved in excess hydrochloric acid. The gas is collected at room temperature and pressure. The molar volume is 24 litres.

The equation for the reaction is:

$CaCO_3(s) + 2HCl(aq) \rightarrow CaCl_2(aq) + CO_2(g) + H_2O(l)$

1 mol $CaCO_3$ (100 g) gives 1 mol CO_2 (24 litres)

Hence, 0.4 g of $CaCO_3$ gives

$\frac{0.4}{100} \times 24$ litres

= 0.096 litres (or 96 cm^3) of CO_2.

2 Comparing volumes of gases

Worked Example 6.6

Calculate **a)** the volume of oxygen required for the complete combustion of 200 cm^3 of butane
b) the volume of each product.

(All volumes are measured at 150°C and 1 atmosphere pressure.)

	$2C_4H_{10}(g)$ +	$13O_2(g)$ →	$8CO_2(g)$ +	$10H_2O(g)$
Mole ratio of reactants and products:	2	13	8	10
Volume ratio of reactants and products (at same T and P):	2	13	8	10
Simplified volume ratio:	1	6.5	4	5

Hence 200 cm^3 of butane **a)** requires 1300 cm^3 of oxygen for complete combustion
b) produces 800 cm^3 of $CO_2(g)$ and 1000 cm^3 of $H_2O(g)$.

- In many reactions which involve gases one or more of the reactants or products may be a liquid or a solid. The volume of liquid or solid is so small compared with that of the gas or gases that it can be regarded as negligible.

If the reaction in Worked Example 6.6 is carried out when all the volumes are measured at room temperature and pressure the products include water. Since water is a liquid its volume can be ignored.

Gas volume calculations involving excess reactant

Worked Example 6.7 illustrates the application of excess reactant to the previous type of calculation. It is necessary to find out which reactant is in excess before calculating the volume of gaseous product.

Worked Example 6.7

A mixture of 20 cm^3 of methane and 50 cm^3 of oxygen was ignited and allowed to cool. Calculate the volume and composition of the resulting gaseous mixture. All volumes are measured under the same conditions of room temperature and pressure.

	$CH_4(g)$	+	$2O_2(g)$	→	$CO_2(g)$	+	$2H_2O(l)$
Mole ratio:	1		2		1		2
Volume ratio:	1		2		1		negligible

According to the equation,

20 cm^3 of methane requires 2×20 cm^3 of oxygen, i.e. 40 cm^3.

Hence, oxygen is present in excess since its initial volume is 50 cm^3.

Volume of excess oxygen $= (50 - 40) = 10$ cm^3

Volume of carbon dioxide formed $= (1 \times 20) = 20$ cm^3

Therefore the resulting gas mixture consists of 10 cm^3 of O_2 and 20 cm^3 of CO_2.

Questions

SECTION A:

The Avogadro Constant ($L = 6.02 \times 10^{23}$)

1 The Avogadro constant is the same as the number of

A molecules in 1 mol of iodine crystals
B atoms in 1 mol of chlorine gas
C ions in 1 mol of KBr crystals
D protons in 1 mol of helium gas.

2 60 g of carbon contain as many atoms as

A 190 g of fluorine
B 100 g of calcium
C 60 g of cobalt
D 20 g of helium.

3 The number of moles of ions in 1 mol of aluminium nitrate is

A 1 B 2 C 3 D 4

4 The Avogadro constant is the same as the number of

A molecules in 14 g of nitrogen gas
B ions in 500 cm^3 of 1 mol l^{-1} KI(aq)
C atoms in 2 g of hydrogen gas
D molecules in 200 g of $C_7H_{16}(l)$.

Questions 5, 6 and **7** relate to the following quantities

A 12 g of CH_4 B 6 g of H_2O
C 5 g of C_2H_6 D 17 g of NH_3

5 Which of the above quantities contains 6.02×10^{23} molecules?

6 Which of the above quantities contains L atoms?

7 Which of the above quantities contains one mole of hydrogen atoms?

8 200 cm^3 of 3 mol l^{-1} $ZnCl_2$ solution contains the same number of chloride ions as

A 100 cm^3 of 2 mol l^{-1} $FeCl_3$
B 200 cm^3 of 3 mol l^{-1} NaCl
C 800 cm^3 of 1.5 mol l^{-1} HCl
D 400 cm^3 of 2 mol l^{-1} $MgCl_2$

9 The number of sodium ions in 212 g of sodium carbonate is approximately

A 6.0×10^{23} B 1.2×10^{24}
C 2.4×10^{24} D 3.6×10^{24}

10 A mixture of sodium chloride and sodium carbonate contains 2 mol of chloride ions and 1 mol of carbonate ions. How many moles of sodium ions are present?

A 1 B 2 C 3 D 4

11* A mixture of magnesium bromide and magnesium sulphate is known to contain 3 mol of magnesium and 4 mol of bromide ions. How many moles of sulphate ions are present?

A 1 B 2 C 3 D 4

12 How many hydrogen atoms are there in 0.25 mol of methane? ($L = 6.02 \times 10^{23}$)

A 0.25 L B 0.5 L C 1.0 L D 2.0 L

13 Identify the quantity of substance containing the greatest number of atoms.

A 4g H_2
B 12g C
C 8g O_2
D 16g CH_4

14 0.2 mol of $CuSO_4$ and 0.1 mol of Na_2SO_4 were dissolved in water and the solution made up to 500cm^3. Which is the true statement?

A The solution contained 0.1 mol of sodium ions.
B The solution contained equal numbers of positive and negative ions.
C The concentration of copper(II) ions in the solution is 0.4 mol l^{-1}.
D The concentration of sulphate ions in the solution is 0.8 mol l^{-1}.

15 The value for the Avogadro Constant is 6.02×10^{23} mol^{-1}. Identify the true statement.

A 24g of carbon contains 6.02×10^{23} atoms.
B 500cm^3 of 2.0 mol l^{-1} sodium hydroxide solution contains 6.02×10^{23} sodium ions.
C 9g of water contains 6.02×10^{23} atoms.
D 44g of carbon dioxide contains 6.02×10^{23} oxygen atoms.

16 If the Avogadro Constant is identified by L, which is the true statement?

A There are L atoms in 0.5 mol of neon gas.
B There are L electrons in 0.5 mol of hydrogen gas.
C There are L oxide ions in 0.5 mol of potassium oxide.
D There are L carbon atoms in 0.5 mol of carbon dioxide.

17 Hydrazine, N_2H_4
Hydrogen sulphide, H_2S Benzene, C_6H_6

For each of the above compounds calculate the mass that contains

a) 6.02×10^{23} molecules
b) 6.02×10^{23} atoms of hydrogen.

18 a) Using L to represent the Avogadro constant, calculate the number of

i) positive ions in 82 g of sodium phosphate, Na_3PO_4
ii) negative ions in 328 g of calcium nitrate, $Ca(NO_3)_2$
iii) ions in total in 42 g of ammonium dichromate, $(NH_4)_2Cr_2O_7$

b) What volume of 2 mol l^{-1} sulphuric acid contains

i) L positive ions
ii) L negative ions?

SECTION B: Gas volumes

1 The density of oxygen gas at 50°C and 1 atmosphere pressure was found to be 1.2 g l^{-1}.
Under these conditions, the molar volume in litres is

A 13.3 B 24.0 C 26.7 D 32.0

2 $2NO(g) + O_2(g) \rightarrow 2NO_2(g)$

How many litres of nitrogen dioxide could be obtained by mixing 1.5 litres of nitrogen monoxide and 0.5 litres of oxygen? (All volumes are measured under the same conditions of temperature and pressure.)

A 0.5 B 1.0 C 1.5 D 2.0

3 Which of the following has the same volume as 16 g of oxygen gas? (All volumes are measured under the same conditions of temperature and pressure.)

A 4 g of helium gas
B 16 g of methane gas
C 2 g of hydrogen gas
D 14 g of carbon monoxide gas

4 Using the density given in the SQA Data Book how many moles of chlorine molecules are there in a 2 litre container?

A 0.09 B 0.18 C 0.29 D 0.38

Questions 5 and **6** relate to the decomposition of sodium hydrogencarbonate when heated, as shown in the following equation.

$$2NaHCO_3(s) \rightarrow Na_2CO_3(s) + CO_2(g) + H_2O(l)$$

In an experiment 0.1 mol of sodium hydrogencarbonate was heated until no further change took place.

5 The volume of gas obtained after cooling to room temperature was (V_{mol} = 24 litres mol^{-1})

A 0.6 litres B 1.2 litres
C 1.8 litres D 2.4 litres

6 The mass of remaining solid was

A 5.3 g B 8.4 g C 10.6 g D 13.7 g

7 In which reaction is the volume of the product equal to the volume of the reactants?

A $2H_2(g) + O_2(g) \rightarrow 2H_2O(g)$
B $3H_2(g) + N_2(g) \rightarrow 2NH_3(g)$
C $N_2(g) + O_2(g) \rightarrow 2NO(g)$
D $2SO_2(g) + O_2(g) \rightarrow 2SO_3(g)$

8 The equation for the complete combustion of ethene is

$$C_2H_4(g) + 3O_2(g) \rightarrow 2CO_2(g) + 2H_2O(l)$$

50 cm^3 of ethene is mixed with 200 cm^3 of oxygen and the mixture is ignited. What is the volume of the resulting gas mixture? (All volumes are measured under the same conditions of temperature and pressure.)

A 100 cm^3 B 150 cm^3
C 200 cm^3 D 250 cm^3

Questions 9 and **10** refers to the reaction between hydrogen gas and iodine vapour as shown in the following equation.

$$H_2(g) + I_2(g) \rightarrow 2HI(g)$$

1.2 mol of hydrogen gas and 1.0 mol of iodine vapour were mixed and allowed to react. After 5 minutes, 0.6 mol of iodine remained.

9 The number of mol of hydrogen which remained at this time was

A 1.0 B 0.8 C 0.6 D 0.4

10 The number of mol of hydrogen iodide produced after 5 minutes was

A 1.6 B 1.2 C 0.8 D 0.4

11 A mixture of 20 cm^3 of methane and 30 cm^3 of oxygen at 130°C was ignited.

$$CH_4 + 2O_2 \rightarrow CO_2 + 2H_2O$$

Which is the true statement about this experiment, assuming that all volumes were measured at the same temperature and pressure?

A Oxygen was in excess by 10 cm^3.
B Methane was in excess by 5 cm^3.
C 20 cm^3 of carbon dioxide were produced.
D 40 cm^3 of water vapour were produced.

12 A mixture containing 0.2 mol magnesium and 0.1 mol magnesium carbonate was added to 200 cm^3 of 5.0 mol l^{-1} hydrochloric acid. Gas volumes were measured at room temperature and pressure (V_{mol} = 24 litres mol^{-1}). Identify the true statement about this experiment.

A There was 0.4 mol of excess acid.
B 1.2 litres of hydrogen gas were produced.
C 1.2 litres of carbon dioxide were produced.
D 0.6 mol of magnesium chloride were produced.

13 Hydrogen sulphide gas reacts with zinc(II) nitrate solution, forming a precipitate of zinc(II) sulphide and a solution of nitric acid. The equation for the reaction is:

$$Zn(NO_3)_2(aq) + H_2S(g) \rightarrow ZnS(s) + 2HNO_3(aq)$$

Hydrogen sulphide gas was passed through 200 cm^3 of 0.5 mol l^{-1} zinc(II) nitrate solution until no further reaction occurred. The precipitate was removed by filtration, washed, dried and weighed.

a) Hydrogen sulphide is a highly poisonous gas. What safety precaution should be taken during this experiment?
b) Calculate the volume of hydrogen sulphide required for complete reaction. (V_{mol} = 24 litres mol^{-1})
c) Calculate the mass of precipitate obtained.
d) Assuming that there is no loss of water during the experiment, what will be the concentration of the filtrate?

14 Silver(I) carbonate decomposes when heated according to the following equation.

$$2Ag_2CO_3 \rightarrow 4Ag + 2CO_2 + O_2$$

In an experiment, 2.758 g of silver(I) carbonate was heated as shown in the diagram until decomposition was complete.

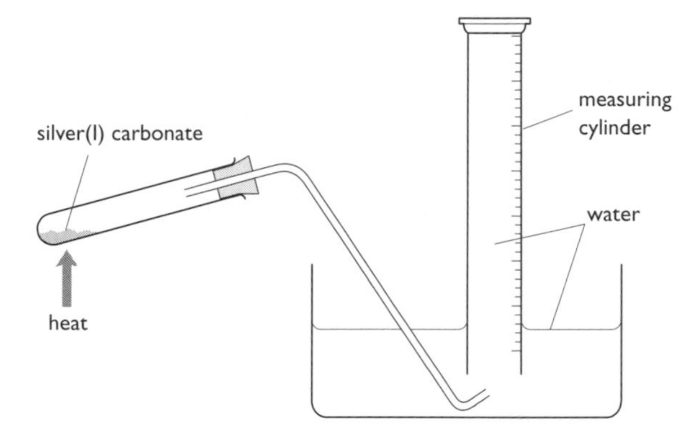

a) Calculate the mass of silver obtained.
b) Calculate the volume of gas collected. (V_{mol} = 24 litres mol^{-1})
c) If sodium hydroxide solution was then added to the water, how would the volume of gas be affected? Explain why this happens.

15 Methoxymethane, CH_3OCH_3, is a gas at room temperature. The equation for the complete combustion of methoxymethane is shown below.

$$CH_3OCH_3 + 3O_2 \rightarrow 2CO_2 + 3H_2O$$

a) A mixture of 25 cm^3 of methoxymethane and 100 cm^3 of oxygen was ignited. Assume that all volumes are measured at room temperature and pressure.
   i) Show by calculation which reactant is in excess and by how much.
   ii) Calculate the volume of the gaseous product of this reaction.
b) Methoxymethane and ethanol (boiling point 78°C) are isomers.
   i) Explain the meaning of '*isomers*'.
   ii) Suggest why methoxymethane has a much lower boiling point than ethanol.

16 Lime water (calcium hydroxide solution) turns cloudy at first when carbon dioxide is passed into it. When more carbon dioxide is added, the mixture turns clear again due to the production of calcium hydrogencarbonate which is soluble in water. The equation for the overall reaction is as follows.

$$Ca(OH)_2(aq) + 2CO_2(g) \rightarrow Ca(HCO_3)_2(aq)$$

Carbon dioxide was passed into a solution of calcium hydroxide containing 1.48 g l^{-1}.

a) Calculate the concentration of calcium hydroxide in moles per litre.
b) Calculate the volume of carbon dioxide which can react with 250 cm^3 of this solution according to the equation shown above (V_{mol} = 24 litres mol^{-1}).

17 Ammonia burns in air which is enriched with extra oxygen. The balanced equation for this reaction is:

$$4NH_3 + 3O_2 \rightarrow 2N_2 + 6H_2O$$

a) Assuming that all volumes are measured at room temperature and pressure, calculate
   i) the volume of oxygen needed to burn 600 cm^3 of ammonia, and
   ii) the volume of gas produced.
b) Write the balanced equation for the catalytic oxidation of ammonia in which nitrogen monoxide is formed instead of nitrogen.

18 Using the apparatus shown in the diagram, 0.12 mol of copper(II) oxide was completely reduced by ammonia according to the following equation.

$$3CuO(s) + 2NH_3(g) \rightarrow 3Cu(s) + N_2(g) + 3H_2O(g)$$

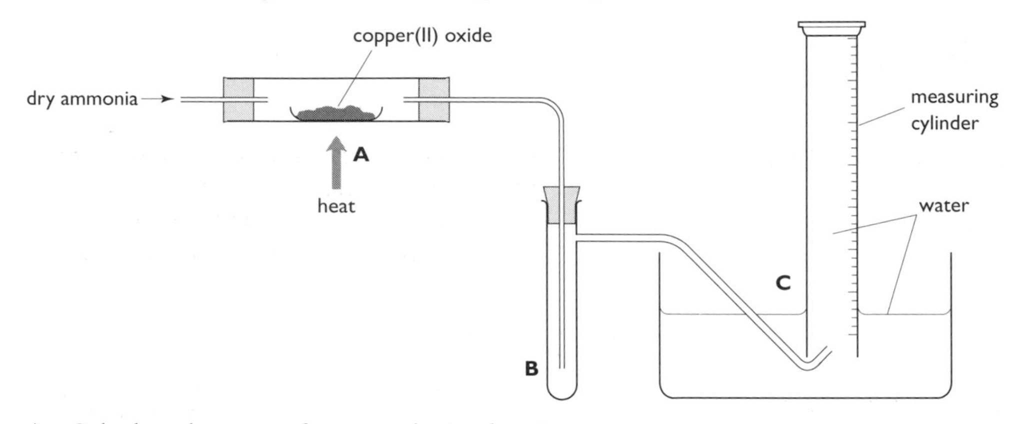

a) Calculate the mass of copper obtained at **A**.

b) Assume that all the steam produced in this experiment condenses in tube **B**. What volume of water, to the nearest cm^3, will be collected in tube **B**? (density of water = 1.0 g cm^{-3})

c) Calculate the volume of nitrogen collected over water at **C**. (V_{mol} = 24 litres mol^{-1})

d) To ensure complete reduction of the copper oxide, excess ammonia is used. What effect, if any, will this have on

i) the pH of the water at **C**

ii) the volume of gas collected?

Questions 1, 2, and 3 on pages 120–1 are based on the Unit 1 Prescribed Practical Activities.

THE WORLD OF CARBON

7 Fuels

Fuels are substances which release energy on burning.

For Higher Chemistry, detailed additional knowledge of only one crude-oil derived fuel, petrol, is required.

Petrol

❍ In addition to production from the gasoline fraction of the distillation of crude oil, petrol can also be made by **reforming** the naphtha fraction.

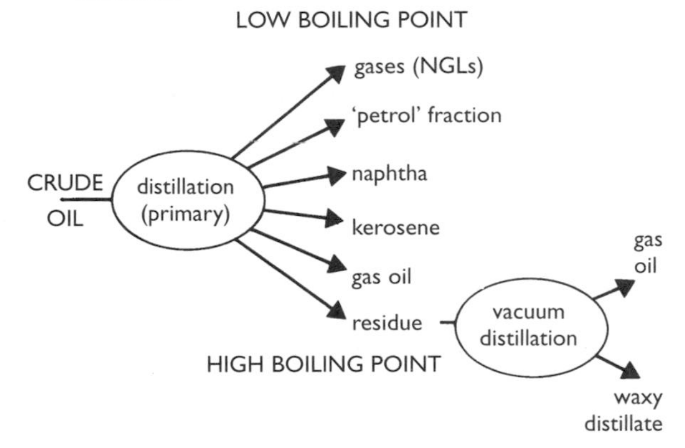

Figure 1 Fractions from crude oil

❍ Reforming alters the arrangement of atoms in molecules. It *may* also alter the number of carbon and/or hydrogen atoms per molecule.

$$C_7H_{16} \rightarrow C_7H_8 + 4H_2$$

heptane **methylbenzene**

$$CH_3CH_2CH_2CH_2CH_2CH_2CH_2CH_3 \rightarrow$$

octane

$$CH_3C(CH_3)_2CH_2CH(CH_3)CH_3$$

2,2,4-trimethylpentane

❍ After adding the products of reforming, petrol contains branched-chain alkanes, cycloalkanes, and aromatic hydrocarbons as well as straight-chain hydrocarbons. (Aromatic hydrocarbons are dealt with in Chapter 8.)

❍ To suit prevailing temperatures, any petrol is a blend of hydrocarbons of different volatilities, i.e. for low temperatures more high-volatility components are included.

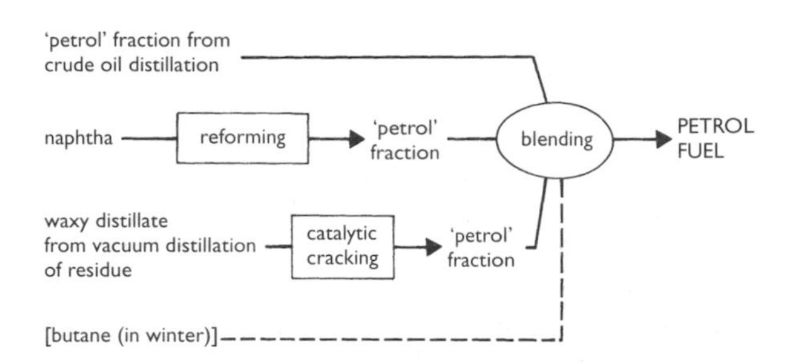

Figure 2 Making petrol

❍ In a petrol engine, the petrol is mixed with air, and at the correct instant just before the end of the compression stroke of the 'four-stroke' cycle it is ignited by an electrical spark.

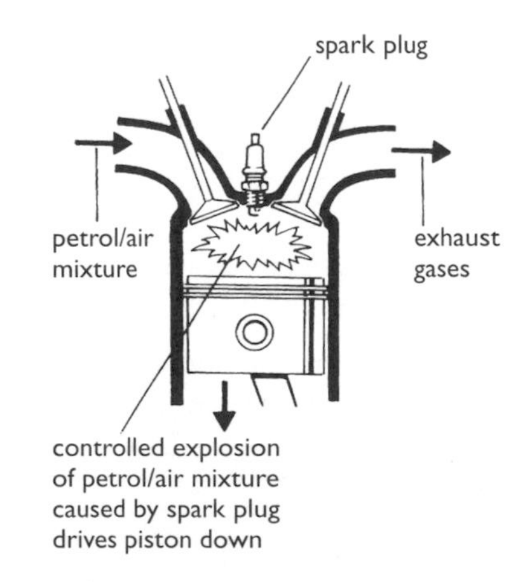

Figure 3 Petrol engine

❍ 'Knocking', or 'pinking', is caused by auto-ignition, i.e. the fuel-air mixture ignites in the hot engine before the spark occurs.

- Auto-ignition can be reduced by adding lead compounds to petrol, but leaded petrol is being phased out.
- In unleaded petrol, more highly branched and aromatic compounds are used instead of lead compounds to reduce auto-ignition.

Alternative fuels

Ethanol

Ethanol can be mixed with petrol. The ethanol can be obtained by fermenting sucrose from surplus sugar-cane production in, for example, Brazil. The sugar cane is a **renewable** source of fuel. The equation for the combustion of ethanol is:

$$C_2H_5OH + 3O_2 \rightarrow 2CO_2 + 3H_2O$$

Methanol

Methanol is an alternative to petrol. For its manufacture, see Chapter 11, page 71.

Advantages as a fuel
Virtually complete combustion, less carbon monoxide than from petrol.
Contains no aromatic carcinogens.
Cheaper to produce than primary fractionation gasoline.
Less volatile, and explosive, than petrol.
The car engine requires little modification.
Disadvantages as a fuel
Difficulty of mixing methanol and petrol without an extra cosolvent.
Methanol absorbs water, tending to form immiscible layers, becomes corrosive to car engine.
Methanol is toxic.
Less energy produced, volume for volume than from petrol. Bigger car fuel tanks needed.
Methanol is made from synthesis gas made from fossil methane.
Increases 'greenhouse' gases, unless 'biogas' is used to make synthesis gas.

Methane

Organic waste (e.g. animal manure, domestic refuse and human sewage) can be fermented anaerobically (in the absence of oxygen) to produce methane. Methane can be used as a domestic gaseous fuel or, from high pressure cylinders, as a motor fuel.

$$CH_4 + 2O_2 \rightarrow CO_2 + 2H_2O$$

Hydrogen

Electrolysis of water using solar energy which is a renewable resource could produce hydrogen to be used as a means of storing and distributing energy. If safety fears can be overcome, the hydrogen could be used as fuel in internal combustion engines, replacing petrol, and 'slowing' the increase of carbon dioxide in the atmosphere. The only product of hydrogen's combustion is water.

$$2H_2 + O_2 \rightarrow 2H_2O$$

Questions

Questions 1 and **2** relate to the following hydrocarbons.

A C_2H_6 **B** C_4H_{10}
C C_8H_{18} **D** $C_{16}H_{34}$

1 Which hydrocarbon is most likely to be found in petrol?

2 Which hydrocarbon is added to petrol in winter to make it more volatile?

3* Which of the following occurs when crude oil is distilled?

A Covalent bonds break and form again.
B Covalent bonds break and van der Waals' bonds form.
C Van der Waals' bonds break and covalent bonds form.
D Van der Waals' bonds break and form again.

4 Biogas is mainly a mixture of

A hydrogen and nitrogen
B methane and carbon dioxide
C hydrogen and carbon dioxide
D methane and nitrogen.

5 Which of the following substances would **not** decrease the pre-ignition of petrol?

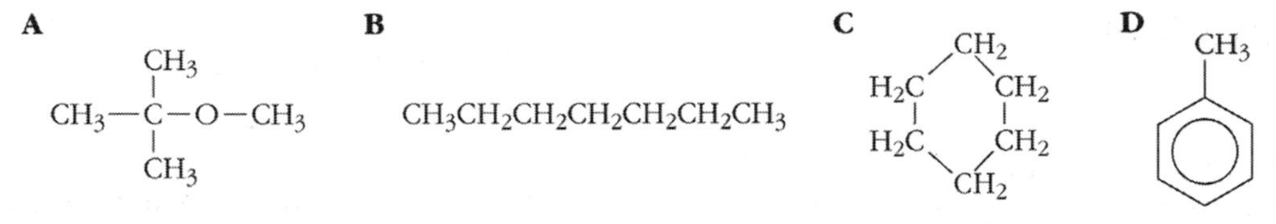

6 Which of the following substances is least likely to be present in petrol?

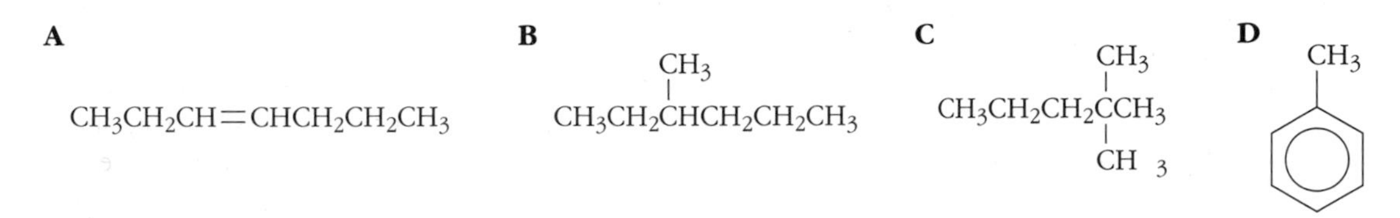

Questions 7 and **8** refer to the following compounds

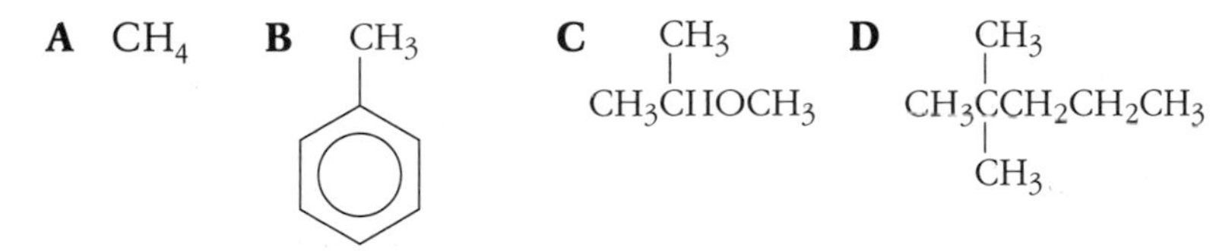

7 Which compound is a branched alkane?

8 Which compound is the main component of biogas?

9 Which of the following is a renewable source of motor fuel?

A Ethanol
B Octane
C Methanol
D Benzene

10 Which of the following substances when burned would **not** contribute to the 'greenhouse effect'?

A Methane
B Cyclohexane
C Hydrogen
D Ethanol

11 Prior to the year 2000 petrol contained a compound called lead tetraethyl, $Pb(C_2H_5)_4$.

a) Why was this compound added to petrol?
b) Why was its use discontinued?
c) Calculate the percentage by mass of lead in lead tetraethyl.

12

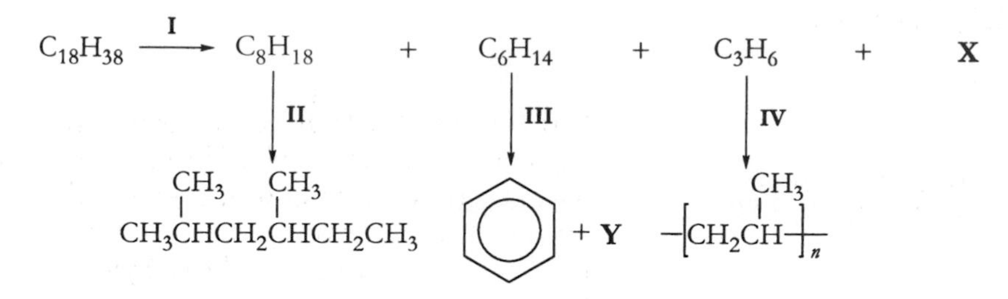

Reaction **I** shows one possible outcome of cracking octadecane, $C_{18}H_{38}$.

Reactions **II**, **III** and **IV** refer to subsequent changes involving the products of cracking.

a) From the following list choose **two** terms for **each** of the reactions **II**, **III** and **IV**.

addition dehydrogenation isomerisation polymerisation reforming

b) Which reaction produces an aromatic compound?
c) What structural feature of the product of reaction **II** makes it suitable for use in petrol?
d) Name the carbon compounds produced in **i)** reaction **III** **ii)** reaction **IV**.
e) What substance is **X** and why is it a problem in catalytic cracking?

13 a) Place the following fuels in order of **increasing** enthalpy of combustion.

Ethanol Hydrogen Methane

b) Calculate the quantity of energy released per **gram** of each of these fuels and again place them in ascending order.
c) How does the position of hydrogen in the **second** list account for its use as a rocket fuel?

8 Hydrocarbons

Homologous Series

- There are different series of hydrocarbons such as **alkanes**, **alkenes** and **cycloalkanes**.
- Alkanes and alkenes have carbon atoms linked in chains, while cycloalkanes have a ring structure.
- Alkanes and cycloalkanes are **saturated** as all of their carbon–carbon bonds are single bonds. Alkenes are **unsaturated** since they contain carbon–carbon double bonds.
- The carbon–carbon double bond is an example of what is called a **functional group** since it has a major influence on the chemical behaviour of the compound.
- Whether a hydrocarbon is saturated or unsaturated can be shown by testing it with bromine water. Alkenes rapidly decolourise the bromine water, while alkanes and cycloalkanes do not do so.
- It is also possible to have a carbon–carbon triple bond in a molecule. The **alkynes** are a series of unsaturated hydrocarbons which possess this functional group. The first member of this series is called **ethyne** (acetylene).

Members of the three homologous series – alkanes, alkenes and alkynes – are listed below in Table 1.

- Each of these series is an **homologous series**. Members of a given series, known as **homologues**, have the following characteristics:

  1 Physical properties show a gradual change from one member to the next.

  2 Chemical properties and methods of preparation are very similar.

  3 Successive members differ in formula by $-CH_2-$ and in molecular mass by 14.

Alkanes	Alkenes	Alkynes
Methane CH_4 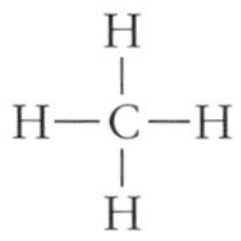		
Ethane C_2H_6 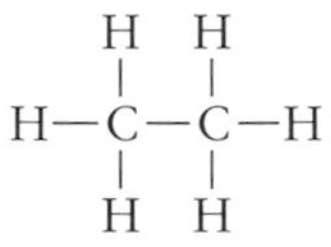	Ethene C_2H_4 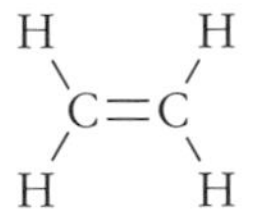	Ethyne C_2H_2 $H-C\equiv C-H$
Propane C_3H_8 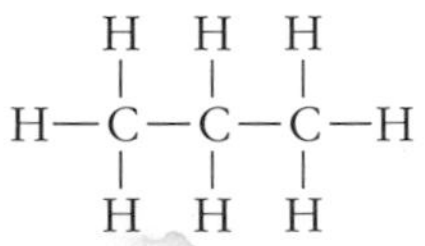	Propene C_3H_6 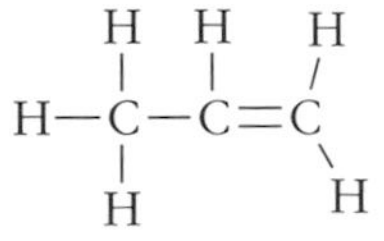	Propyne C_3H_4 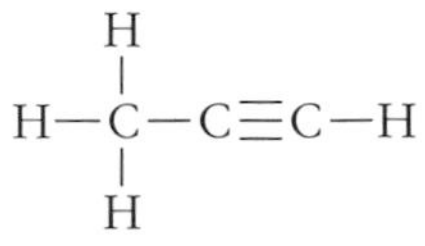
Butane C_4H_{10} Pentane C_5H_{12} Hexane C_6H_{14} Heptane C_7H_{16} Octane C_8H_{18} General formula: C_nH_{2n+2}	Butene C_4H_8 Pentene C_5H_{10} Hexene C_6H_{12} Heptene C_7H_{14} Octene C_8H_{16} General formula: C_nH_{2n}	Butyne C_4H_6 Pentyne C_5H_8 Hexyne C_6H_{10} Heptyne C_7H_{12} Octyne C_8H_{14} General formula: C_nH_{2n-2}

4 They can be represented by a **general formula**.

5 Members of a series possess the same **functional group**, i.e. a certain group of atoms or type of bond which is mainly responsible for the characteristic chemical properties of that homologous series.

Naming isomers of alkanes

- Isomers have the same molecular formula but different structural formulae.
- The first three members of the alkane series – methane, ethane and propane – do not have isomers.
- Butane, C_4H_{10}, has two isomers. Their full and shortened structural formulae are shown below.

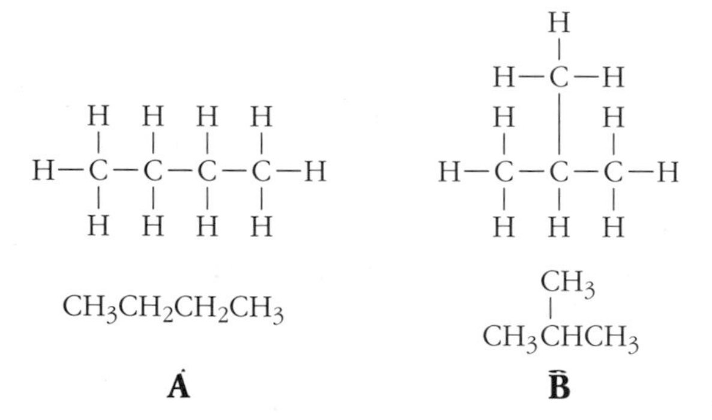

Compound A is called butane and is an example of a **straight-chain hydrocarbon.** Compound B is a **branched** hydrocarbon and its name is 2-methylpropane.

The method of naming isomers operates as follows:

1 Select and name the longest chain of carbon atoms in the molecule.

2 Find out which atom the branch is attached to by giving each carbon in the longest chain a number. Begin numbering from the end which is nearer the branch.

3 Identify the branch, e.g. as a methyl group (CH_3–), an ethyl group (C_2H_5–) etc. Figure 1 illustrates how this system applies to the above example.

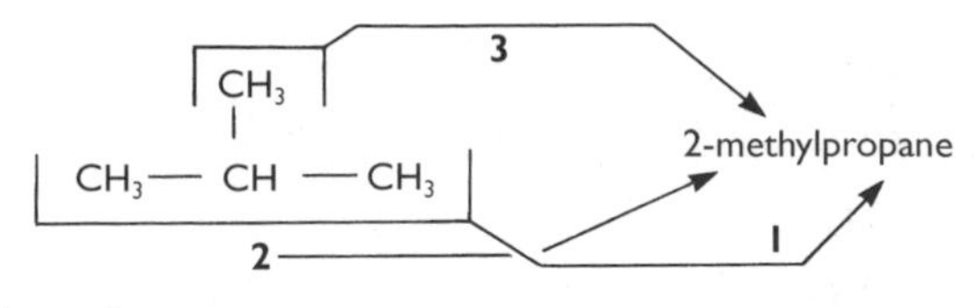

Figure 1

- Pentane, C_5H_{12}, has three isomers as follows.

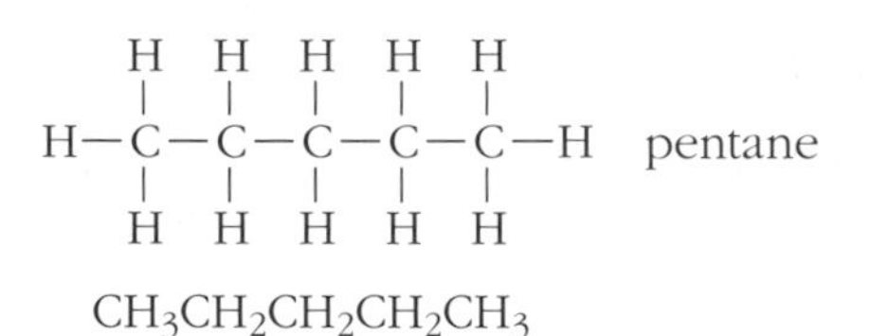

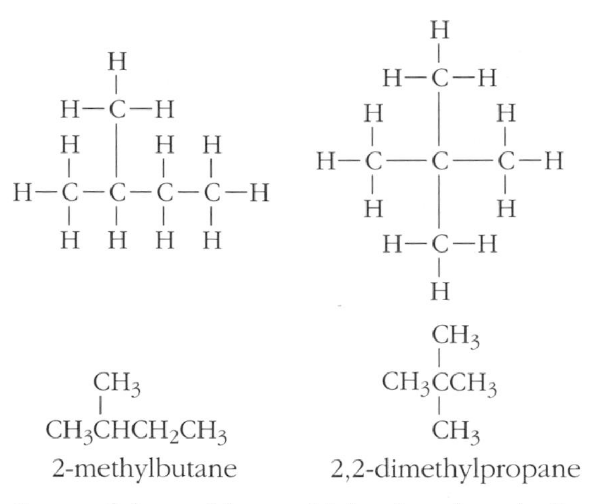

Some of the problems which arise when dealing with structural formulae are shown below. Four other ways of representing 2-methylbutane, different from the structural formula given above, are shown.

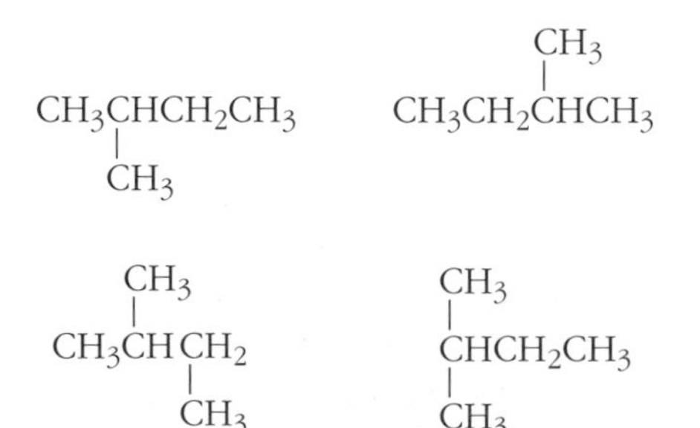

The system of naming can be applied to more complicated molecules as shown by the following two examples.

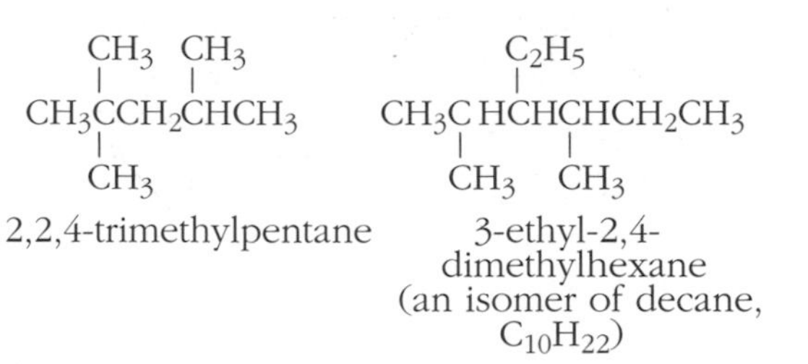

Naming isomers of alkenes and alkynes

- Ethene has no isomers. Propene has no isomers which are alkenes but it is isomeric with cyclopropane.
- Isomers of alkenes can arise for two reasons, since
  1 the position of the double bond in the chain can vary and
  2 the chain can be straight or branched.

  This is illustrated by the isomers of butene, C_4H_8.

```
H  H  H   H          H  H  H  H
|  |  |  /           |  |  |  |
H—C—C—C=C          H—C—C=C—C—H
|  |     \           |        |
H  H      H          H        H
```

$CH_3CH_2CH{=}CH_2$ but-1-ene $CH_3CH{=}CHCH_3$ but-2-ene

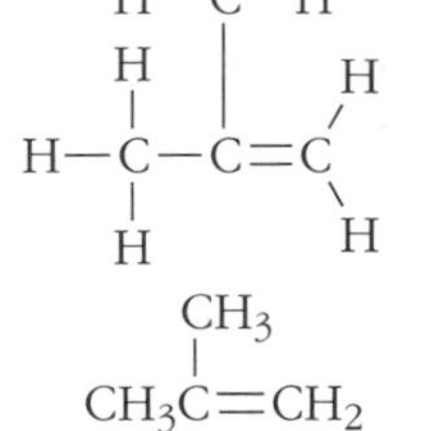

$CH_3C(CH_3){=}CH_2$

2-methylpropene

- Where necessary, the name shows the position of the double bond. Thus but-2-ene has the double bond between the second and third carbon atoms in the chain.
- When naming a branched alkene the position of the double bond is more important than the position of the branch. The name of the alkene whose structure is shown here is 3-methylbut-1-ene (and **not** 2-methylbut-3-ene).

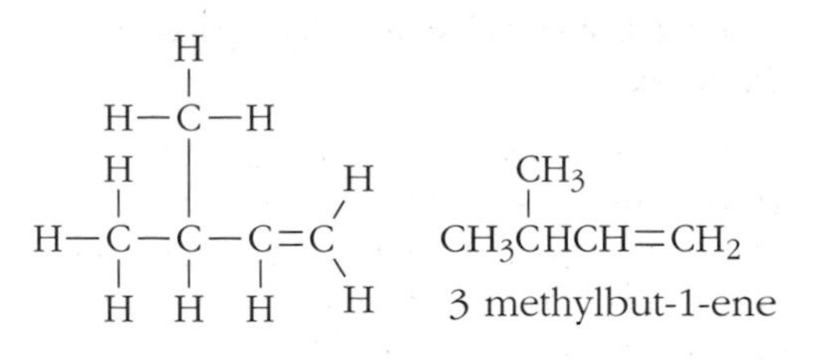

- Isomers of alkynes (from butyne) are named in a similar way to alkenes.

Addition reactions of alkenes

- Alkenes can undergo addition reactions because they contain C=C bonds. Alkanes are unable to react by addition since they contain only single bonds.

Addition of hydrogen (H_2) – hydrogenation

An ALKENE reacts with HYDROGEN to form an ALKANE.

e.g.

$$CH_3CH_2CH{=}CH_2 + H_2 \xrightarrow[\text{Ni catalyst}]{200^\circ C} CH_3CH_2CH_2CH_3$$

but-1-ene → **butane**

Addition of halogens (Br_2, Cl_2)

An ALKENE reacts with BROMINE to form a DIBROMOALKANE.

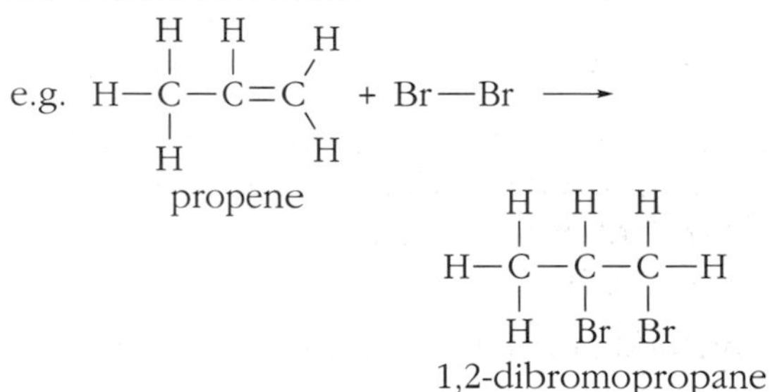

Similarly, an ALKENE reacts with CHLORINE to form a DICHLOROALKANE.

e.g.

$$CH_3CH_2CH{=}CH_2 + Cl_2 \rightarrow CH_3CH_2CHClCH_2Cl$$

but-1-ene → **1,2-dichlorobutane**

Addition of hydrogen halides (HX, where X = F, Cl, Br or I)

An ALKENE reacts with a HYDROGEN HALIDE to form a HALOGENOALKANE.

e.g. $CH_2{=}CH_2 + HBr \rightarrow CH_3CH_2Br$

ethene → **bromoethane**

Addition of water (H_2O) – Hydration

An ALKENE reacts with WATER (steam) to form an ALKANOL.

$$CH_2{=}CH_2 + H_2O \xrightarrow[\text{phosphoric acid catalyst}]{300°C} CH_3CH_2OH$$

ethene **ethanol**

This method can be applied to the production of other alkanols from alkenes with more carbon atoms, e.g. the catalytic hydration of propene produces propanol.

Note: In each of the addition reactions described above the product is **saturated**. The atoms of the substance being added form bonds with the carbon atoms which were originally held together by the double bond.

Alkenes from dehydration of alkanols

- The reverse process of the addition of water (hydration) is **dehydration** of an alkanol.

This can be easily demonstrated in the laboratory using the apparatus illustrated in Figure 2.

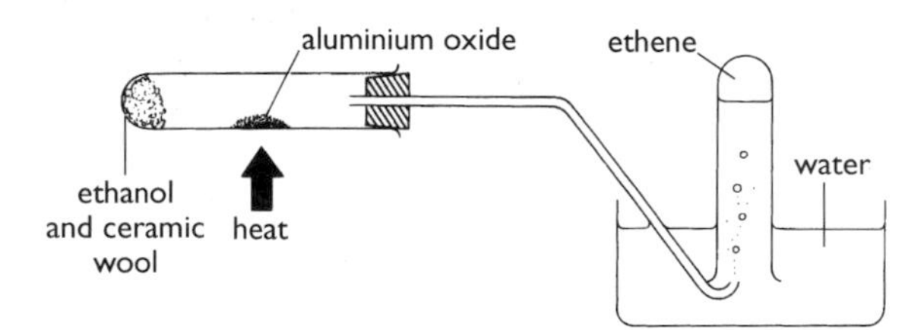

Figure 2 Dehydration of ethanol

- When ethanol vapour is passed over heated aluminium oxide, which acts as a catalyst, dehydration occurs forming ethene.

$$C_2H_5OH \rightarrow CH_2{=}CH_2 + H_2O$$

- Concentrated sulphuric acid has a strong attraction for water and can also be used to dehydrate alkanols.
- Other alkenes can be produced by dehydration of alkanols, e.g. dehydration of propanol produces propene.
- When an alkanol is dehydrated the –OH group is removed along with a hydrogen atom from an adjacent carbon atom.
- Sometimes this produces two different alkenes, e.g. dehydration of butan-2-ol can produce either but-1-ene or but-2-ene as shown below.

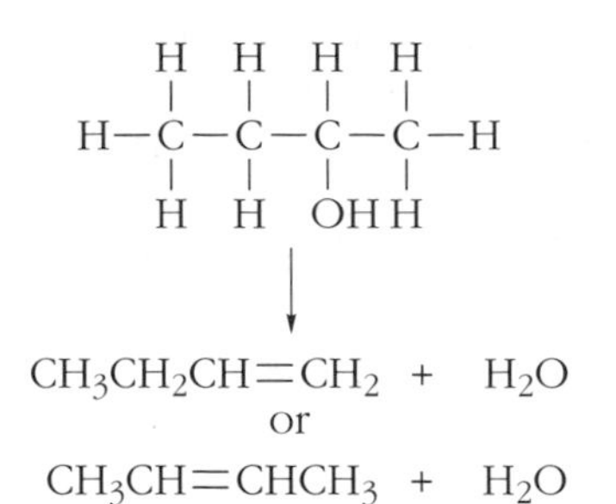

Uses of halogenoalkanes

- Halogenoalkanes have properties which make them suitable for a wide range of consumer products.
- 2-bromo-2-chloro-1,1,1-trifluoroethane is the anaesthetic, 'halothane'.

$CF_3CHClBr$ halothane
CF_2Cl_2 dichlorodifluoromethane
CF_3CH_2F 1,1,1,2-tetrafluoroethane

- Chlorofluorocarbons, or CFCs, were developed for a number of uses where a non-flammable, volatile liquid was required. Dichlorodifluoromethane was important as a refrigerant and as an aerosol propellent. Other CFCs were used as degreasing solvents and 'blowing agents' for making plastic foam, e.g. expanded polystyrene.
- It is believed that CFCs are chiefly responsible for the depletion of the ozone layer in the upper atmosphere which absorbs ultra-violet (UV) radiation from the sun.
- HFAs, i.e. hydrofluoroalkanes, are being developed as alternatives to CFCs, e.g. 1,1,1,2-tetrafluoroethane is a refrigerant. HFAs are more reactive than CFCs and are less likely to persist in the atmosphere sufficiently to affect the ozone layer.

Addition reactions of ethyne

- Ethyne decolourises bromine water which shows that ethyne is unsaturated. It burns with a very sooty flame due to its high carbon content.
- Ethyne undergoes similar addition reactions to ethene when it reacts with hydrogen, hydrogen halides, bromine and chlorine but the addition is a two-stage process. In the first stage one molecule of ethyne combines with one molecule of the addition reagent to form a compound with a C=C bond. In the second stage another molecule of the reagent is needed to form a saturated product.

Figure 3 summarises the reactions of ethyne with hydrogen, hydrogen chloride and bromine.

Aromatic hydrocarbons

- **Aromatic** compounds contain a distinctive ring structure of six carbon atoms.
- The simplest aromatic compound is the hydrocarbon called **benzene**, molecular formula: C_6H_6.
- Aromatic compounds often behave in a markedly different way due to their characteristic chemical structure.

Structure of benzene

- From its formula, C_6H_6, we might expect benzene to be unsaturated but the bromine water test shows that C=C bonds are not present in benzene.
- Benzene has a planar molecule and all its carbon–carbon bonds are the same length, longer than the C=C bond but shorter than the C–C bond.
- Each carbon atom has four outer electrons and in benzene three of these are used in forming bonds with a hydrogen atom and the two adjacent carbon atoms. The six remaining electrons, one from each carbon atom, occupy electron clouds which are **delocalised**.

Figure 4 illustrates the structure of benzene.

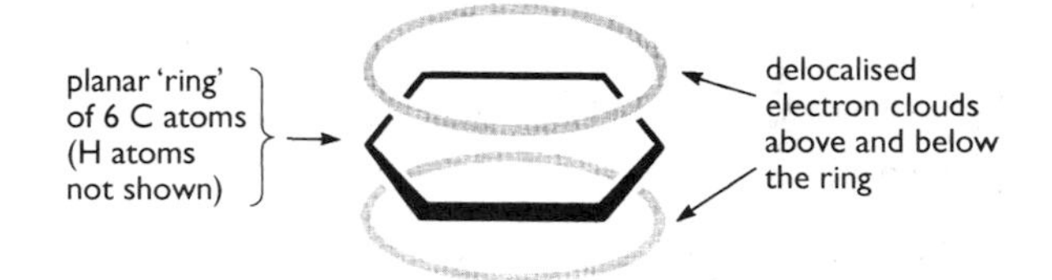

Figure 4

- The structure of benzene is usually represented as a regular hexagon suggesting a single bond framework (Figure 5). The circle inside the hexagon indicates the additional bonding due to the delocalised electrons. Each corner of the hexagon represents a carbon atom with one hydrogen atom attached.

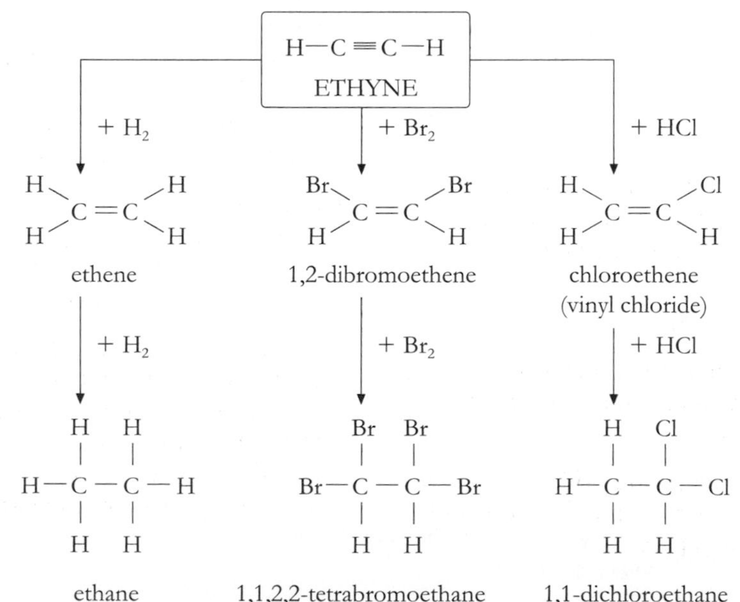

Figure 3

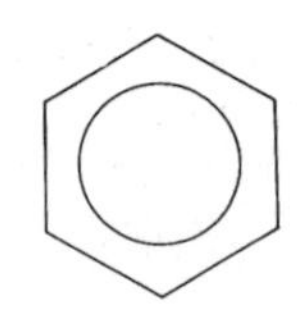

Figure 5

- All of the above explains why benzene is reluctant to react by addition.

Aromatic compounds

- Aromatic compounds have at least one of the hydrogen atoms on the benzene ring replaced by another atom or group of atoms.

For example, if one hydrogen atom is replaced, or substituted, by a methyl group, methylbenzene (toluene), $C_6H_5CH_3$, is produced.

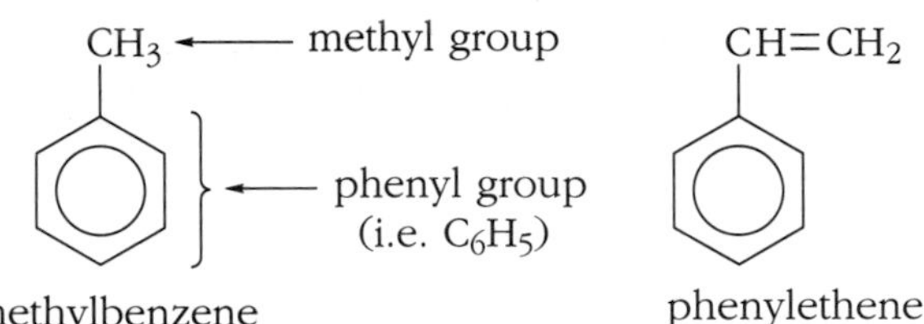

- The **phenyl group** consists of a benzene ring minus one hydrogen atom, i.e. C_6H_5– and does not exist on its own but must be attached to another atom or group of atoms.

- The prefix 'phenyl' is sometimes used in naming aromatic compounds.

For example, phenylethene is the systematic name for styrene which is the monomer for producing poly(phenylethene) or polystyrene as shown in Figure 6.

- It is possible to have more than one of the hydrogen atoms in benzene replaced by methyl groups, e.g. 1,4 dimethylbenzene (para-xylene), $C_6H_4(CH_3)_2$.

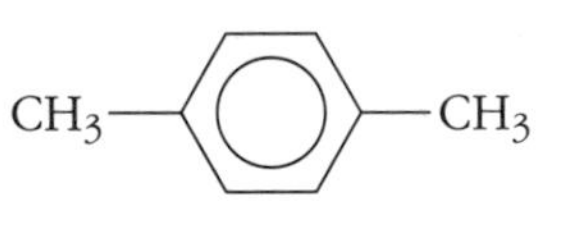

$CH_3C_6H_4CH_3$

1,4-dimethylbenzene

Some examples of important monosubstituted aromatic compounds are given below.

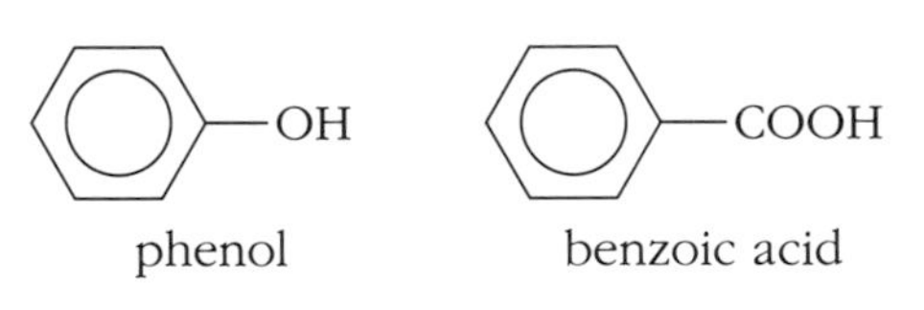

phenol benzoic acid

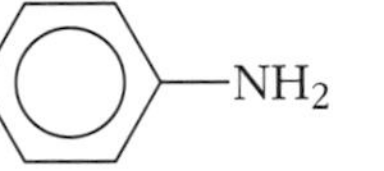

phenylamine (aniline)

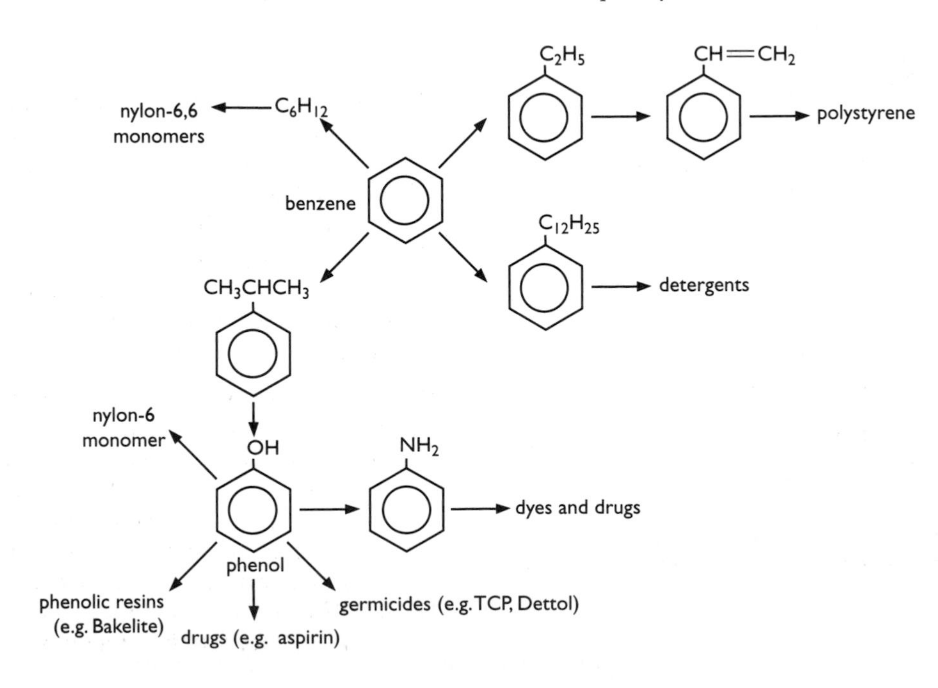

 Figure 6

- Many aromatic compounds, including benzene itself, are important as feedstocks for the production of everyday materials.

Some of these feedstocks and their products are illustrated in Figure 6 as a summary of synthetic routes emanating from benzene.

Questions

1 What kind of reaction occurs when propanol vapour is passed over hot aluminium oxide?

A Dehydration B Hydrolysis
C Dehydrogenation D Hydration

2 One mole of a hydrocarbon combines with two moles of hydrogen gas to form a saturated hydrocarbon. The hydrocarbon could be

A benzene B butene
C cyclopentane D propyne.

3

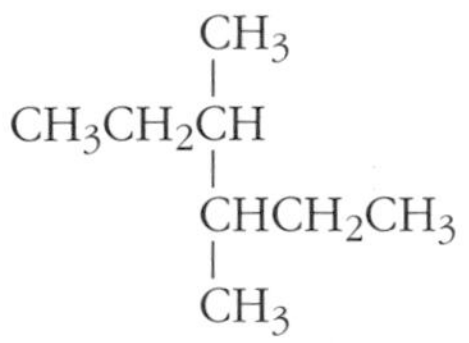

$$\begin{array}{l} \quad\quad\quad\quad\; CH_3 \\ \quad\quad\quad\quad\;\; | \\ CH_3CH_2CH \\ \quad\quad\quad\quad\;\; | \\ \quad\quad\quad\quad\; CHCH_2CH_3 \\ \quad\quad\quad\quad\;\; | \\ \quad\quad\quad\quad\; CH_3 \end{array}$$

The name of this compound is

A 2,3-diethylbutane
B 2-ethyl-3-methylpentane
C 3,4-dimethylhexane
D 3-ethyl-2-methylpentane

4 When but-2-ene reacts with bromine the formula of the product is

A $CH_3CHBrCHBrCH_3$
B $CH_3CHBrCH_2CH_2Br$
C $CH_3CBr_2CH_2CH_3$
D $CH_2BrCHBrCH_2CH_3$

5 A bottle is labelled with the formula C_5H_{10}. The liquid in the bottle is **not**

A pent-1-ene B pent-2-yne
C cyclopentane D 2-methylbut-2-ene.

6 Draw the full structural formula of each of the following compounds.
a) 3-ethyl-2-methylpentane
b) 1,3-dimethylcyclobutane
c) pent-2-ene
d) 2-methylbut-1-ene
e) hex-3-yne
f) 3-methylbut-1-yne
g) ethylbenzene
h) 1,3,5-trimethylbenzene

7 Draw the shortened structural formula of each of the following compounds.
a) 3,3-diethylpentane
b) 2,2,6-trimethylheptane
c) 3-methylhex-3-ene
d) 3-ethylpent-1-ene
e) 4,4-dimethylhex-2-yne
f) 1,3-dichloropropane

8 Name the following compounds

a)
$$\begin{array}{l} \quad\quad\quad\; CH_3 \quad\; CH_3 \\ \quad\quad\quad\quad | \quad\quad\;\; | \\ CH_3CH_2CHCH_2CCH_3 \\ \quad\quad\quad\quad\quad\quad\;\; | \\ \quad\quad\quad\quad\quad\;\; CH_3 \end{array}$$

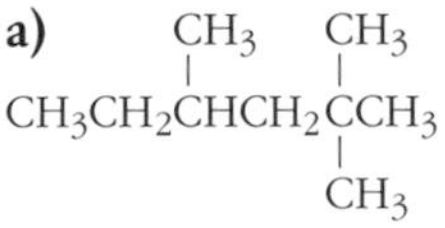

b)
$$\begin{array}{l} CH_3CH{=}CHCHCH_2CH_3 \\ \quad\quad\quad\quad\quad\;\; | \\ \quad\quad\quad\quad\; C_2H_5 \end{array}$$

c)
$$\begin{array}{l} \quad\quad\quad\quad CH_3 \\ \quad\quad\quad\quad\;\; | \\ CH_3 - CH \quad\; CH \quad\; CH_2 \\ \quad\quad\;\; | \quad\quad\quad\quad\quad | \\ \quad\quad CH_2 - CH_2 \end{array}$$

(ring: CH–CH(CH₃)–CH₂–CH₂–CH₂, with CH_3 on the first CH)

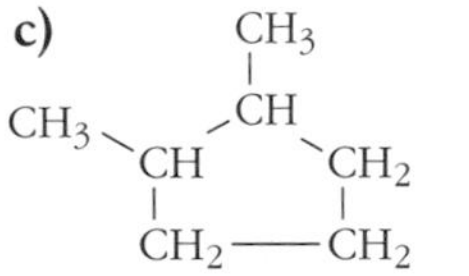

d)
$$\begin{array}{l} \quad\quad\quad\quad Br \\ \quad\quad\quad\quad\; | \\ CH_3CH_2CCH_3 \\ \quad\quad\quad\quad\; | \\ \quad\quad\quad\quad Br \end{array}$$

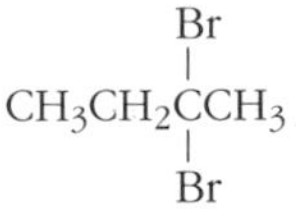

e)
$$\begin{array}{l} CH_3C{\equiv}CCHCH_3 \\ \quad\quad\quad\quad\;\; | \\ \quad\quad\quad\quad CH_3 \end{array}$$

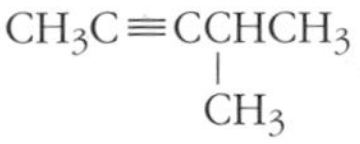

f) Benzene ring with CH_3 at position 1 and CH_3 at position 3

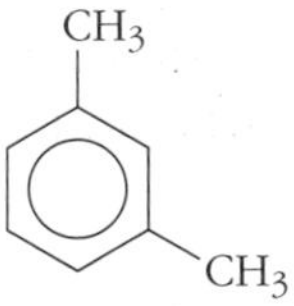

9 For each of the following reactions, draw the full structural formula and give the systematic name of the carbon compound produced.
a) But-2-ene + hydrogen chloride
b) Ethene + chlorine
c) 2-methylpropene + hydrogen
d) Ethyne + chlorine [1:2 mol ratio]
e) But-2-yne + HBr [1:1 mol ratio]
f) Propyne + hydrogen [1:1 mol ratio]

10 Iodine monochloride, ICl, reacts with alkenes in a similar manner to bromine.
a) What type of reaction occurs when an alkene reacts with iodine monochloride?
b) Draw the structural formulae of the two possible products formed when propene reacts with iodine monochloride.
c) Why is there only one product when but-2-ene reacts with iodine monochloride?

11 In the Wurtz reaction, iodoalkanes react with sodium to form alkanes. The reaction is used to make alkanes which have an even number of carbon atoms, e.g. iodomethane and sodium react to produce ethane as shown below.

$$CH_3 - I + 2Na + I - CH_3 \rightarrow CH_3{-}CH_3 + 2NaI$$

a) Name the iodoalkane required to make octane.

b) Draw the structural formula of 2-iodopropane and the product when it undergoes the Wurtz reaction.

c) Alkanes such as pentane, which have odd numbers of carbon atoms per molecule, can be produced by the Wurtz reaction.

$$C_3H_7I + 2Na + C_2H_5I \rightarrow C_5H_{12} + 2NaI$$

However, **two** other alkanes are formed in this reaction. Name these alkanes.

12 The following diagram shows some reactions involving propene and propyne.

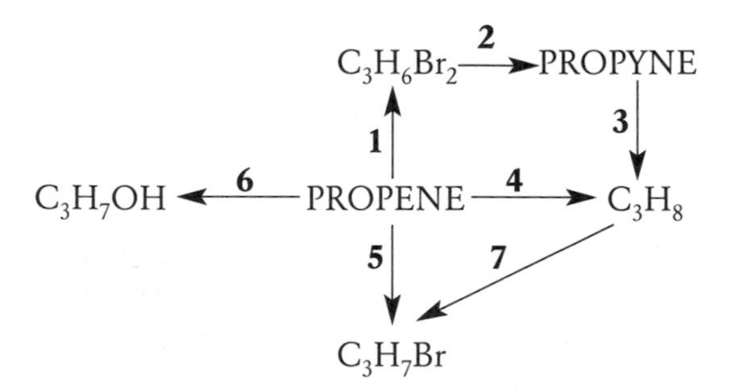

a) Which of the reactions **1** to **7** are examples of hydrogenation?

b) Which of the reactions **1** to **7** are **not** addition reactions?

c) Reaction **6** can be reversed in a laboratory experiment.

i) What type of reaction is this reverse reaction?

ii) Describe briefly with the aid of a diagram how it can be carried out in the laboratory.

d) Reaction **7** occurs when a mixture of propane and bromine vapour is exposed to sunlight. An acidic compound is also produced in the reaction. Work out an equation for this reaction using molecular formulae.

13 Aromatic compounds which have hydroxyl groups attached to the benzene ring are known collectively as phenols. The names and structural formulae of three examples are shown below.

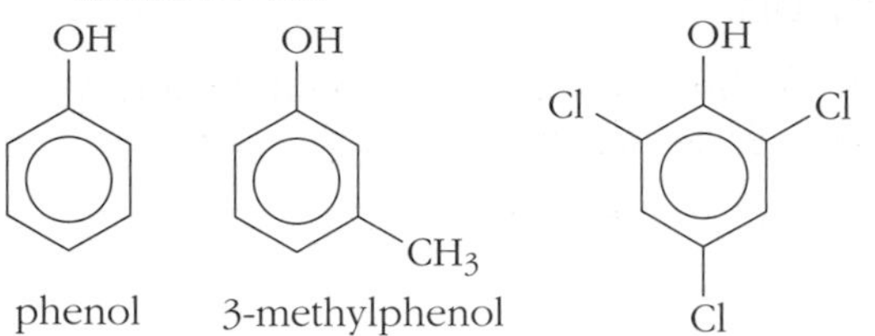

a) Give the molecular formula of each of these phenols.

b) Phenol is slightly soluble in water forming a weak acid known as carbolic acid. Benzene is not acidic.
Write the formulae of both ions present in carbolic acid solution.

c) The name 'trichlorophenol' does not fully describe its structure. Work out the full name of this compound.

14 a) Draw the shortened structural formula of pent-2-yne.

b) Name an isomer of pent-2-yne which has

i) a straight-chain structure

ii) a branched-chain structure

iii) a ring structure.

15*

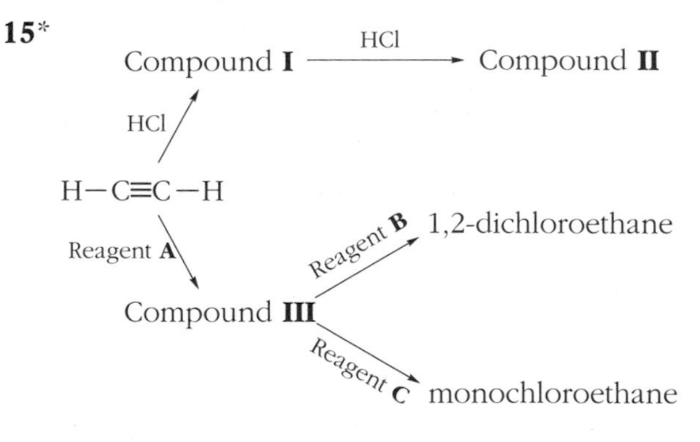

All compounds in the above diagram have different structures, and the reagent and compound are in the ratio of 1 mole : 1 mole in every case.

a) Draw the full structural formulae for compounds **I**, **II** and **III**.

b) Name reagents **A**, **B** and **C**.

9 Alcohols, Aldehydes and Ketones

Alcohols

- An alcohol is a carbon compound containing the hydroxyl functional group, –OH.
- An alcohol name ends in -ol.

Alkanols

- Alkanols are a homologous series of alcohols.
- An alkanol is a substituted alkane in which one of the hydrogen atoms has been replaced by the **hydroxyl** group.
- The name of an alkanol is obtained by replacing the final letter of the corresponding alkane by the name ending '-ol'.

Alkanes		Alkanols	
Methane	CH_4	Methanol	CH_3OH
Ethane	C_2H_6	Ethanol	C_2H_5OH
Propane	C_3H_8	Propanol	C_3H_7OH
Butane	C_4H_{10}	Butanol	C_4H_9OH
Pentane	C_5H_{12}	Pentanol	$C_5H_{11}OH$
General formula: C_nH_{2n+2}		General formula: $C_nH_{2n+1}OH$ or $C_nH_{2n+2}O$	

Table 1

Structural formulae and isomers

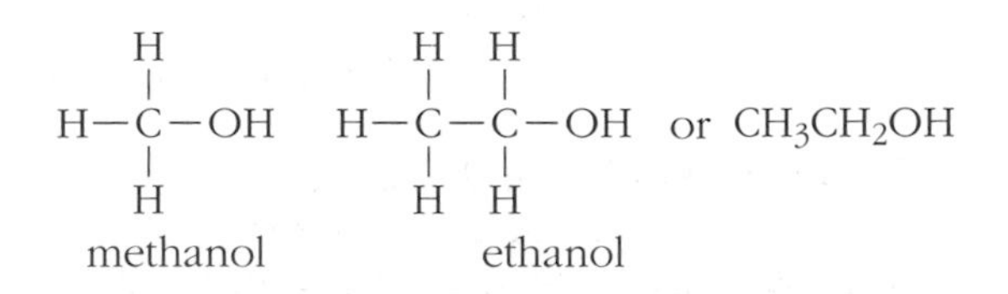

- Methanol has no isomers. Ethanol has no isomer which is an alkanol.
- Propanol has two isomers since the hydroxyl group can be attached either to a carbon atom at the end of the chain or to the carbon atom in the middle of the chain.

```
   H  H  H              H  H  H
   |  |  |              |  |  |
H—C—C—C—OH          H—C—C—C—H
   |  |  |              |  |  |
   H  H  H              H  OH H
```

$CH_3CH_2CH_2OH$ — propan-1-ol

CH_3CHCH_3 (with OH on the middle carbon) or $CH_3CH(OH)CH_3$ — propan-2-ol

- Butanol has more isomers since the chain may be branched. Three isomers of butanol are shown here.

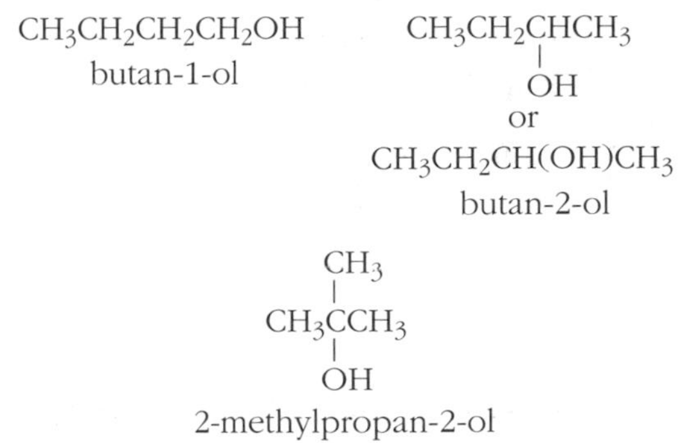

- When an alkanol contains three or more carbon atoms per molecule it is necessary when naming it to specify the carbon atom to which the –OH group is attached.
- When naming alkanols which have branched chains the position of the hydroxyl group takes precedence over the position of the branch. Thus the alkanol shown below is called 3-methylbutan-2-ol, not 2-methylbutan-3-ol.

$CH_3CHCHCH_3$ (with CH_3 on C-3 and OH on C-2) or $CH_3CHCH(OH)CH_3$ (with CH_3 on C-3)

Types of alcohol

- Alcohols fall into three different types depending on the position of the hydroxyl group (Table 2).
- Methanol and ethanol are examples of primary alkanols.

Type	Primary	Secondary	Tertiary
Position of –OH group.	Joined to the **end** of the carbon chain **R**–CH_2OH e.g. propan-1-ol butan-1-ol	Joined to an **intermediate** carbon atom. **R**–CH–**R′** (with OH on the CH) e.g. propan-2-ol butan-2-ol	Joined to an **intermediate** carbon atom which also has a **branch** attached. **R**–C–**R′** (with **R″** above and OH below the C) e.g. 2-methylpropan-2-ol

(The symbols **R**, **R′** and **R″** stand for alkyl groups, e.g. methyl, CH_3–, ethyl, C_2H_5–, etc.)

Table 2

Oxidation of alcohols

- Oxidation occurs when a substance combines with oxygen.
- Combustion is an extreme example of oxidation. Complete combustion of an alcohol produces carbon dioxide and water. e.g.

$$C_2H_5OH(l) + 3O_2(g) \rightarrow 2CO_2(g) + 3H_2O(l)$$

- Other oxidation reactions cause less drastic changes to the structure of the alcohol.
- Primary and secondary alcohols can be oxidised by various oxidising agents but tertiary alcohols do not readily oxidise.
- Acidified potassium dichromate solution is a suitable oxidising agent. The dichromate ions gain electrons from the alcohol which, therefore, has been oxidised. The equation for this reaction is:

$$Cr_2O_7^{2-} + 14H^+ + 6e^- \xrightarrow{heat} 2Cr^{3+} + 7H_2O$$

dichromate ions – orange; chromium(III) ions – blue–green

- Oxidation also results from passing the alcohol vapour over heated copper(II) oxide as shown in Figure 1.

The oxide is reduced to copper. Ethanol is oxidised to form ethanal. The equation for the reaction is:

$$CH_3CH_2OH + CuO \rightarrow CH_3CHO + Cu + H_2O$$

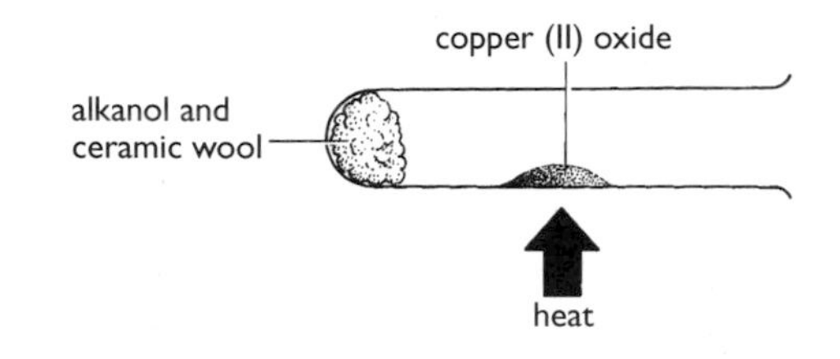

Figure 1 Oxidation of an alkanol

- **Primary alcohols** are oxidised to produce **aldehydes**,

e.g. propan-1-ol → PROPANAL

$$CH_3CH_2CH_2OH \rightarrow CH_3CH_2CHO$$

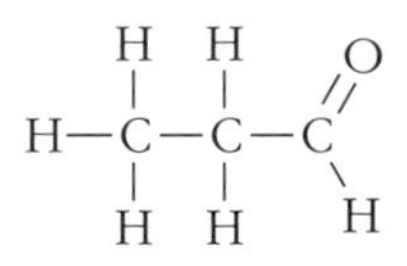

- **Secondary alcohols** are oxidised to produce **ketones**.

e.g propan-2-ol → PROPANONE

$$CH_3CH(OH)CH_3 \rightarrow CH_3COCH_3$$

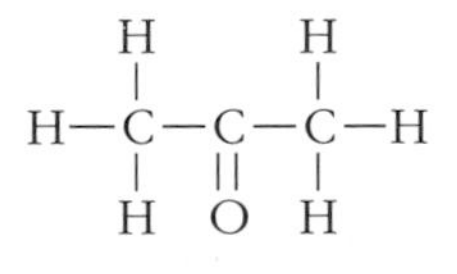

- Either of the oxidising agents, $Cr_2O_7^{2-}/H^+$ (aq) or CuO(s) can be used for these reactions.
- When primary alcohols are oxidised to aldehydes and secondary alcohols are oxidised to ketones, two hydrogen atoms are removed, namely the hydrogen atom of the –OH group and a hydrogen on the adjacent carbon atom. A tertiary alcohol cannot be similarly oxidised since it does not have a hydrogen atom attached to the carbon atom adjacent to the hydroxyl group.

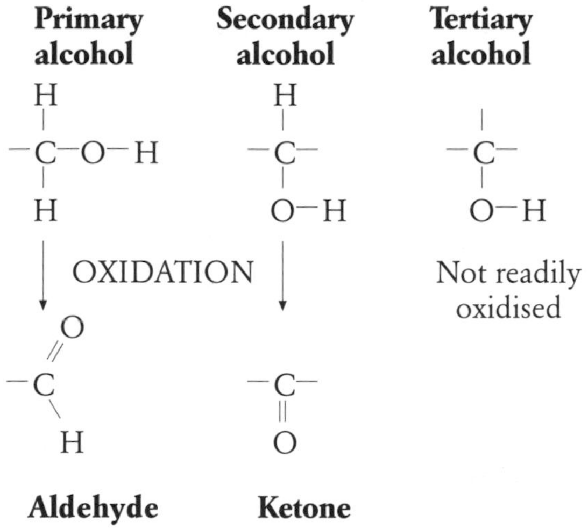

Primary alcohol | **Secondary alcohol** | **Tertiary alcohol**

OXIDATION

Not readily oxidised

Aldehyde | **Ketone**

Figure 2

Aldehydes and Ketones

- Aldehydes and ketones contain a carbon–oxygen double bond, C=O, called a **carbonyl** group. In an **aldehyde** the carbonyl group is at the **end** of the carbon chain and has a hydrogen atom attached to it. In a **ketone** the carbonyl group is **joined** to two other carbon atoms.
- There is a homologous series of aldehydes called **alkanals**, and a homologous series of ketones called **alkanones**. These are shown in Table 3.
- Note that in the shortened structural formula of an alkanal the functional group is written as –CHO and not –COH.
- Branched alkanals and alkanones do exist. When naming these the position of the functional group takes precedence over the position of the branch, as shown in the following examples. There is no need to indicate a number for the functional group when naming an alkanal as it must always be at the end of the chain but when naming an alkanone it is usually necessary to specify the number of the carbonyl group.

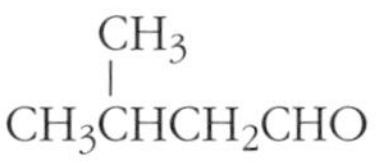

$CH_3CH(CH_3)CH_2CHO$ — 3-methylbutanal

$CH_3COCH(CH_3)CH_2CH_3$ — 3-methylpentan-2-one

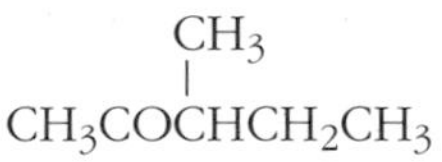

Alkanals		Alkanones	
Methanal	HCHO		
Ethanal	CH_3CHO		
Propanal	CH_3CH_2CHO	Propanone	CH_3COCH_3
Butanal	$CH_3CH_2CH_2CHO$	Butanone	$CH_3CH_2COCH_3$
General formula:	$C_nH_{2n}O$ [$n \geq 1$]		$C_nH_{2n}O$ [$n \geq 3$]
Functional group:	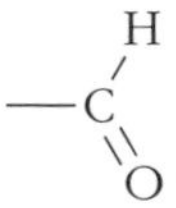 –C(H)=O or –CHO		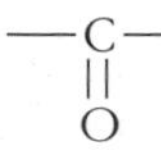 –C(=O)– or –CO–

Table 3

Oxidation of aldehydes

- Aldehydes and ketones are similar in that both types of compound contain the carbonyl group and are obtained by oxidation of alcohols.
- Aldehydes are readily oxidised while ketones are not.

Prescribed Practical Activity

Several oxidising agents can be used to distingush aldehydes from ketones:

1. Acidified potassium dichromate solution, $Cr_2O_7^{2-}(aq) + H^+(aq)$
2. Benedict's solution or Fehling's solution, $Cu^{2+}(aq)$ in alkaline solution
3. Tollen's reagent, i.e. ammoniacal silver(I) nitrate solution, $Ag^+(aq) + NH_3(aq)$.

In the experiment two samples of each oxidising agent are poured into separate test tubes. A few drops of an aldehyde are added to one set of each oxidising agent and a few drops of a ketone are added to the other set. The test tubes are placed in a hot water bath for several minutes (Figure 3).

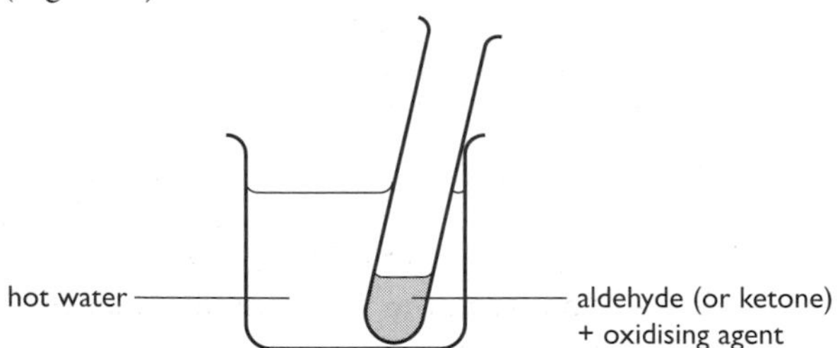

Figure 3

- Acidified potassium dichromate solution oxidises an aldehyde to form a **carboxylic acid**. If the aldehyde is an alkanal, it is oxidised to form an **alkanoic acid**, e.g.

$$CH_3CH_2CH_2CHO \rightarrow CH_3CH_2CH_2COOH$$

butanal → **butanoic acid**

- The other two oxidising agents are alkaline and as a result oxidise an alkanal to give the appropriate alkanoate ion, e.g.

$$CH_3CH_2CHO \rightarrow CH_3CH_2COO^-$$

propanal → **propanoate ion**

- The change that occurs to the functional group when an aldehyde is oxidised is shown below.

- This shows that during the oxidation of an aldehyde the C–H bond next to the carbonyl group is broken and the hydrogen atom is replaced by a hydroxyl group. This cannot happen with a ketone since it does not have a hydrogen atom attached to its carbonyl group.

Table 4 summarises the results for the reactions of an aldehyde, ketones do **not** react.

Oxidising Agent	Observations	Explanation
Acidified potassium dichromate solution	Orange solution → blue-green solution	$Cr_2O_7^{2-}(aq)$ reduced to $Cr^{3+}(aq)$
Benedict's solution or Fehling's solution	Blue solution → orange-red ppt.	$Cu^{2+}(aq)$ reduced to $Cu_2O(s)$ [copper(I) oxide] i.e. $Cu^{2+} + e^- \rightarrow Cu^+$
Tollen's reagent	Colourless solution → silver mirror	$Ag^+(aq)$ reduced to $Ag(s)$, i.e. $Ag^+ + e^- \rightarrow Ag$

Table 4

Summary of the oxidation of alcohols

	PRIMARY ALCOHOL	→	ALDEHYDE	→	CARBOXYLIC ACID
	Primary Alkanol	→	**Alkanal**	→	**Alkanoic Acid**
e.g.	CH_3CH_2OH	→	CH_3CHO	→	CH_3COOH
	ethanol		ethanal		ethanoic acid
	SECONDARY ALCOHOL	→	KETONE		not readily oxidised further
	Secondary Alkanol	→	**Alkanone**		
e.g.	$CH_3CH(OH)CH_3$	→	CH_3COCH_3		
	propan-2-ol		propanone		

TERTIARY ALCOHOLS are not readily oxidised.

In each of the examples above, oxidation of an alkanol or an alkanal has resulted in an increase in the oxygen to hydrogen ratio. The reverse of these reactions would be reductions and would involve a decrease in the oxygen to hydrogen ratio.

- **Oxidation** occurs when there is an **increase** in the oxygen to hydrogen ratio.
- **Reduction** occurs when there is a **decrease** in the oxygen to hydrogen ratio.

Questions

1 Which process occurs when ethanol is converted to ethanal?

A Reduction **B** Hydrogenation
C Oxidation **D** Dehydration

2* Which is true of a compound with the following formula?

$$CH_3CH(OH)CH_3$$

A It is a primary alcohol.
B It can be oxidised to an aldehyde.
C It is a tertiary alcohol.
D It can be oxidised to a ketone.

3 Which compound is formed by the oxidation of butan-2-ol?

A $CH_3CH_2CH_2CHO$

B $CH_3CH_2COCH_3$

C $CH_3CH_2CH_2COOH$

D $CH_3CH{=}CHCH_3$

4 Oxidation of butanal to butanoic acid results in the compound

A gaining 2 g per mole
B losing 16 g per mole
C gaining 16 g per mole
D losing 2 g per mole.

5* The dehydration of butan-2-ol can produce two isomeric alkenes, but-1-ene and but-2-ene. Which one of the following alkanols can similarly produce, on dehydration, a pair of isomeric alkenes?

A propan-2-ol **B** pentan-3-ol
C hexan-3-ol **D** heptan-4-ol

6 Which alkanol can be oxidised to produce an alkanone?

A 2-methylpropan-1-ol
B 3-methylbutan-2-ol
C 2,2-dimethylpropan-1-ol
D 2-methylbutan-2-ol

7 Which of the following statements is **not** true about aldehydes?

A They are formed by oxidation of secondary alcohols.
B They can be oxidised to produce carboxylic acids.
C Their molecules contain a carbonyl group of atoms.
D They will reduce acidified potassium dichromate solution.

8 Draw a structural formula and name the type of each of the following alcohols.
a) hexan-3-ol
b) 2-methylbutan-2-ol
c) 3,3-dimethylbutan-1-ol
d) 3-ethylpentan-2-ol

9 Give the systematic name and indicate the type of each of the following alcohols.

a) $CH_3CH_2CH(OH)CH_2CH_3$

b)
```
        C2H5
        |
CH3CH2CCH2CH3
        |
        OH
```

c) $CH_3CH(CH_3)CH_2CH_2OH$

d)
```
                  CH3
                  |
HOCH2CH2CH2CCH3
                  |
                  CH3
```

10 Butanone is an important solvent which can be made from butene in a two-stage process outlined below.

$$\text{BUTENE} \xrightarrow{H_2O(g)} \text{Compound } \mathbf{X} \xrightarrow{\text{oxidation}} \text{BUTANONE}$$

a) Draw the full structural formula of butanone.
b) Give the systematic name of compound **X**.
c) Name the type of reaction occurring in the first stage.
d) Name a suitable oxidising agent for the second stage.

11 The bromoalkane shown below can react with potassium hydroxide under different conditons to give two other carbon compounds, **Y** and **Z**.

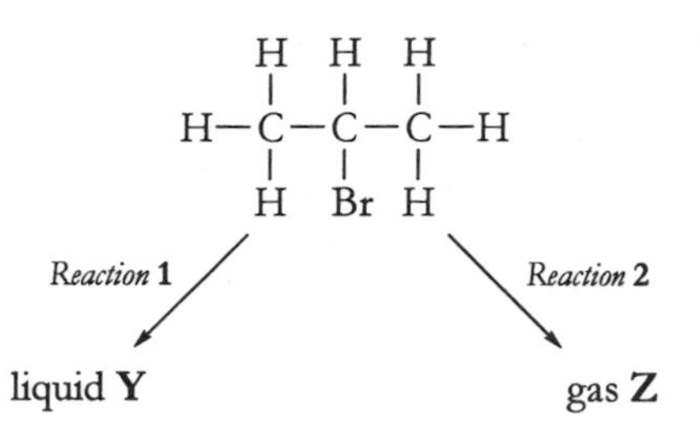

In reaction **1** the bromine atom is replaced by a hydroxyl group.
In reaction **2** the bromine atom and a hydrogen atom from an adjacent carbon atom are removed.

a) Name **Y** and **Z**.
b) **Y** and **Z** can be distinguished by chemical tests.
   i) Describe one test (with result) which would give a positive result with **Y**.
   ii) Describe one test (with result) which would give a positive result with **Z**.

12

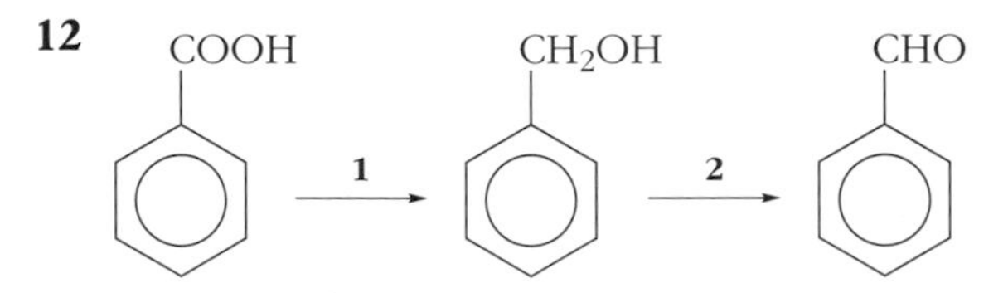

Benzoic acid can be converted into two other aromatic compounds by the reaction sequence shown above.
Work out the oxygen:hydrogen ratio for each compound and decide whether reactions **1** and **2** occur by oxidation or reduction.

13 When an alkene reacts with ozone, the molecule is split in two at the double bond to form carbonyl compounds. An example is shown below.

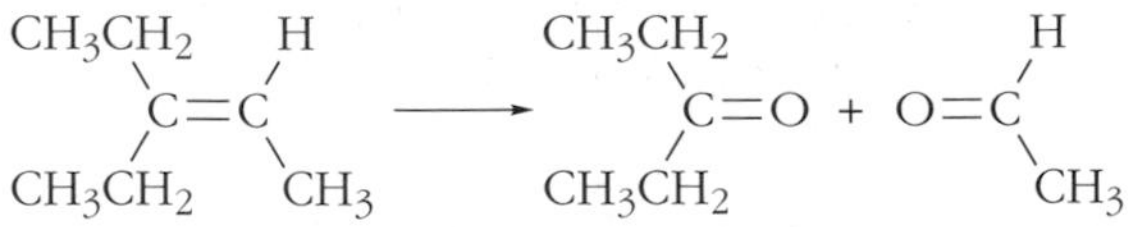

$$(CH_3CH_2)_2C{=}C(H)CH_3 \longrightarrow (CH_3CH_2)_2C{=}O + O{=}C(H)CH_3$$

a) Name the alkene shown above.
b) Name the alkanone produced in the reaction shown above.
c) Describe a chemical test which would distinguish between the carbonyl compounds produced in the above reaction.
d) When hex-3-ene reacts with ozone there is only one compound produced. Explain why.
e) Draw a structural formula of the alkene which reacts with ozone to form only propanone.

Questions 14 and **15** refer to the following reaction sequence:

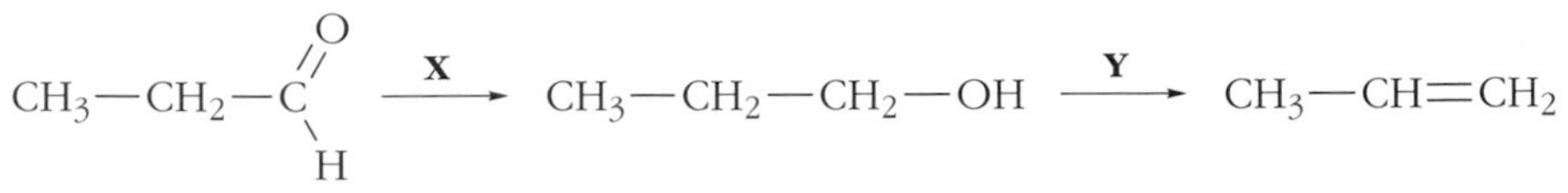

$$CH_3{-}CH_2{-}C(=O)H \xrightarrow{\textbf{X}} CH_3{-}CH_2{-}CH_2{-}OH \xrightarrow{\textbf{Y}} CH_3{-}CH{=}CH_2$$

14 Reaction **X** involves reduction and this is confirmed by

A a decrease in the carbon-hydrogen ratio
B an increase in the number of atoms
C a decrease in the oxygen-hydrogen ratio
D an increase in the number of single bonds.

15 The type of reaction occurring in step **Y** is

A condensation
B dehydration
C dehydrogenation
D hydrolysis.

16 The following list contains carbon compounds:

Butan-1-ol	Propanone	Pentan-3-ol
Propan-2-ol	Butanal	2-Methylpropan-2-ol.

a) Which compound listed above is
i) a tertiary alcohol?
ii) formed when $CH_3CH{=}CH_2$ is hydrated?
iii) an isomer of methoxyethane, $CH_3OCH_2CH_3$?

b) Which pair of compounds listed above
i) contain a carbonyl group?
ii) are secondary alcohols?
iii) are isomers?

c) One compound in the list can be produced by oxidation and can also itself be oxidised.
i) Which compound behaves in this way?
ii) Give the formula of the product when this compound is oxidised.
iii) What colour change occurs when acidified potassium dichromate is used as the oxidising agent in these reactions?

10 Carboxylic Acids and Esters

Carboxylic Acids

- A carboxylic acid is characterised by the **carboxyl** functional group, and by its name ending, '-oic acid'.

  The carboxyl functional group:

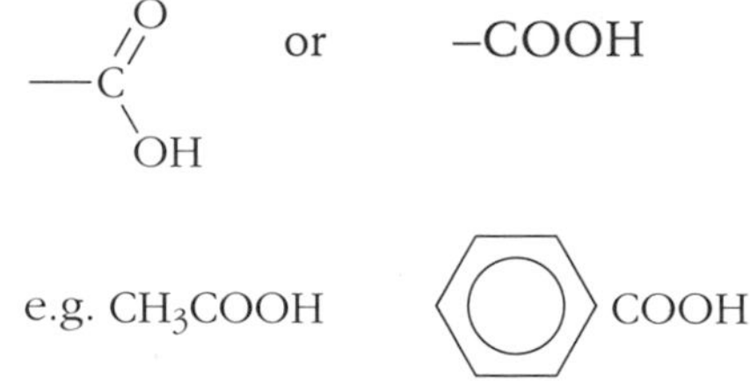

- There is a homologous series of carboxylic acids called alkanoic acids which are based on the corresponding parent alkanes. This is shown in Table 1.
- Branched-chain alkanoic acids can also be obtained and these are named with the functional group taking precedence. There is no need to give a number for the functional group as it must be at the end of the chain, i.e. the carbon atom of the –COOH group is the first carbon atom in the chain.

e.g.

$$CH_3CH(CH_3)CH_2COOH$$

3-methylbutanoic acid

Uses of carboxylic acids

- Ethanoic acid prevents bacterial and fungal growth. Dilute solutions, vinegar, are used for pickling food.
- Ethanoic acid is a feedstock for the production of, for example, ethenyl ethanoate (vinyl acetate), which is polymerised to give the plastic component of vinyl emulsion paints, and cellulose ethanoate, which is used to produce films and lacquers.
- Benzene-1,4-dicarboxylic acid, is a diacid used to make polyester and hexanedioic acid is a monomer used in the production of nylon.
- Benzoic acid, C_6H_5COOH, is the food additive, E210, acting as a preservative and antioxidant.

$CH_3COOCH{=}CH_2$ — ethenyl ethanoate

$HOOC{-}C_6H_4{-}COOH$ — benzene-1,4-dicarboxylic acid

$HOOC(CH_2)_4COOH$ — hexanedioic acid

Name of acid	Structural formulae	
Methanoic acid	$HCOOH$	H–C(=O)–OH
Ethanoic acid	CH_3COOH	H–C(H)(H)–C(=O)–OH
Propanoic acid	CH_3CH_2COOH	
Butanoic acid	$CH_3CH_2CH_2COOH$	
Pentanoic acid	$CH_3CH_2CH_2CH_2COOH$	
General formula:	$C_nH_{2n}O_2$	

 Table 1

Esters

- When a carboxylic acid reacts with an alcohol a new type of carbon compound, called an **ester**, is formed.

A convenient method of preparing an ester is described below.

Prescribed Practical Activity

About 2 cm^3 of an alcohol and an equal volume of a carboxylic acid are mixed together and a few drops of concentrated sulphuric acid are added. This mixture is heated as shown in Figure 1 for several minutes.

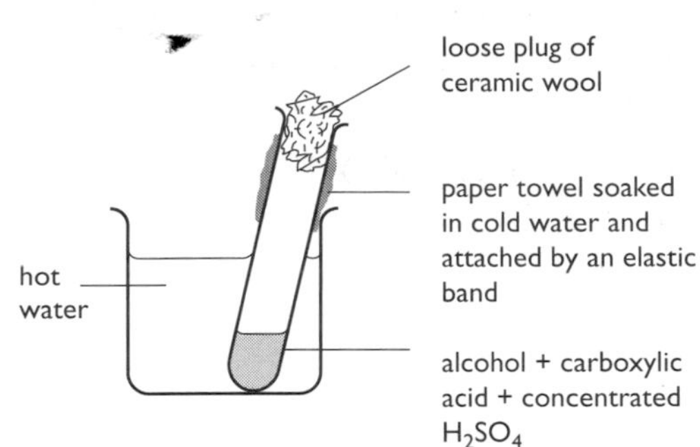

Figure 1

The test tube is removed from the water bath and its contents added to about 20 cm^3 of sodium hydrogencarbonate solution as shown in Figure 2. This solution neutralises the sulphuric acid as well as any unreacted carboxylic acid releasing carbon dioxide in the process. The ester appears as an immiscible layer on the surface of the solution. The ester can also be detected by its characteristic smell.

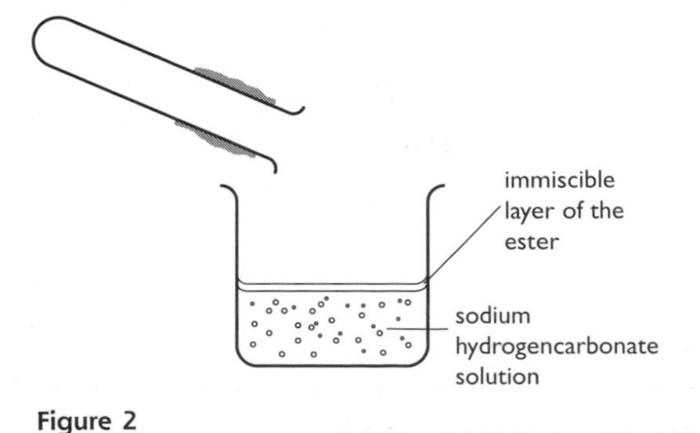

Figure 2

- The functional groups present in the reactants interact to form an ester link and, at the same time, produce a water molecule, i.e. the process of making an ester is an example of a **condensation** reaction (Figure 3).

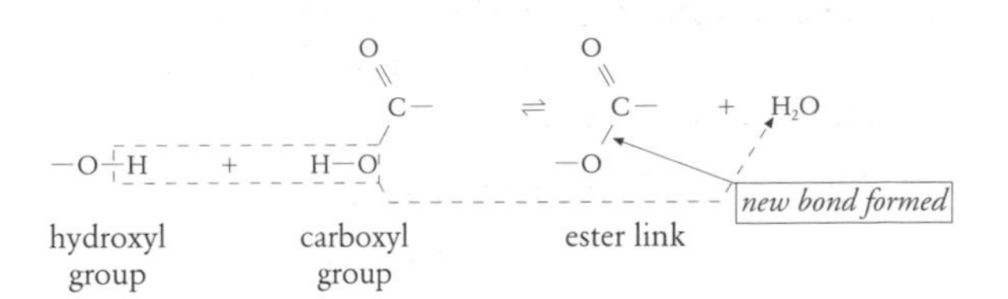

Figure 3

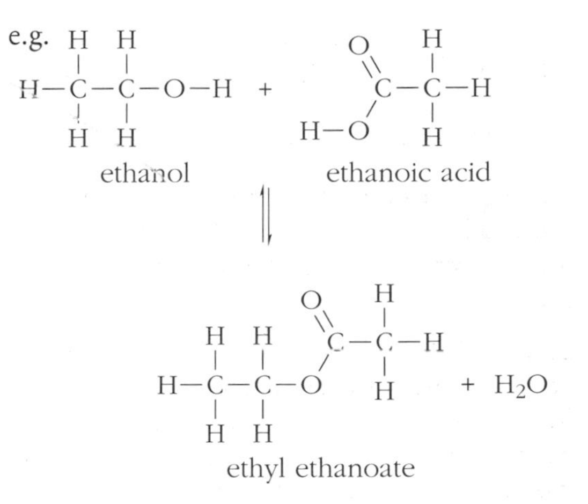

This equation can be rewritten using shortened structural formulae:

$$CH_3CH_2OH + HOOCCH_3 \rightleftharpoons CH_3CH_2OOCCH_3 + H_2O$$

To show the reaction between the functional groups, in the example above the formula of the carboxylic acid has been reversed. The equation can, of course, be written with the formula of the carboxylic acid shown first and the formula of the alcohol written in reverse.

i.e.

$$\underset{\text{ethanoic acid}}{CH_3COOH} + \underset{\text{ethanol}}{HOCH_2CH_3} \rightleftharpoons \underset{\text{ethyl ethanoate}}{CH_3COOCH_2CH_3} + H_2O$$

- Concentrated sulphuric acid provides hydrogen ions which catalyse the reaction. It also helps to increase the yield of ester by absorbing water, the other product. (See Chapter 15.)

Naming esters

- The name of an ester depends on which alcohol and which carboxylic acid have been used in preparing it. The first part of the name, which ends in '-yl', comes from the alcohol and the second part,which ends in '-oate', comes from the acid.

An example is illustrated in Figure 4. The formula of the ester is shown twice since there are two ways of drawing it as explained above.

- Esters prepared from alkanols and alkanoic acids which have the same total number of carbon atoms will have the same molecular formula.

For example, ethyl ethanoate and methyl propanoate are isomers with molecular formula $C_4H_8O_2$. Butanoic acid, C_3H_7COOH, is also an isomer but it belongs to a different homologous series.

Uses of esters

- Different esters differ widely in smell. They are used in perfumes and as artificial flavourings in food, e.g. pentyl ethanoate has a pear-like flavour. Artificial perfumes or flavourings have several natural and synthetic materials blended together.
- The principal use of esters is as solvents, e.g. in car body paints, radiator enamels, adhesives and cosmetic preparations. Rapid evaporation of the solvent requires esters with few carbon atoms, e.g. ethyl ethanoate, as they tend to be more volatile.
- Esters also have medicinal uses, e.g. methyl salicylate or 'oil of wintergreen' is used as a liniment.

Hydrolysis of Esters

- The formation of an ester, **esterification**, is a reversible reaction. The reverse process in which the ester is split by reaction with water to form an alcohol and a carboxylic acid is an example of **hydrolysis**. Hydrolysis is the opposite of condensation. The C–O bond formed when an ester is made is the bond which is broken when the ester is hydrolysed.
- Hydrolysis of an ester can be carried out by heating in the presence of a dilute acid, e.g. HCl(aq) or H_2SO_4(aq), to provide hydrogen ions to catalyse the reaction.

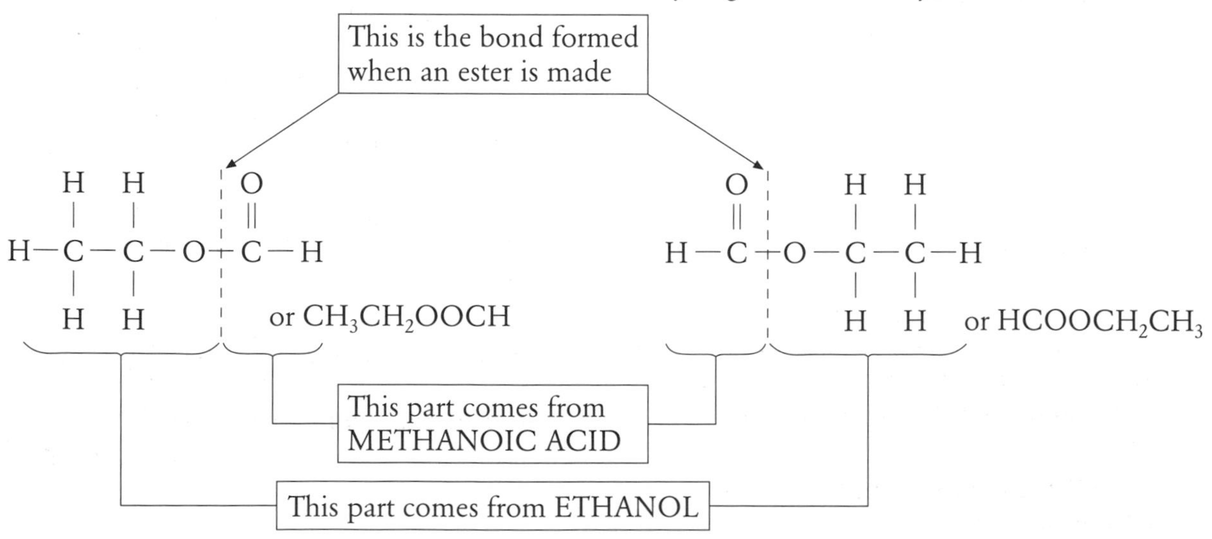

 Figure 4

The equation for the acid-catalysed hydrolysis of ethyl ethanoate is as follows:

$$CH_3CH_2OOCCH_3 + H_2O \rightleftharpoons CH_3CH_2OH + HOOCCH_3$$

ethyl ethanoate → **ethanol**, **ethanoic acid**

- Acid-catalysed hydrolysis is reversible and is incomplete.
- Complete hydrolysis can be achieved by adding the ester to a strong alkali, e.g. NaOH(aq) or KOH(aq), and heating under reflux for about 30 minutes as shown in Figure 5.
- The products are the alcohol and the salt of the carboxylic acid and these can be separated by distillation. The alcohol is distilled off and the salt of the carboxylic acid is left behind in the aqueous solution. When this solution is acidified with hydrochloric acid the salt is converted to the carboxylic acid.

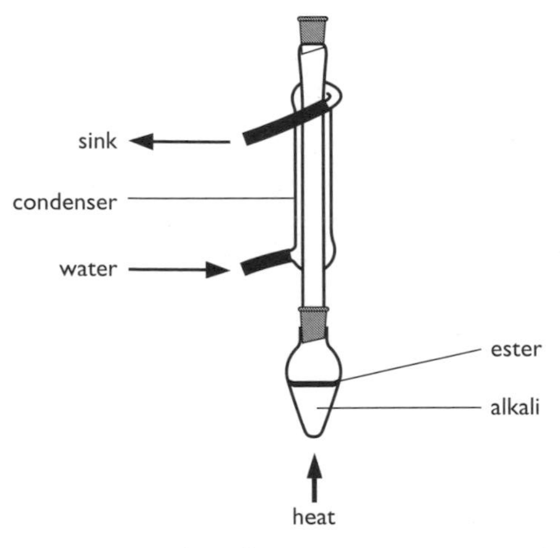

Figure 5 Heating under reflux

The equations for **a)** the hydrolysis of ethyl ethanoate by sodium hydroxide solution and **b)** subsequent acidification of the salt solution (after removal of ethanol by distillation) are as follows:

a) $CH_3COOCH_2CH_3 + Na^+ + OH^- \rightarrow CH_3COO^- + Na^+ + HOCH_2CH_3$

ethyl ethanoate → sodium ethanoate, ethanol

b) $CH_3COO^- + Na^+ + H^+ + Cl^- \rightarrow CH_3COOH + Na^+ + Cl^-$

sodium ethanoate → ethanoic acid

- Ester hydrolysis, especially by alkali, is an important process in the manufacture of soap from fats and oils. (See Chapter 12.)

Percentage yield

- The expected quantity of product from a known mass of reactant can be calculated from the balanced equation. This is called the **theoretical yield**.
- The **actual yield** is usually less than the theoretical yield, particularly so when carrying out a reaction involving carbon compounds because:

  – the reaction may not go to completion, e.g. preparation of an ester,

  – other reactions may also occur which 'compete' with the main reaction,

  – separation of the desired product may be difficult,

  – the product may be impure and some of it lost during purification.

- In industrial processes a high percentage yield as well as high purity of product is desirable. Unconverted reactants are frequently recycled for further reaction, e.g. making ammonia by the Haber process and making ethanol by hydration of ethene.

- The **percentage yield** of product can calculated by using the following relationship:

$$\text{Percentage yield} = \frac{\text{Actual yield}}{\text{Theoretical yield}} \times 100\%$$

Also: Actual yield = Percentage yield × Theoretical yield

Worked Example 10.1

A sample of methyl ethanoate weighing 6.9 g was obtained from a reaction mixture containing 9.0 g of ethanoic acid, excess methanol and a small volume of concentrated sulphuric acid. Calculate the percentage yield of ester using the following equation.

$$CH_3OH + HOOCCH_3 \rightleftharpoons CH_3OOCCH_3 + H_2O$$

1 mol = 60 g (ethanoic acid); 1 mol = 74 g (methyl ethanoate)

In theory, 60 g of ethanoic acid should yield 74 g of methyl ethanoate. Hence, 9.0 g of ethanoic acid should yield $9.0 \times \frac{74}{60}$ g = 11.1 g of methyl methanoate.

Thus, the theoretical yield is 11.1 g, while the actual yield is 6.9 g.

Percentage yield = $\frac{6.9}{11.1} \times 100\ \% = 62.2\ \%$

Summaries of reactions in Carbon chemistry are provided on pages 142–3.

Questions

1* Propanoic acid is reacted with ethanol. The formula for one of the products is

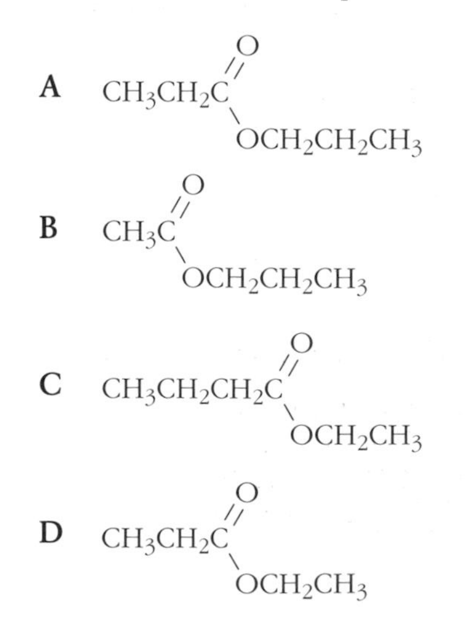

2 The compound with formula CH_3OOCCH_3 can be hydrolysed to give

A ethanol and methanoic acid
B methanol and ethanoic acid
C ethanol and ethanoic acid
D methanol and methanoic acid.

3* Aspirin is one of the most widely used pain relievers in the world. It has the structure:

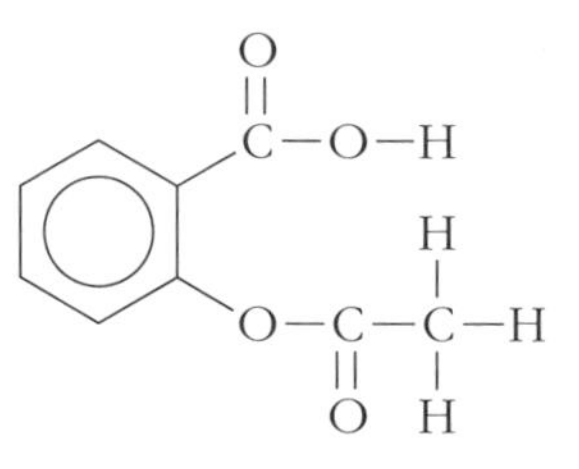

Which two functional groups are present in an aspirin molecule?

A Hydroxyl and carboxyl
B Aldehyde and ketone
C Carboxyl and ester
D Ester and aldehyde

4* Ethyl butanoate is used in pineapple flavouring. The formulae for the molecules from which it is made are

A C_3H_7OH and CH_3COOH
B C_2H_5OH and C_2H_5COOH
C C_3H_7COOH and C_2H_5OH
D C_2H_5COOH and C_3H_7OH

5 The compound with the formula $CH_3CH_2CH_2CH_2COOH$ can be obtained when

A pentan-1-ol is oxidised
B methyl butanoate is hydrolysed
C 3-methylbutan-1-ol is oxidised
D pentanal is reduced.

6 Draw a structural formula for each of the following compounds.

a) Hexanoic acid
b) 2,3-dimethylbutanoic acid
c) 3-ethyl-4-methylpentanoic acid
d) Butyl methanoate
e) Propyl propanoate

7 Name the following compounds.

a)

$$\begin{array}{c} \quad\quad C_2H_5 \\ \quad\quad | \\ CH_3CH_2CHCH_2CH_2COOH \end{array}$$

b) CH_3OOCCH_3

c)

$$\begin{array}{c} CH_3 \\ | \\ CH_3CCH_2COOH \\ | \\ CH_3 \end{array}$$

d) $HCOOCH_2CH_2CH_3$

8

$$\begin{array}{c} CH_3 \\ | \\ CH_3-C-OH \\ | \\ CH_3 \end{array} \quad \text{Compound } \mathbf{Y}$$

5.0 g of compound **Y** was reacted with a slight excess of propanoic acid by heating for several minutes.

a) What other chemical should be added to the mixture before heating?
b) Describe briefly the method of heating that should be used.
c) What type of reaction occurs between compound **Y** and propanoic acid?
d) Draw a structural formula for the carbon compound produced in this reaction.
e) How could you show that a new substance had been produced?
f) Calculate the percentage yield if the actual yield of ester (molecular formula, $C_7H_{14}O_2$) was 6.2 g.

9

$$\begin{array}{c} \;\; CH_3 \qquad\quad CH_3 \\ \;\; | \qquad\qquad\quad | \\ CH_3CHCOOCHCHCH_3 \\ \qquad\quad | \\ \qquad\quad CH_3 \end{array}$$

10.0 g of the compound shown above were added to 50 cm^3 of dilute hydrochloric acid and heated under reflux for about 30 minutes.

a) What type of reaction occurred?
b) Draw a labelled diagram of an apparatus which could be used in this experiment showing clearly what is meant by the phrase 'heated under reflux'.
c) Draw the structural formulae and give the names of the two carbon compounds produced in this reaction.
d) Calculate the mass of alcohol obtained if the percentage yield is 65%.

10* Esters are important and useful compounds. They occur in nature and can also be made in the laboratory.

a) An ester can be made from ethanol and methanoic acid. Draw the full structural formula for this ester.
b) Name the catalyst used in the laboratory preparation of an ester.
c) How can this ester be separated from unreacted ethanol and methanoic acid?

11* Esters are a widely used class of organic compounds.

a) Draw a labelled diagram to show how to prepare an ester from an alkanol and an alkanoic acid.
b) State any safety precaution that would be taken (apart from wearing eye protection) and give a reason for it.
c) Give a use for an ester.
d) Draw a structural formula for the ester produced in the reaction between methanol and ethanoic acid.

Questions 12 to **21** relate to work covered in **Chapters 8, 9** and **10.**

Questions 12 and **13** refer to the following compounds:

A CH_3CH_2CHO
B CH_3CH_2COOH
C CH_3COCH_3
D CH_3OOCCH_3.

12 Which compound is an aldehyde?

13 Which compound can be hydrolysed by sodium hydroxide solution?

14 Which of the following compounds is an isomer of ethyl butanoate?

A hexan-2-ol.
B 2,3-dimethylbutanal.
C 2-methylpentanoic acid.
D propyl ethanoate.

15 Propanoic acid can be produced when

A $CH_3CH{=}CH_2$ is hydrated
B CH_3CH_2CHO is reduced
C CH_3OOCCH_3 is hydrolysed
D $CH_3CH_2CH_2OH$ is oxidised.

16 Partial dehydration of ethanol produces a compound called ethoxyethane, which is commonly called ether. The equation for this reaction is shown below.

$$2C_2H_5OH \rightarrow C_2H_5OC_2H_5 + H_2O$$

High yields of ether are difficult to achieve for various reasons, such as

i) complete dehydration of some of the ethanol
ii) the high volatility of ether.

a) What is the product of complete dehydration of ethanol?

b) Draw the full structural formula of ethoxyethane and suggest why it is so volatile when compared with ethanol.

c) In an experiment, 4.5 g of ethoxyethane were obtained from 20.0 g of ethanol. Calculate the percentage yield of ether.

17 Phenol reacts with chlorine to produce the important antiseptic, trichlorophenol [TCP] according to the following equation.

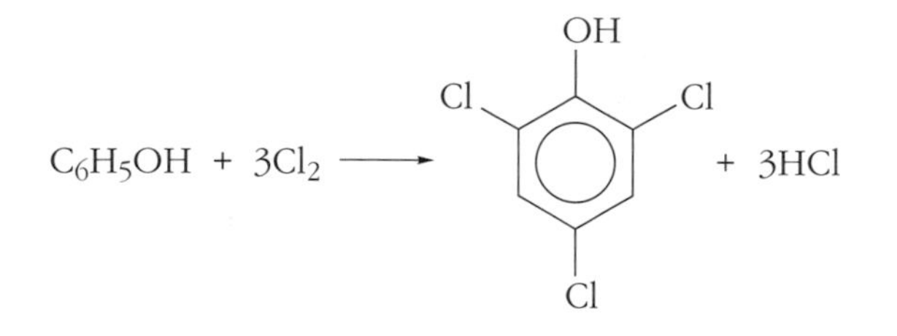

a) Give the molecular formula of TCP.

b) In an experiment to make some TCP in the lab, 28 g of phenol was used. Calculate

i) the number of moles of phenol used
ii) the volume of chlorine required at room temperature and pressure (V_{mol} = 24 litres)
iii) the mass of TCP (1 mol = 197.5 g) obtained if the percentage yield is 83%.

18

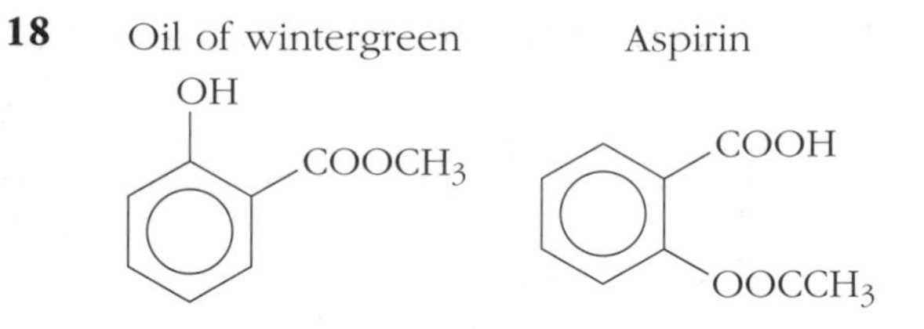

a) What is the molecular formula of aspirin?

b) Which of the following functional groups is/are present in oil of wintergreen?

Carbonyl **Carboxyl** **Ester** **Hydroxyl**

c) When hydrolysed both compounds yield a compound called salicylic acid.

i) Suggest an alternative name for oil of wintergreen.
ii) Work out the structural formula of salicylic acid.

d) Theoretically one mole of salicylic acid (138 g) produces one mole of aspirin (180 g). In an experiment 9.2 g of aspirin was obtained from 9.6 g of salicylic acid. Calculate the percentage yield of aspirin.

19

ETHENE —**I**→ C_2H_5OH
ETHENE —**II**→ C_2H_5CN
C_2H_5CN —**III**→ C_2H_5COOH
C_2H_5OH + C_2H_5COOH —**IV**→ Compound **Q**

a) In reaction **II**, ethene undergoes addition with hydrogen cyanide (structural formula: H–C≡N). Draw the full structural formula of ethyl cyanide, the product of reaction **II**.

b) Reaction **I** is also an addition reaction. What other name can be used to describe this reaction?

c) In reaction **III**, ethyl cyanide reacts with water and HCl(aq). Ammonium chloride is also produced in this reaction. Write a balanced equation for reaction **III**.

d) In reaction **IV**, the products of reactions **I** and **III** react to form another carbon compound, **Q**.

i) Name compound **Q** and draw its full structural formula.
ii) Describe how reaction **IV** can be carried out in the laboratory.

20 The structural formulae of six aromatic compounds are shown below.

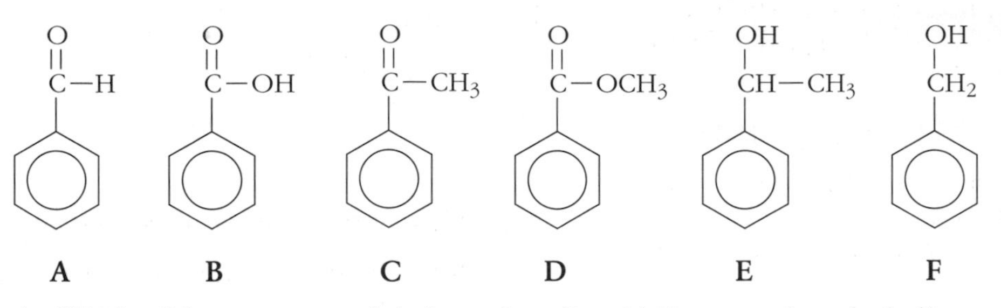

a) Which of these compounds is **i)** a carboxylic acid **ii)** a secondary alcohol?
b) Classify each of the four remaining compounds.
c) Which of these compounds is formed when
  i) **A** is oxidised
  ii) **A** is reduced
  iii) **E** is oxidised
  iv) **B** reacts with methanol.
d) What is the molecular formula of **D**?
e) Which two compounds react with NaOH(aq)? In each case name the type of reaction occurring.
f) Describe a chemical test, with result, which would distinguish **A** and **C**.

21 In the following flow diagram, compound **X** is converted into two other compounds, one of which is then converted into various products as shown.

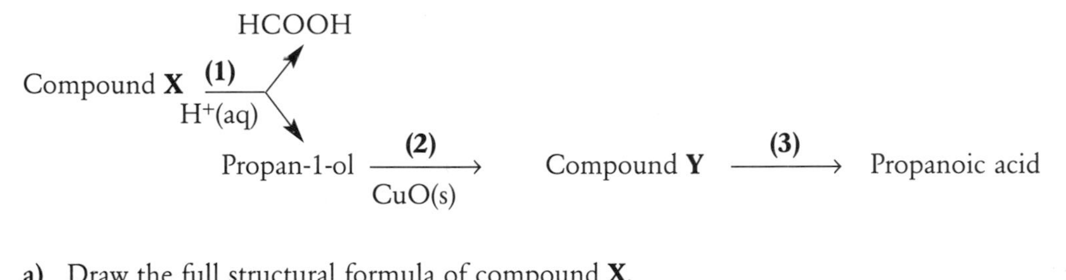

a) Draw the full structural formula of compound **X**.
b) Name compound **Y**.
c) Suggest why **X** and **Y** have lower boiling points than the other carbon compounds shown in the diagram.
d) Alkanols behave differently when in contact with different metal oxides.
  i) What happens to CuO(s) in reaction **(2)**?
  ii) What type of reaction occurs when propan-1-ol vapour is passed over hot Al_2O_3(s)? Name the carbon compound produced.

11 Polymers

Polymers are very large molecules, macromolecules, made by joining small molecules, monomers, in long chains or networks.

Addition polymers

- Addition polymers are made from **unsaturated monomers**. The polymers are saturated.
- Ethene is the starting material for many products of the chemical industry, especially for addition polymers. Ethene can be made by cracking ethane from natural gas or the gas fraction of crude oil or by cracking the naphtha fraction of crude oil. It can then be polymerised:

$$n\,CH_2{=}CH_2 \rightarrow \left[CH_2{-}CH_2\right]_n$$

ethene **poly(ethene)**

- Propene is made by cracking propane from the gas fraction of crude oil or by cracking naphtha. It can be polymerised into poly(propene) as shown in Figure 1.

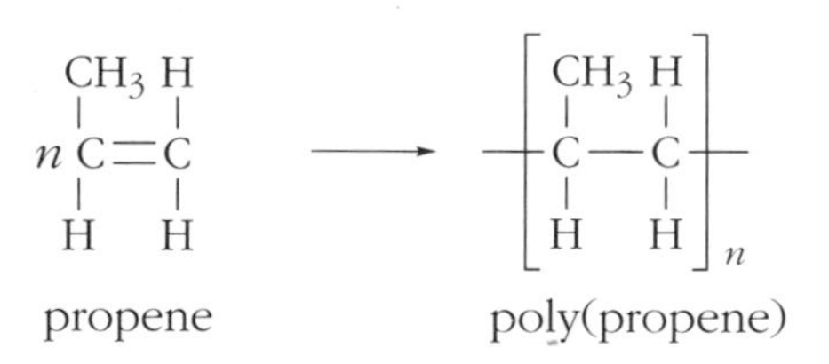

Figure 1

Condensation polymers

- Condensation polymers are made from monomers with two functional groups per molecule. The monomers and repeating units or structure of some condensation polymers are shown below. In each case, water is formed as a by-product.

Polyester

- Polyesters are synthesised by condensation of two monomers. One having two hydroxyl groups is called a **diol** and the other, with two carboxyl groups, is a **diacid**. Their structures are shown in Figure 2, and the structure of polyester, obtained by condensing two diol molecules and two diacid molecules, is shown in Figure 3.

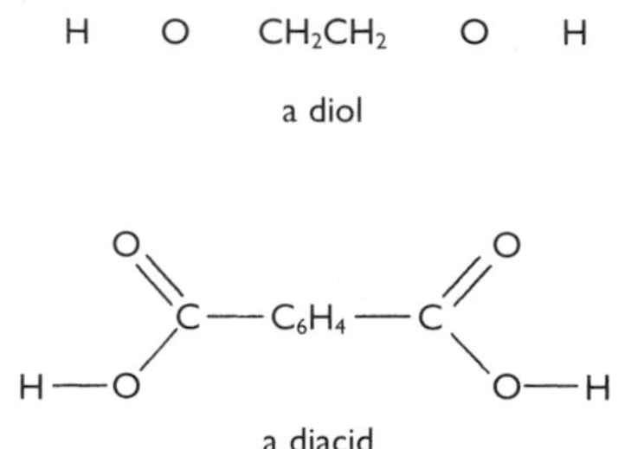

Figure 2

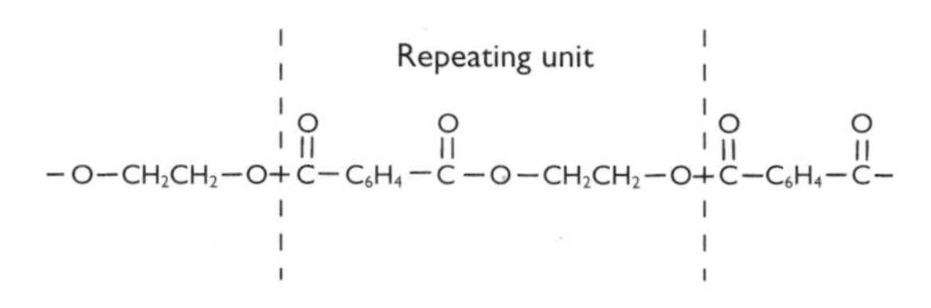

Figure 3

The dotted lines indicate the **repeating unit**.

- Polyesters with a linear structure are used for fibres.
- Polyesters which are cross-linked to produce a network, 3-dimensional structure are used for moulding resins.

Polyamide

- An amide link is formed by the reaction of an amine functional group with a carboxyl group.

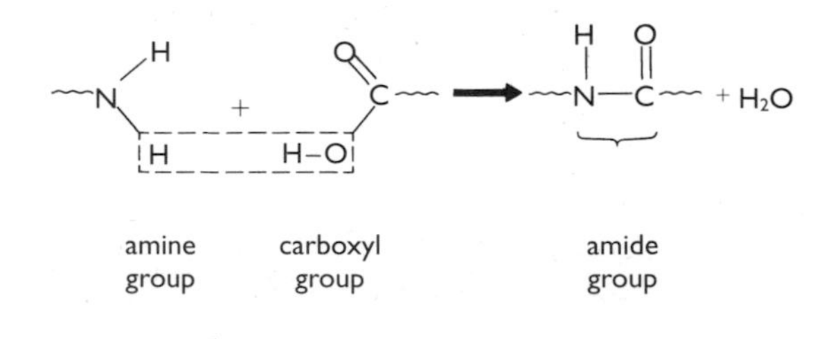

Figure 4

- Nylon is an example of a polyamide. It is used as a fibre and as an engineering plastic.
- The strength of nylon is related to hydrogen bonding between its chains.

Figure 5 shows the structures of typical monomers for making a polyamide, and Figure 6 shows the resulting polyamide made from two molecules of each monomer. Once again the dotted lines show the repeating unit, starting from the middle of the -NH-CO- amide linkage.

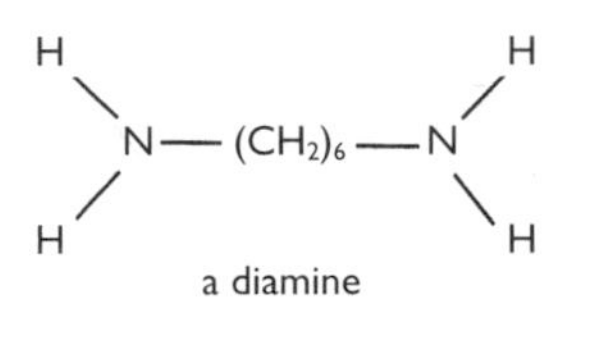

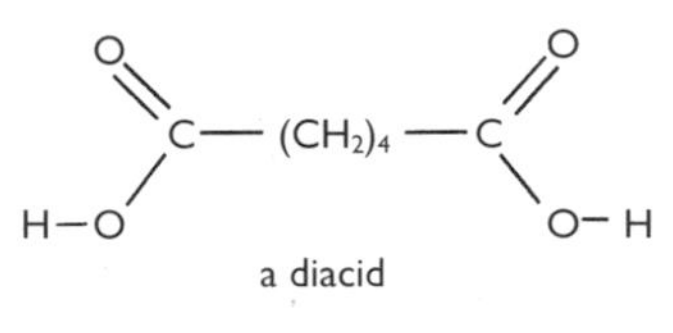

Figure 5

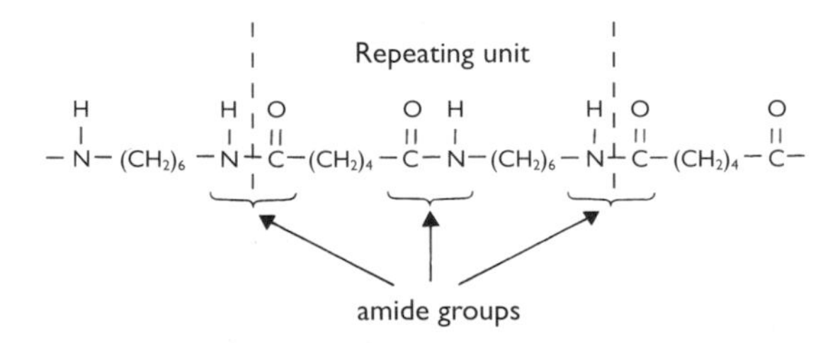

Figure 6

Methanal-based thermosetting polymers

- Thermosetting polymers harden on heating. They cannot be softened by warming or be remoulded. (Polymers which can be remoulded after warming are thermoplastic.)
- Examples of methanal-based polymers are urea-methanal and bakelite (Figure 7).

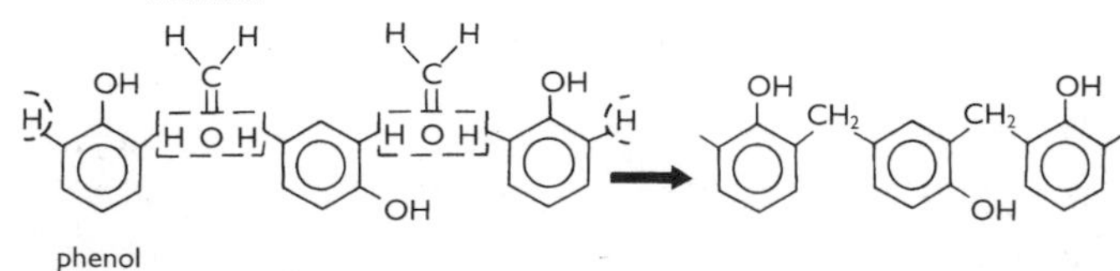

Figure 7 Bakelite

- A major use of both polymers is for electrical fittings because of their good insulating properties.
- Methane or coal is steam reformed to give synthesis gas, a mixture of carbon monoxide and hydrogen.

$$CH_4(g) + H_2O(g) \rightarrow CO(g) + 3H_2(g)$$
(synthesis gas)

- The synthesis gas is the feedstock for the production of methanol, which, in turn, is the feedstock for producing methanal.

$$CO(g) + 2H_2(g) \rightarrow CH_3OH(g)$$
(methanol)

$$CH_3OH(g) + \tfrac{1}{2}O_2(g) \rightarrow HCHO(g) + H_2O(g)$$
(methanal)

Recent developments

Kevlar

- Kevlar is an aromatic polyamide.
- It is very strong because of the way the rigid linear molecules are packed together.

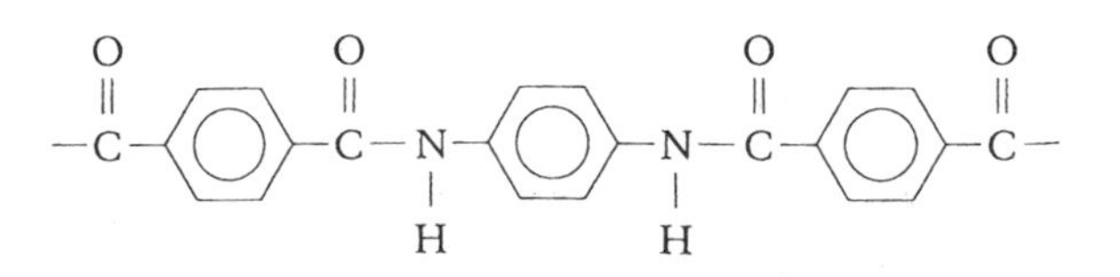

Figure 8 Kevlar

- Kevlar is used in sail fabric, crash helmets and in body armour.

Poly(ethenol)

- Poly(ethenol) is a water-soluble plastic.
- It is made from another polymer by 'ester exchange'

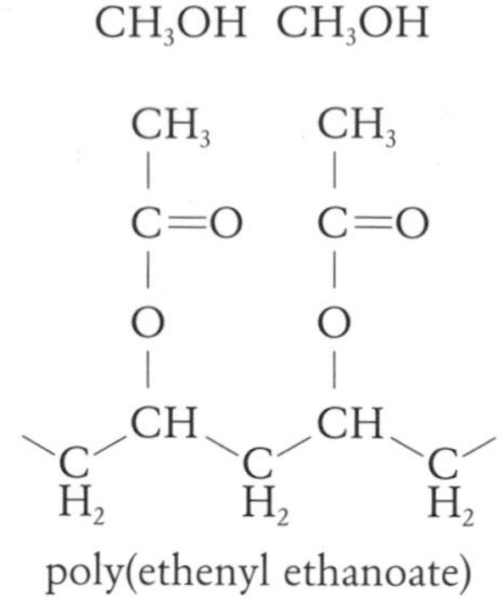

$CH_3O-C(=O)-CH_3$ $\quad$ $CH_3O-C(=O)-CH_3$

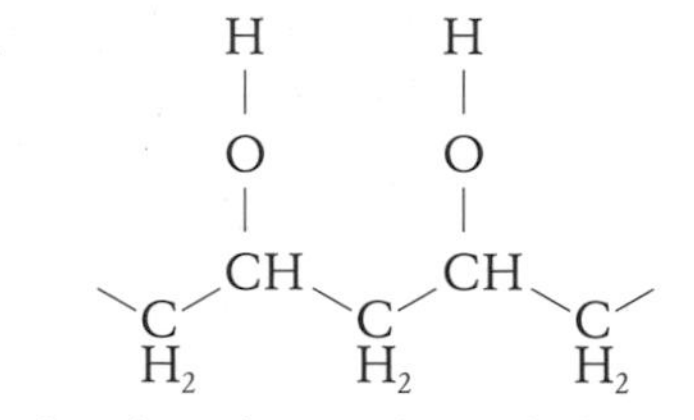

poly(ethenyl ethanoate)
Reacts with methanol

poly(ethenol) or polyvinyl alcohol
Ester groups on the side chains of the polymer are removed and new ester groups form in methyl ethanoate

Figure 9 Formation of poly(ethenol)

- The proportion of acid groups removed in the production process determines the strength of the intermolecular forces upon which the solubility depends.
- Poly(ethenol) is used to make adhesives, water-soluble protective coatings for cars in storage and soluble hospital laundry bags.

Poly(ethyne)

- Poly(ethyne) is a polymer which can be made to conduct electricity.

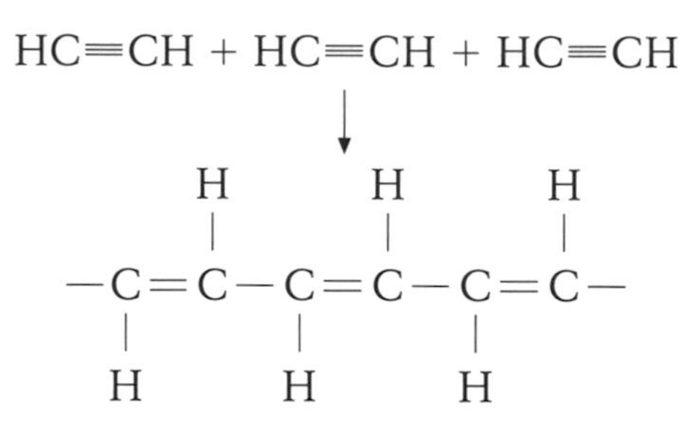

Figure 10

- Poly(ethyne)'s conductivity depends on delocalised electrons. It is used to make loud speaker membranes.

Poly(vinylcarbazole)

- Poly(vinylcarbazole) is able to conduct electricity when exposed to light. It is used in photocopiers.

Biopol

- Biopol is a biodegradable polymer. It is not proving to be popular because it is not in keeping with the present move to encourage recycling.
- Biopol included with other polymers can make the mixture unsuitable for recycling.

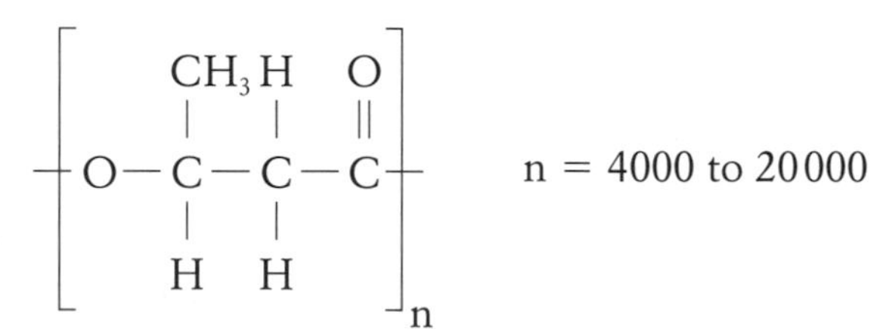

Figure 11 The structure of Biopol

Photodegradable LDPE

- Low density poly(ethene) can be made photodegradable i.e. to break up under the action of light.

Questions

1* Part of a polymer is shown.

$$
\begin{array}{cccccccc}
H & C_2H_5 & H & H & H & C_2H_5 & H & H \\
| & | & | & | & | & | & | & | \\
-C- & C- & -C- & C- & C- & C- & -C- & C- \\
| & | & | & | & | & | & | & | \\
H & H & H & H & H & H & H & H
\end{array}
$$

Which pair of alkenes was used as monomers?

A Ethene and propene **B** Ethene and but-1-ene
C Propene and but-1-ene **D** Ethene and but-2-ene

2* Part of a polyester chain is shown below.

$$-O-\overset{\overset{\displaystyle O}{\|}}{C}-(CH_2)_4-\overset{\overset{\displaystyle O}{\|}}{C}-O-(CH_2)_6-O-\overset{\overset{\displaystyle O}{\|}}{C}-(CH_2)_4-\overset{\overset{\displaystyle O}{\|}}{C}-O-(CH_2)_6-O-$$

Which compound, when added to the reactants during polymerisation, would stop the polyester chain from getting too long?

A $HO-\overset{\overset{\displaystyle O}{\|}}{C}-(CH_2)_4-\overset{\overset{\displaystyle O}{\|}}{C}-OH$

B $HO-(CH_2)_6-OH$

C $HO-(CH_2)_5-\overset{\overset{\displaystyle O}{\|}}{C}-OH$

D CH_3-OH

3* Polyester fibres and cured polyester are both very strong.
What kinds of structure do their molecules have?

	Fibre	Cured resin
A	cross-linked	cross-linked
B	linear	linear
C	cross-linked	linear
D	linear	cross-linked

Questions 4 and **5** refer to the following polymers.

A Biopol **B** Poly(ethenol)
C Kevlar **D** Poly(ethyne)

4 Which polymer can conduct electricity?

5 Which polymer may be used for water-soluble laundry bags?

6 The structural formula of biopol is:

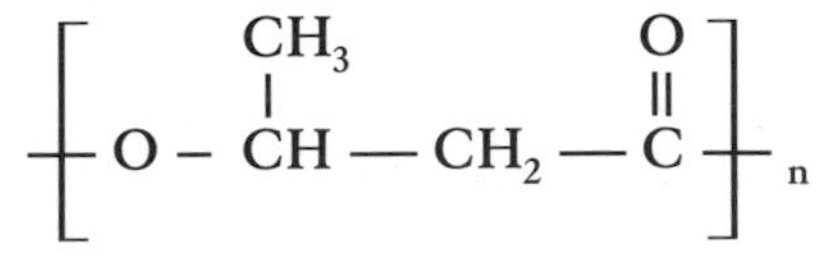

Which of the following statements about biopol is **not** true?

A It is an addition polymer. **B** It is biodegradable.
C It is a linear polymer. **D** It is a polyester.

7 Compounds **A** to **F** are important monomers in manufacturing plastics.

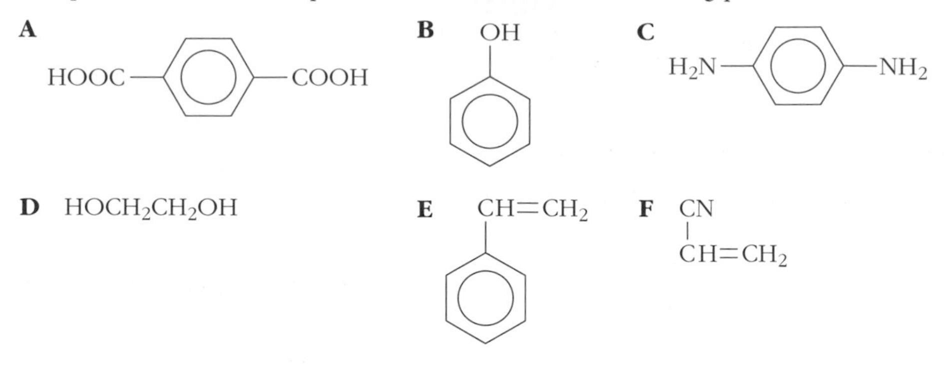

a) Kevlar is a polyamide made from two of the monomers shown above.
   i) Which **two** monomers are used to make kevlar?
   ii) Draw a structural formula to show the repeating unit of kevlar.
   iii) Describe one use of kevlar.

b) Terylene is a polyester made from two of the monomers shown above.
   i) Which **two** monomers can be used to make terylene?
   ii) Draw a structural formula to show the repeating unit of terylene.

c) SAN is a plastic made by co-polymerising compounds **E** and **F**. Draw a section of this polymer's structure in which two molecules of compound **E** have joined one on each side of a molecule of compound **F**.

d) Bakelite is a condensation thermosetting polymer made from compound **B** and methanal. Explain the meaning of the terms '*condensation*' and '*thermosetting*'.

8

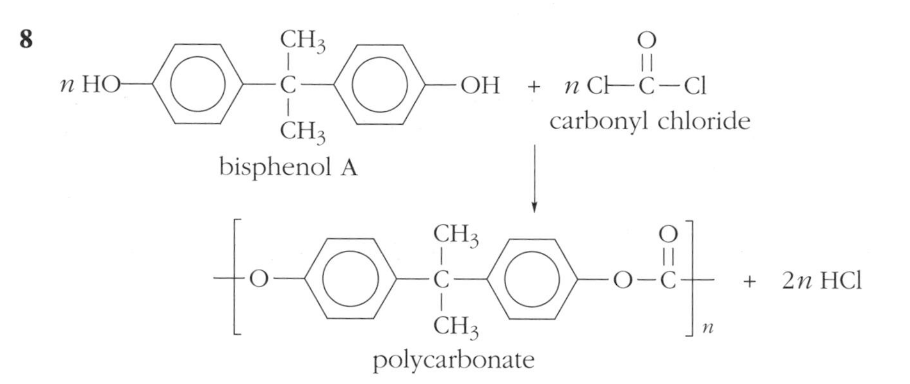

The condensation reaction shown above is used to manufacture 'polycarbonates'. Polycarbonates are transparent, strong and have good resistance to heat and chemicals. They are thermoplastic and are used in making compact discs, safety glass and feeding bottles.

a) How does the above reaction differ from more familiar condensation reactions?

b) What part of the structure gives rise to the name 'polycarbonate'?

c) In the manufacturing process, bisphenol A is dissolved in NaOH(aq) and carbonyl chloride is bubbled through the solution. Sodium hydroxide solution is a better solvent for the monomer than water alone. Suggest another reason why the presence of NaOH is helpful.

d) The polymer is insoluble in NaOH(aq) and is removed by dissolving it in dichloromethane, CH_2Cl_2, which has to be stirred continuously and vigorously with the NaOH(aq). What property of CH_2Cl_2 and water makes this stirring necessary?

9* The compound C_2H_2 can be used to make various plastics:

$$C_2H_2 \begin{cases} \nearrow \text{but-1-ene-3-yne} \xrightarrow{\mathbf{A}} \text{synthetic rubbers} \\ \searrow \text{chloroethene} \longrightarrow \text{poly(chloroethene)} \end{cases}$$

a) To which homologous series does C_2H_2 belong?
b) Draw the structure of but-1-ene-3-yne.
c) Which type of polymerisation occurs in reaction **A**?
d) Draw the structure of part of poly(chloroethene) showing at least three monomer units linked together.

10 The 'polyurethane' group of polymers are made by the following reaction.

$$\underset{\text{a diol}}{\text{HO}-[\mathbf{X}]-\text{OH}} \quad + \quad \underset{\text{a diisocyanate}}{\text{O}=\text{C}=\text{N}-[\mathbf{Y}]-\text{N}=\text{C}=\text{O}} \quad + \quad \text{HO}-[\mathbf{X}]-\text{OH}$$

$$\downarrow$$

$$-\text{O}-[\mathbf{X}]-\text{O}-\underset{\substack{\| \\ \text{O}}}{\text{C}}-\underset{\substack{| \\ \text{H}}}{\text{N}}-[\mathbf{Y}]-\underset{\substack{| \\ \text{H}}}{\text{N}}-\underset{\substack{\| \\ \text{O}}}{\text{C}}-\text{O}-[\mathbf{X}]-\text{O}-$$

This reaction is exothermic.

a) The 'urethane link' is similar to but not identical to the amide link. Copy the polymer structure and mark on it the urethane link.
b) Explain why the polymerisation shown above is neither
  i) a typical addition process, nor
  ii) a typical condensation process.
c) Name and give the structural formula of a diol containing three carbon atoms per molecule.
d) In what way would the presence of some monomer molecules with a third –OH group or a third –NCO group alter the structure of the polymer?
e) When polyurethane was made, an inert gas, CFC 11, was sometimes passed into the reaction mixture. What would the function of CFC 11 have been?
f) CFC 11 is now being replaced by HFAs similar to CH_2FCF_3. What environmental reason is there for this change?

12 Natural Products

Fats and oils

- Natural fats and oils can be of animal, vegetable or marine origin.
- Oils are liquid at room temperature because they contain more unsaturated molecules (i.e. carbon–carbon double bonds) than fats which are solid. The lower melting point of oils is caused by smaller van der Waals' forces of attraction between their molecules. This is in turn caused by the distorted structure of the unsaturated oils which prevent close packing.

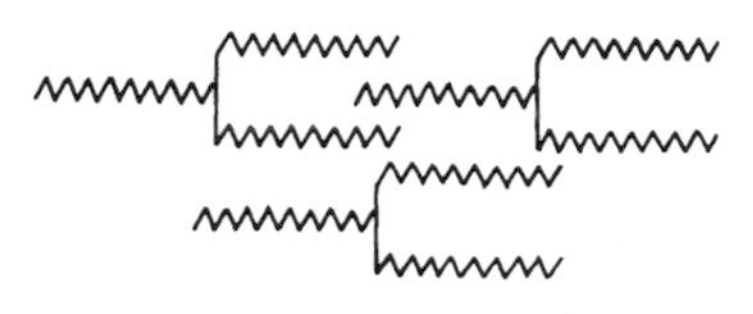

Figure 1 Diagrammatic representation of the structure of fat molecules

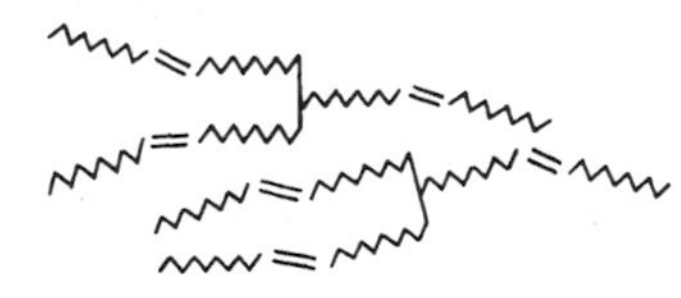

Figure 2 An exaggerated picture of oil molecules

- Oils can be converted into solid fats by the addition of hydrogen which decreases the unsaturation.

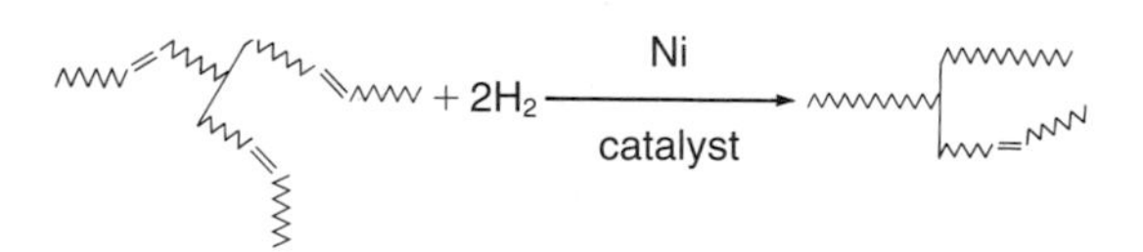

Figure 3 Partial hydrogenation of an oil molecule

- Fats and oils in the diet are a source of energy. They supply more energy per gramme than carbohydrates, but the energy is released more slowly.
- Fats and oils are esters. When hydrolysed, they produce three moles of 'fatty' acids to one mole of glycerol.
- Glycerol is a trihydric alcohol i.e. it has three –OH groups per molecule.
  It is propan-1,2,3-triol.

Foodstuff	Energy yield kJ/100g
Sucrose ('sugar' – carbohydrate)	1672
Sunflower oil	3700
Cooking margarine	3006
'Low fat' spread	1569
Butter	3140
White bread (carbohydrate)	961

Table 1

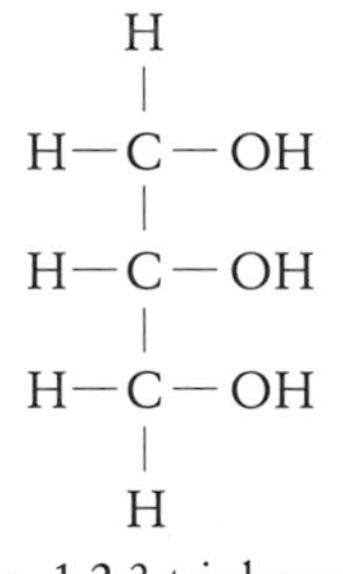

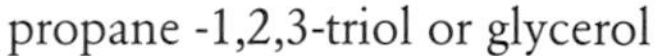

propane -1,2,3-triol or glycerol

Figure 4

- Fatty acids are straight-chain carboxylic acids, either saturated or unsaturated, containing between 4 and 24 carbon atoms per molecule. The commonest are C_{16} and C_{18}.

e.g. Palmitic acid $CH_3(CH_2)_{14}COOH$
Stearic acid $CH_3(CH_2)_{16}COOH$
Oleic acid
$CH_3(CH_2)_7CH = CH(CH_2)_7COOH$
Linoleic acid
$CH_3(CH_2)_3(CH_2CH{=}CH)_2(CH_2)_7COOH$
Ricinoleic acid
$CH_3(CH_2)_5CH(OH)CH_2CH{=}CH(CH_2)_7COOH$

- Fats and oils are largely mixtures of molecules in which three molecules of fatty acid are joined, with the loss of water, to one molecule of glycerol. This is called a triglyceride. The fatty acid molecules can be saturated or unsaturated and may or may not be identical.

```
                 O
                 ||
CH3(CH2)14C—O—CH2       O
                    |           ||
             O  H—C—O—C(CH2)14
             ||      |
CH3(CH2)16C—O—CH2
```

Figure 5 A triglyceride

- Hydrolysis of fats and oils using sodium hydroxide produces soaps, the sodium salts of carboxylic acids.

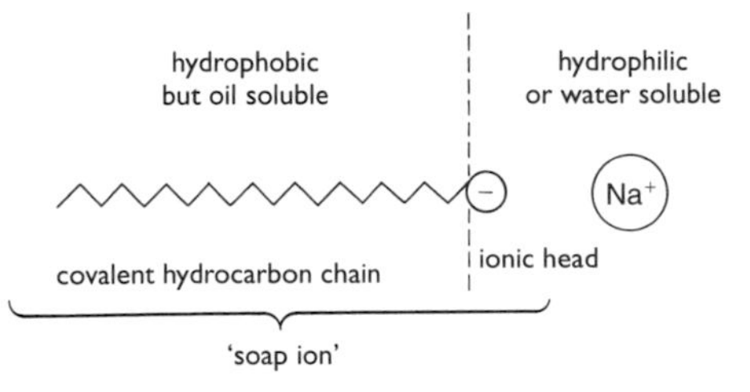

Figure 6 The structure of a 'soap ion'

Proteins

- All proteins contain nitrogen. It is essential for their formation by plants and animals.
- Proteins are condensation polymers made by many amino acid molecules linking together. An amine group of one molecule condenses with the carboxyl group of another molecule to form an amide or peptide link.
- Proteins specific to the body's needs are built up within the body from the appropriate sequence of amino acids.
- The body cannot make all the amino acids required to build proteins. It relies on dietary protein for the supply of some amino acids, these are known as essential amino acids.
- Digestion of protein involves hydrolysis to amino acids.

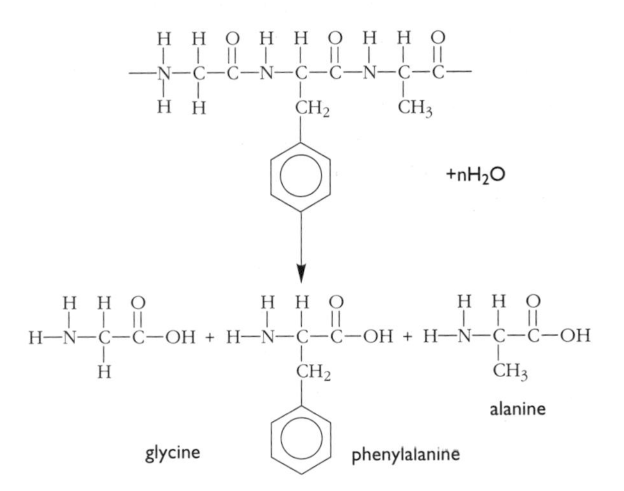

Figure 8 Hydrolysis of a protein to amino acids

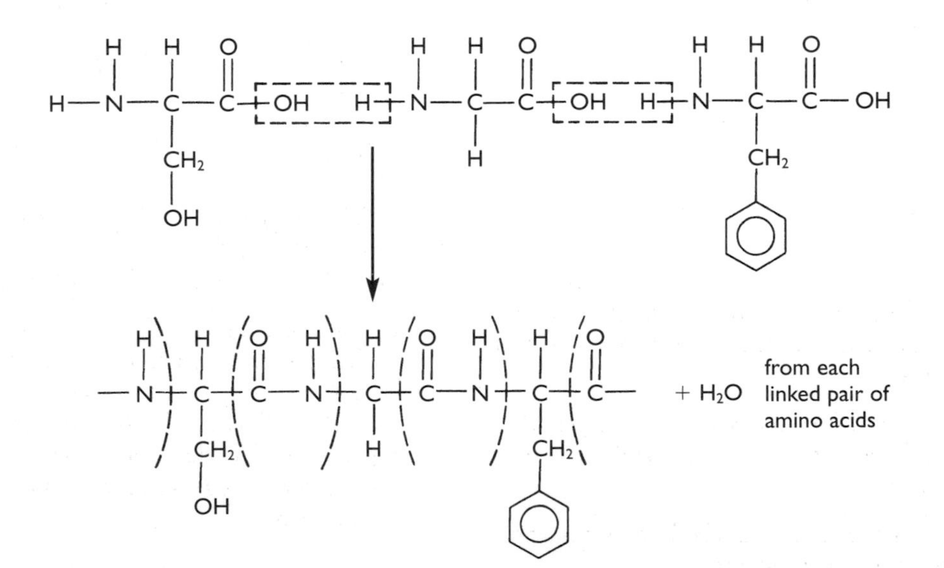

Figure 7 Condensation of amino acids to form amide links in a protein. The amide links are shown in brackets.

Classification of proteins

- Fibrous proteins, the major structural materials of animal tissue, are long and thin.
- Globular proteins are the proteins involved in the regulation of life processes, e.g. haemoglobin, enzymes, hormones like insulin. Their amino acid chains form helices which are then folded into compact units.

Enzymes

- Enzyme function is related to the molecular shapes of proteins. The enzyme and its substrate fit together on the 'lock and key' principle.

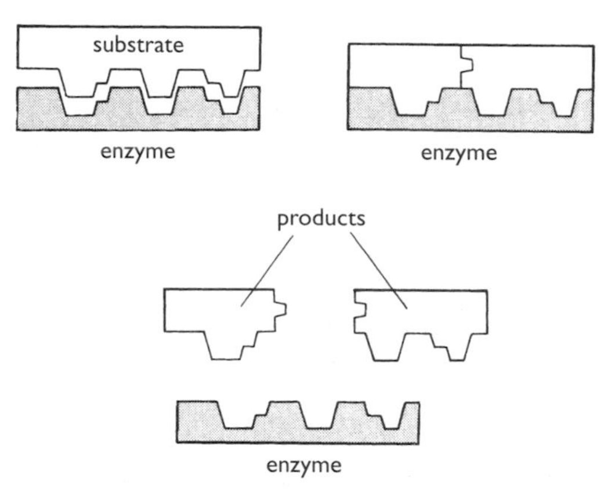

Figure 9 The 'lock and key' principle. A complex molecule being split by an enzyme

- All proteins can be denatured by changes in temperature or pH which bring about physical changes in the molecules. Enzymes are therefore affected by changes in pH and temperature and operate best within a narrow range of each.

Prescribed Practical Activity

It is possible to investigate the factors affecting the activity of **catalase**, an enzyme which catalyses the breakdown of hydrogen peroxide into water and oxygen, with the apparatus shown.

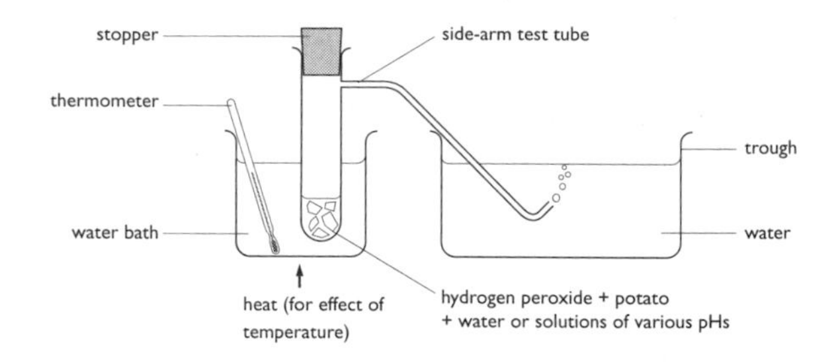

Figure 10

The basic procedure is to place three discs of potato (a source of catalase) in the side-arm tube with 5 cm^3 of water. 1 cm^3 of 30 volume hydrogen peroxide solution is added with a syringe and the tube is stoppered. The number of bubbles of oxygen emerging from the delivery tube in three minutes is counted.

To investigate the effect of temperature, the experiment is carried out at temperatures close to 20°C, 30°C, 40°C, 50°C and 60°C, by heating the water bath. Fresh potato, water and hydrogen peroxide are used each time. The potato and water are left for a few minutes to reach the chosen temperature before the peroxide is added.

To investigate the effect of pH, the water is replaced in turn by buffer solutions of pH 4, 7, and 10 and by 0.1 mol l^{-1} hydrochloric acid (pH 1) and 0.1 mol l^{-1} sodium hydroxide solution (pH 13). This time the water bath is kept at room temperature.

Questions

SECTION A: Fats and Oils

1* Which of the following decolourises bromine solution **least** rapidly?

A Palm oil
B Hex-1-ene
C Cod liver oil
D Mutton fat

2* The conversion of linoleic acid, $C_{18}H_{32}O_2$ into stearic acid, $C_{18}H_{36}O_2$, is likely to be achieved by

A hydrogenation
B hydrolysis
C hydration
D dehydrogenation.

3* The production of fatty acids and glycerol from fats in foods is an example of

A hydrolysis
B hydrogenation
C dehydration
D dehydrogenation.

4* In the formation of 'hardened' fats from vegetable oils, the hydrogen

A causes cross-linking between chains
B causes hydrolysis to occur
C increases the carbon chain length
D reduces the number of carbon–carbon double bonds.

5*

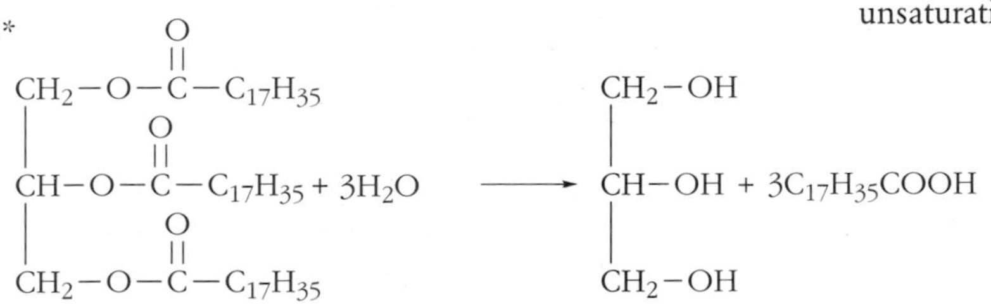

$$
\begin{array}{l}
CH_2-O-\overset{\overset{\displaystyle O}{||}}{C}-C_{17}H_{35} \\
| \\
CH-O-\overset{\overset{\displaystyle O}{||}}{C}-C_{17}H_{35} + 3H_2O \\
| \\
CH_2-O-\overset{\overset{\displaystyle O}{||}}{C}-C_{17}H_{35}
\end{array}
\longrightarrow
\begin{array}{l}
CH_2-OH \\
| \\
CH-OH + 3C_{17}H_{35}COOH \\
| \\
CH_2-OH
\end{array}
$$

Which process is represented by the equation?

A Condensation B Hydrolysis C Oxidation D Dehydration

6 What is the structural formula of glycerol?

A $\underset{\displaystyle OH}{\underset{|}{CH_2}}-CH_2-\underset{\displaystyle OH}{\underset{|}{CH_2}}$

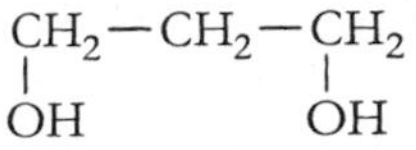

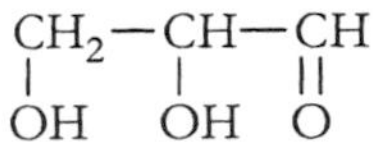

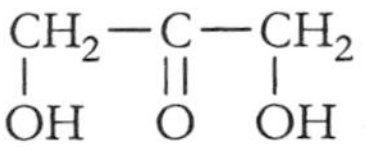

B $\underset{\displaystyle OH}{\underset{|}{CH_2}}-\underset{\displaystyle OH}{\underset{|}{CH}}-\underset{\displaystyle O}{\underset{||}{CH}}$

C $\underset{\displaystyle OH}{\underset{|}{CH_2}}-\underset{\displaystyle O}{\underset{||}{C}}-\underset{\displaystyle OH}{\underset{|}{CH_2}}$

D $\underset{\displaystyle OH}{\underset{|}{CH_2}}-\underset{\displaystyle OH}{\underset{|}{CH}}-\underset{\displaystyle OH}{\underset{|}{CH_2}}$

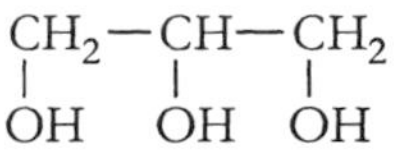

7* Which type of reaction is involved in the conversion of vegetable oils into 'hardened' fats?

A Condensation
B Hydration
C Hydrogenation
D Polymerisation

8 Which statement is true?

A Fats and oils are a less concentrated source of energy than carbohydrates.
B Melting points of fats are likely to be relatively low compared to oils.
C Molecules in fats are more closely packed together than molecules in oils.
D Fats are likely to have a higher degree of unsaturation than oils.

9* Triglycerides are important in our diet. Three are shown below.

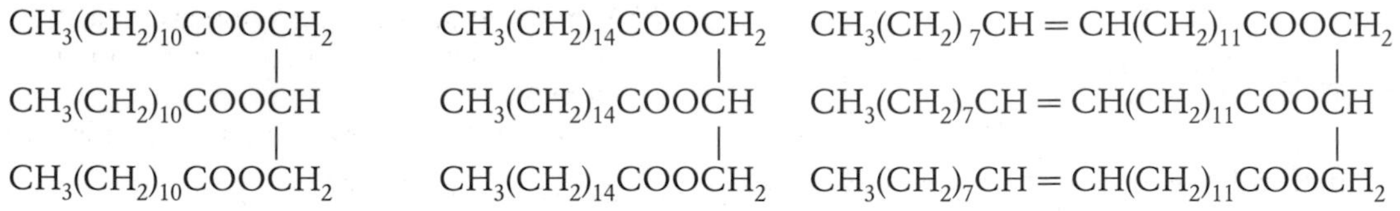

$$\begin{array}{ccc} CH_3(CH_2)_{10}COOCH_2 & CH_3(CH_2)_{14}COOCH_2 & CH_3(CH_2)_7CH=CH(CH_2)_{11}COOCH_2 \\ | & | & | \\ CH_3(CH_2)_{10}COOCH & CH_3(CH_2)_{14}COOCH & CH_3(CH_2)_7CH=CH(CH_2)_{11}COOCH \\ | & | & | \\ CH_3(CH_2)_{10}COOCH_2 & CH_3(CH_2)_{14}COOCH_2 & CH_3(CH_2)_7CH=CH(CH_2)_{11}COOCH_2 \end{array}$$

glyceryl trilaurate | glyceryl tripalmitate | glyceryl trierucate

a) Why are triglycerides an important part of our diet?
b) Glyceryl trilaurate is a liquid at 25°C, but glyceryl tripalmitate is a solid at the same temperature. Why does the triglyceride with the greater molecular mass have the higher melting point?
c) Explain why glyceryl trierucate is a liquid at 25°C, whereas glyceryl tripalmitate is a solid at that temperature even though it has a smaller molecular mass.

10 The triglyceride shown below can be present in fats and oils.

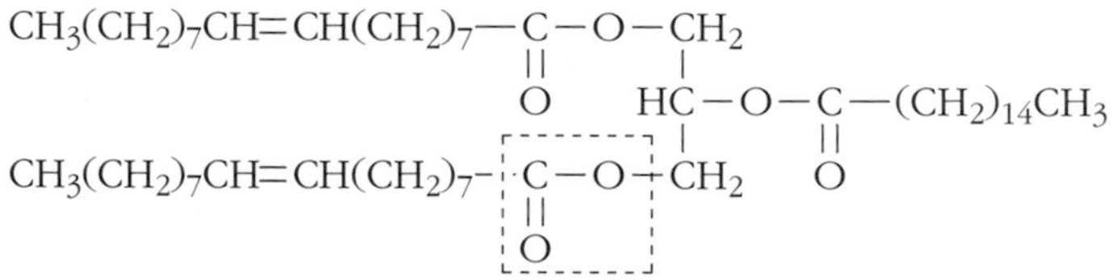

$$\begin{array}{l} CH_3(CH_2)_7CH{=}CH(CH_2)_7{-}\underset{\underset{O}{\|}}{C}{-}O{-}CH_2 \\ \qquad\qquad\qquad\qquad\qquad\qquad\quad | \\ \qquad\qquad\qquad\qquad\qquad\qquad HC{-}O{-}\underset{\underset{O}{\|}}{C}{-}(CH_2)_{14}CH_3 \\ \qquad\qquad\qquad\qquad\qquad\qquad\quad | \\ CH_3(CH_2)_7CH{=}CH(CH_2)_7{-}\boxed{\underset{\underset{O}{\|}}{C}{-}O}{-}CH_2 \end{array}$$

a) Name the functional group inside the dotted lines.
b) Explain the meaning of the word '*triglyceride*'.
c) Give the molecular formula of the unsaturated acid which can be obtained from the triglyceride shown above.

11 Vegetable oils can be converted into important everyday products.

$$\text{Product } \mathbf{X} \xleftarrow[\text{H}_2/\text{Ni/heat}]{\mathbf{I}} \text{VEGETABLE OIL} \xrightarrow[\text{NaOH(aq)/heat}]{\mathbf{II}} \text{SOAP} + \text{Compound } \mathbf{Y}$$

a) Condensation Hydration Hydrolysis Hydrogenation Reforming

Choose one word from the list given above to describe
i) reaction **I**
ii) reaction **II.**

b) What is the function of nickel in reaction **I**?
c) Product **X** is used to make an important substance in the home. What is this substance?
d) Name compound **Y** and draw its structural formula.
e) Oleic acid, $CH_3(CH_2)_7CH{=}CH(CH_2)_7COOH$, is the most common acid obtained from vegetable oils. Give the formula of the soap 'molecule' produced when this acid reacts with NaOH(aq).
f) With the help of a diagram explain how soap cleans a greasy dish.

SECTION B: Proteins

1

$$CH_3-CH(OH)-CH(NH_2)-C(=O)OH$$

The molecule shown above can be classified as

A an enzyme
B an amino acid
C a peptide
D a protein.

2 Which compound could be produced when a protein is hydrolysed?

A $CH_3COOC_2H_5$
B C_2H_5COOH
C $C_3H_7NH_2$
D H_2NCH_2COOH

3* Some amino acids are called α-amino acids because the amino group is on the carbon atom next to the acid group. Which of the following is an α-amino acid?

A $CH_3-CH(CH_2-NH_2)-COOH$

B $CH_2(SH)-CH(NH_2)-COOH$

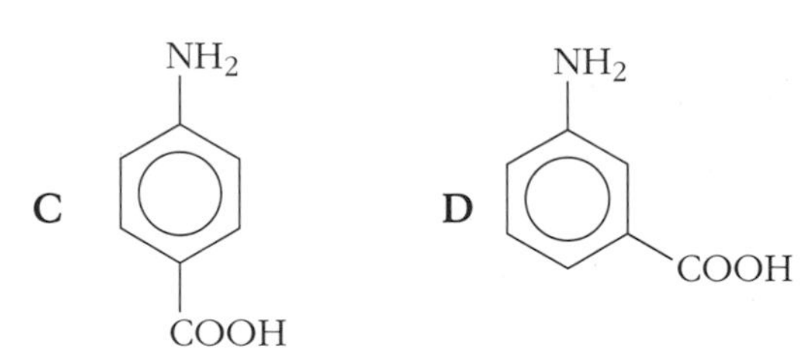

4* The rate of hydrolysis of a protein, using an enzyme, was studied at different temperatures.

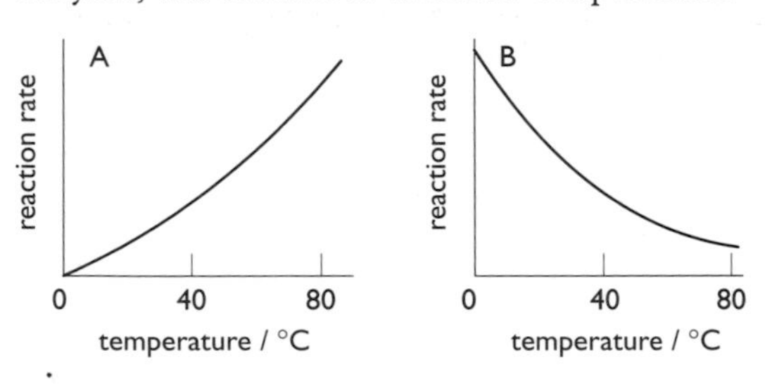

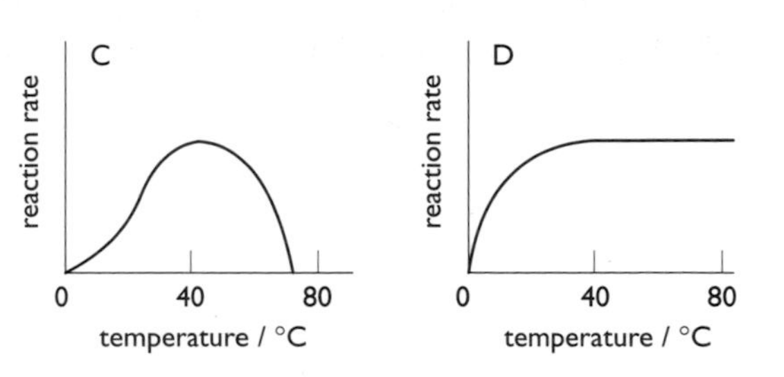

Which graph could be obtained?

5* When two amino acids condense together, water is eliminated and a peptide link is formed.
Which of the following represents this process?

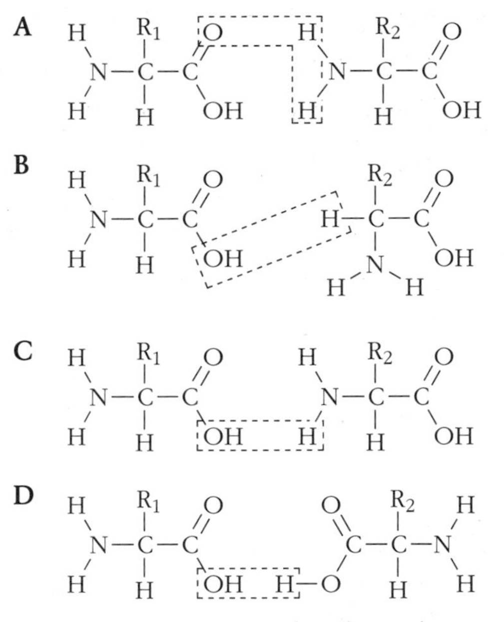

6* Proteins can be denatured under acid conditions. During this denaturing, the protein molecule

A changes shape
B is dehydrated
C is neutralised
D is polymerised.

7* The monomer units used to construct enzyme molecules are

A esters
B amino acids
C fatty acids
D monosaccharides.

8* There are many different enzymes in the human body.

a) Which **four** elements do all enzymes contain?
b) Salivary amylase is an enzyme which can convert starch into maltose. The pH of saliva is about 7, which is close to the optimum pH for that enzyme. Amylase stops functioning when it enters the stomach where the pH is about 2. What happens to the enzyme, on entering the stomach, that would cause it to stop functioning?
c) Many enzymes are specific and can catalyse only one reaction. For example, salivary amylase can catalyse the hydrolysis of starch to maltose, but cannot catalyse the hydrolysis of proteins to amino acids. Give a reason for this.

9 Functional groups play a vital part in the reactions of carbon compounds.

$-CHO$ $-COOH$ $>CHOH$ $-NH_2$ $-NH-CO-$ $-CH_2OH$

a) Which of these is an amide group?
b) Which group (or groups) are present in glycerol?
c) Which groups interact when a protein chain is being produced?
d) Which group (or groups) are produced when a primary alcohol is oxidised?

10* a) A section of a protein chain is shown below.

```
                                      CH3  CH3
                                         \ /
     H   H    O   H    CH3  O  H         CH   O
     |   |    ||  |    |    ||  |        |    ||
   - N - C -  C - N  - C  - C - N  -     C -  C -
         |             |                 |
         H             H                 H
```

i) Draw structural formulae of **two** of the amino acids used to make this section of protein.
ii) What type of polymerisation is involved in the formation of proteins?

b) The body can make some of the amino acids required for growth and repair.
What term is used to describe amino acids which **cannot** be made and which must therefore be included in the diet?

Questions 4, 5 and 6 on pages 121–2 are based on the Unit 2 Prescribed Practical Activities.

Unit 3

CHEMICAL REACTIONS

13 The chemical industry

- The UK chemical industry is a major contributor to our quality of life.

Major products include:

Plastics	Pharmaceuticals
Cosmetics & toiletries	Paints
Aerosols	Disinfectants
Detergents	Fertilisers
Explosives	Pesticides
Adhesives	Herbicides
Veterinary health products	
Inks, dyestuffs & pigments	
Chemicals used in treating water, metals, paper & fabrics	
Domestic polishes & cleaners	
Intermediates for making synthetic fibres	

- The UK chemical industry also plays a vital role in our national economy.
- The chemical industry is one of the UK's largest manufacturing industries. Its share of the total gross value added, i.e. the money raised in converting raw materials into end-products, achieved by all manufacturing industries in the UK was 11.5% in 1995.
- The average growth rate of the chemical industry, 3.0% per year between 1990 and 2000, was nearly five times that of all manufacturing industries in the UK.
- The chemical industry has maintained a positive trade balance, i.e. exports exceeding imports, for many years.
- The UK chemical industry was sixth after USA, Japan, Germany, China and France in 1998 in terms of total sales of chemicals.
- The chemical industry also contributes to our 'invisible' exports. In 1995, income from licensing of chemical processes developed in the UK was, at £1.3 billion; about three and a half times payments made to overseas countries for their technology.

New products

The stages needed before a new product can be manufactured are shown below.

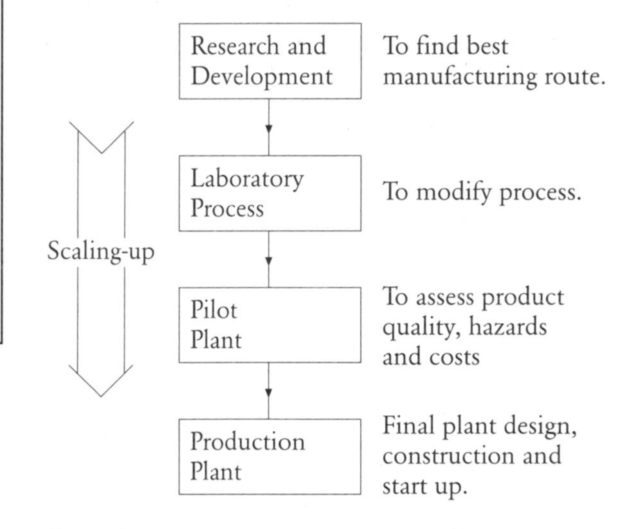

Figure 1

The manufacturing process

- A chemical manufacturing process usually involves a sequence of steps.

For example, sulphuric acid is still a vital substance in the chemical industry. It is manufactured from sulphur by the Contact Process in a series of steps outlined below.

Step 1: Sulphur burning

$$S(l) + O_2(g) \rightarrow SO_2(g) \quad \Delta H = -297\ kJ\ mol^{-1}$$

Step 2: SO_2 conversion

$$SO_2(g) + \tfrac{1}{2}O_2(g) \underset{V_2O_5\ catalyst}{\overset{450°C}{\rightleftharpoons}} SO_3(g) \quad \Delta H = -94\ kJ\ mol^{-1}$$

Step 3: SO_3 absorption

$$SO_3(g) + H_2O(l) \rightarrow H_2SO_4(l)$$

Choosing a manufacturing route

- Several factors may need to be taken into account when choosing a particular route to manufacture a product.

These factors include:

- the cost, availability and suitability of the feedstock(s)
- the yield of product(s)
- whether or not unreacted materials can be recycled
- how marketable are the by-products
- energy consumption
- environmental considerations e.g. emissions to the atmosphere, waste disposal.

Raw materials and feedstocks

- A **feedstock** is a reactant from which other chemicals can be extracted or synthesised. Feedstocks are themselves derived from **raw materials**.

The major raw materials which are used in the chemical industry are:

Crude oil

Crude oil gives naphtha, a fraction obtained by distillation, used as an important feedstock for various chemical processes such as

1 steam cracking to produce ethene and propene for the manufacture of plastics, and

2 reforming to produce aromatic hydrocarbons for the manufacture of dyes, drugs etc.

Ores and minerals

Two examples are:

1 bauxite produces alumina, Al_2O_3, to manufacture aluminium

2 rock salt produces NaOH, Cl_2 & HCl on electrolysis of its aqueous solution.

Air

Air provides:

1 nitrogen for the production of ammonia, and

2 oxygen for the oxidation of
 a) sulphur to sulphur oxides to manufacture sulphuric acid
 b) ammonia to nitrogen oxides to manufacture nitric acid.

Water

Water is a raw material in the steam cracking of naphtha and ethane and in the hydration of ethene to produce ethanol.

Air and water have other functions in the chemical industry. Both can be used as coolants, while water is of course an important solvent.

Batch or continuous process

- In a batch process the reactants are added to the reactor. The reaction is started and its progress carefully monitored. At the end of the reaction the reactor is emptied and the product mixture passes on to the separation and purification stages. A batch reactor is usually a large cylindrical tank.

Continuous process	Batch process
Advantages: Ideal for large quantities of product. Cheaper product if operated to full capacity. Smaller workforce. Good for fast, single-step reactions.	Better suited to small quantities. Plant cheaper to build. More versatile plant. Good for multi-step reactions. Can use reactants in any physical state.
Disadvantages: Can be difficult to use with solid reactants unless 'fluidised'. Can be difficult to control when starting-up, but easier to control when in operation.	Filling and emptying plant increases production time. Can be hard to control if reaction is exothermic.

Table 1

- In a continuous process reactants flow into the reactor at one end and the products flow out at the other end. The design of the reactor varies from one process to another.
- Batch and continuous processes each have their advantages and disadvantages (Table 1).

Economic aspects

- Each industrial chemical process has its characteristic set of conditions under which it operates. These conditions are chosen so as to maximise economic efficiency.
- The chemical industry is research-based.
- The chemical industry can be described as capital intensive rather than labour intensive. The entire chemical industry only employs about 1.5% of the British workforce.
- Manufacturing costs in the chemical industry can be divided into different categories:

Capital costs	Fixed costs	Variable costs
Research & development	Depreciation of plant	Raw materials
Plant construction	Labour	Energy
Buildings	Land purchase or rental	Overheads
Infrastructures	Sales expenses	Effluent treatment or disposal

Table 2

Variable costs relate to the chemical process involved. These costs will not be incurred if production is halted but **fixed costs** will still have to be paid. A company will incur fixed costs whether it manufactures one tonne of product or thousands of tonnes. The effect of the fixed cost on the selling price of the product diminishes as the scale of operation increases.

Capital costs are recovered as depreciation included under fixed costs. Depreciation occurs as chemical plants frequently operate under severe and/or corrosive conditions or become obsolete through technological progress.

The use of energy

- Energy is a major variable cost. At times of international tension, the price of energy from oil has risen very rapidly.
- The chemical industry has responded to the high cost of oil by
  - switching where possible to processes which use less energy
  - saving energy by using heat from exothermic processes elsewhere in the plant
  - using 'waste' heat to generate electricity for the plant
  - selling energy to supply district heating schemes for local housing.
- Apart from its cost, wasted energy is causing needless pollution. When derived from fossil fuels it is contributing to global warming.

The location of chemical industry

Major chemical manufacturing sites have been established as a result of historical and practical considerations as illustrated by the following case study:

Grangemouth chemical works

Grangemouth Works started in 1919 as a dyeworks and has since expanded to produce pharmaceuticals, agrochemicals, pigments and speciality chemicals.

It was sited at Grangemouth for several important practical reasons.

- A large area of flat land was available.
- There was plenty of water for manufacturing processes.
- There were good transport links by railway and by Grangemouth docks to import raw materials (from coal tar) and to export products.
- There was a pool of skilled labour with experience in the chemical industry because of the nearby shale-oil works.
- Effluent could be disposed of to the sea.

Historical reasons for the siting of the Grangemouth Works included the following.

- A shortage of dyes, which had normally

been imported from Germany until the start of the First World War.
– The company's founder wanted to set up the works in his native Scotland!

Safety in the chemical industry

- The chemical industry has, like all industries, a duty to its employees and to the public to operate without causing accidental injury and without causing risks to health.
- Yet disastrous incidents in the industry have occurred.

For example, in 1974 at Flixborough in Lincolnshire a plant making cyclohexane suffered an explosion and fire. Casualties were 28 killed and 104 injured, some in neighbouring housing. Lessons learnt included building new plants away from housing and ensuring that control rooms were fire- and blast-proof.

In 1984 at Bhopal, India, a leak of toxic gas killed several thousand people in the worst chemical plant incident ever. The company had to pay huge sums in compensation, its reputation throughout the world was damaged greatly. Other companies realised the need to avoid similar accidents, not only for economic reasons.

- To put this matter in perspective, deaths on the roads in the UK exceed accidental deaths in the chemical industry by a factor of 1000.
- Chemical industry has a history of causing long-term damage to the health of its workers. Now the rules governing exposure to harmful chemicals are very strict and rigorously enforced.
- In the 1980's the chemical industry recognised that it needed to make a significant improvement in its safety, health and environmental performance. In the UK, the USA and Canada, a programme called Responsible Care has been adopted.

Questions

1* Which of the following substances is a raw material for the chemical industry?
A Benzene
B Methane
C Aluminium
D Iron

2* The costs involved in the industrial production of a chemical are made up of fixed costs and variable costs. Which of the following is most likely to be classed as a variable cost?
A The cost of land rental
B The cost of plant construction
C The cost of labour
D The cost of raw materials

3 Which of the following is a feedstock, but not a raw material, for the chemical industry?
A Crude oil
B Methane
C Naphtha
D Octane

4 In 2004, crude oil prices rose considerably. Suggest different reasons why the retail price of each of the following could increase.
a) Petrol
b) Fertilisers
c) Bottled mineral water

5 Draw up a table with the headings:

Capital Costs, Fixed Costs, Variable Costs.

Place each of the following costs under the appropriate heading.

Staff training
Advertising
Fuel bills
Construction of new plant
Landfill tax
Construction of rail siding,
Rental of land for expansion,
Wages
Development of new product.

6 In February 2000, millions of litres of highly toxic cyanide solution from a gold mine in Romania leaked into a tributary of the River Danube which flows through Hungary, Serbia and Bulgaria into the Black Sea.

What could be the consequences for

a) the aquatic environment
b) the large human populations along the Danube
c) international relations?

7 The technology used in the oil industry had its origins in West Lothian where a rock, oil shale, was mined and then distilled to produce crude oil products. In the nineteenth century, a batch process was used and the main products were lamp oil, candle wax and lubricants. If the remaining deposits of oil shale were mined today, suggest

a) what would be the main products
b) two reasons why a continuous process would be the preferred method of refining
c) a major reason why a continuous process would be difficult to introduce in this case.

8 During the 1970s and the 1980s an aluminium smelter operated at Invergordon on the Cromarty Firth, the site of a former naval base. Alumina (i.e. purified aluminum oxide) was imported and electrolysed using power from the National Grid.

Suggest **two** favourable and **two** unfavourable factors affecting the siting of the industrial plant. You may find it helpful to consult a map of the north of Scotland.

9 Chemical reactions involved in the manufacture of cement include:

$$CaCO_3 \xrightarrow{\text{heat}} CaO + CO_2 \quad \text{and}$$

$$CaO + SiO_2 \xrightarrow{\text{heat}} CaSiO_3$$
[from shale]

a) Use the sketch map below to answer the following:
   i) Cement manufacture requires shale as a raw material.
   What other essential materials are available nearby?
   ii) What other advantages does the specific site for the works have?
b) The cement works takes in quantities of waste paint, resins and solvents for use as fuel.
   How would this benefit
   i) the operating company
   ii) the wider community?
c) In the UK about 10 million car tyres are dumped on landfill sites each year. Rubber consists of hydrocarbon polymers. Reduction in landfill waste would benefit the wider community, but how could the use of 'chipped' tyre rubber benefit the cement company?

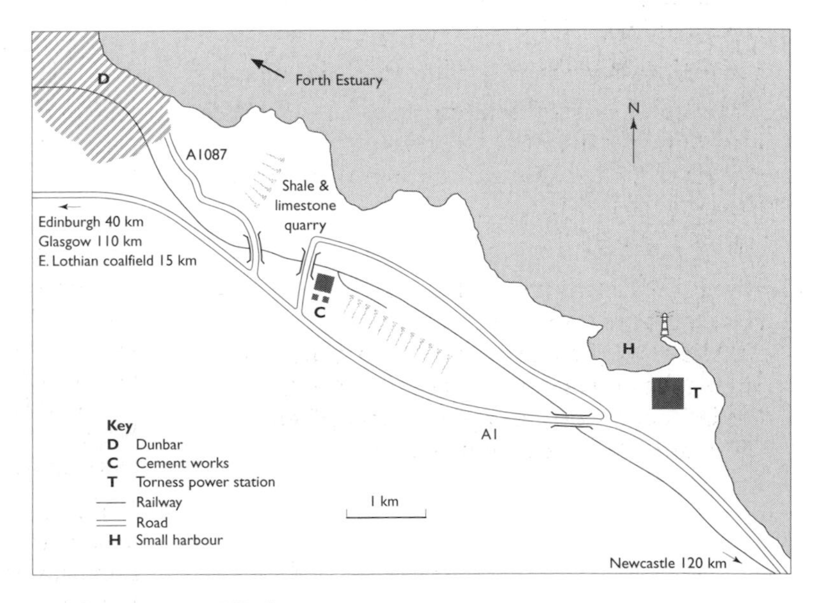

Dunbar Cement Works

10 The following diagram summarises the manufacture of epoxyethane (which itself is used in sterilising and fumigating) and three possible processes for converting it into ethane-1,2-diol.

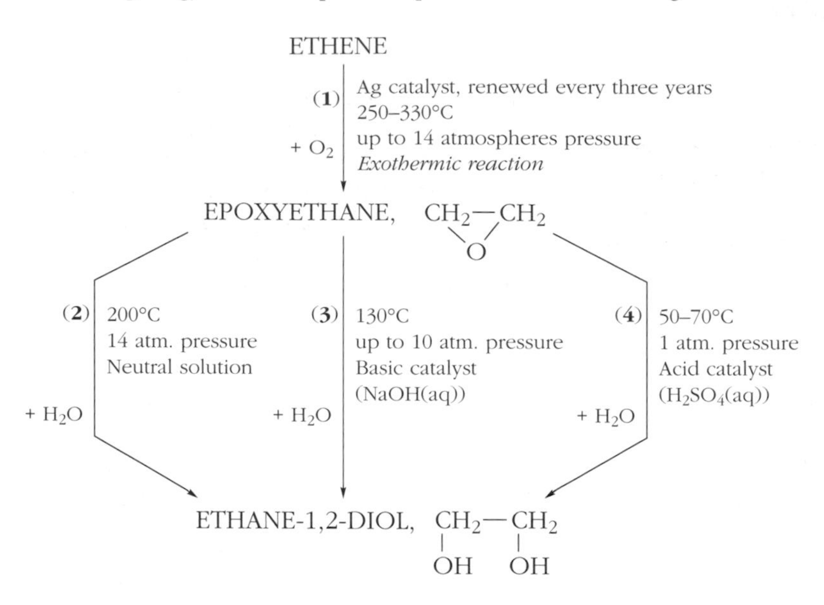

a) Name **two** feedstocks from which ethene can be obtained.
b) Write a balanced equation for reaction **(1)**.
c) In order to prevent another, more common, reaction competing with reaction **(1)** the ethene and oxygen spend less than 4 seconds in the reactor.
What would you expect the products of the competing reaction to be?
d) Give **one** advantage and **one** disadvantage for each of the processes **(2)** and **(4)**.
e) Give **one** important use of ethane-1,2-diol.
f) Explain one factor which has a significant effect on the high **capital** cost of the plant for process **(1)**.
g) Give **two variable** costs in the operating of process **(1)**.

11* In the Downs Process, sodium is extracted from sodium chloride (melting point 801°C) by electrolysis.

a) Suggest why calcium chloride is added to the electrolyte.
b) Is the Downs Process a batch or continuous process?
Explain your answer with reference to the diagram.
c) i) In which state of matter is the sodium collected?
ii) Explain how you arrived at your answer.

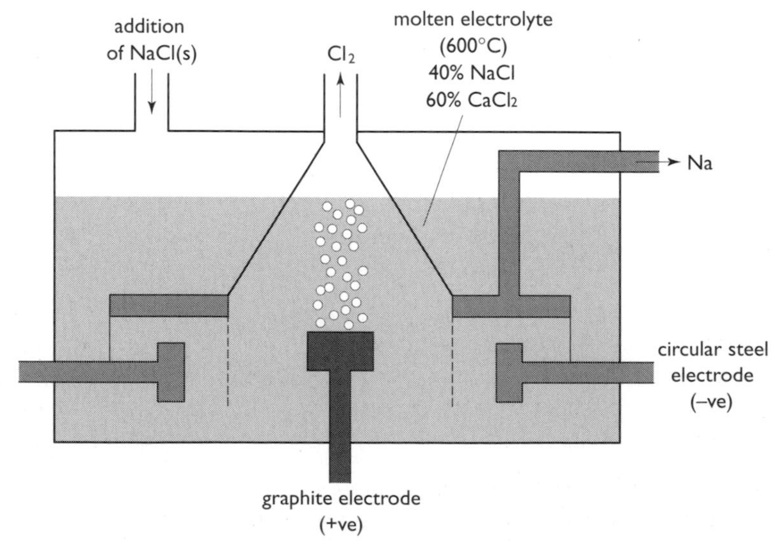

14 Hess's Law

Hess's Law states:

'The enthalpy change of a chemical reaction depends only on the chemical nature and physical states of the reactants and products and is independent of any intermediate steps.'

i.e. the enthalpy change of a chemical reaction does not depend on the route taken during the reaction.

Hess's Law can be tested by experiment as this example shows. Solid potassium hydroxide is converted into potassium chloride solution by two different routes illustrated in Figure 1.

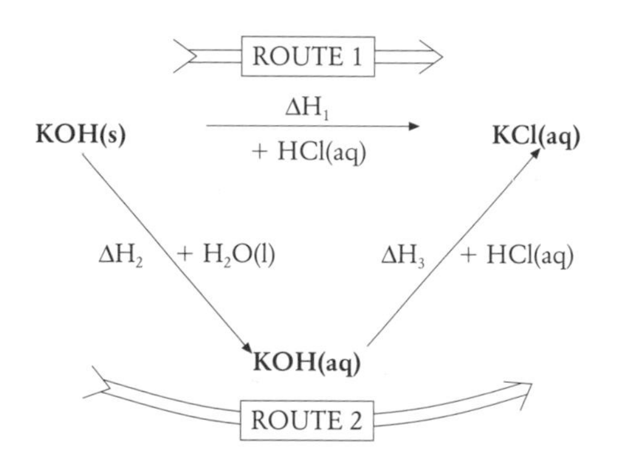

Figure 1

According to Hess's Law: $\Delta H_1 = \Delta H_2 + \Delta H_3$

Prescribed Practical Activity

Route 1: a one-step process. 25 cm^3 of dilute hydrochloric acid is added to a known mass, about 1.2 g, of solid potassium hydroxide in a polystyrene cup to produce potassium chloride solution. The initial temperature of the acid, and the highest temperature of the resulting solution after stirring are measured.

Route 2: *Step 1:* The same mass of solid potassium hydroxide is added to water to form potassium hydroxide solution. The initial temperature of the water, and the highest temperature of the resulting solution after stirring are measured.

Step 2: The temperature of the potassium hydroxide solution from step 1 is remeasured, the temperature of 25 cm^3 of dilute hydrochloric acid is measured and the two temperatures averaged to give a starting temperature. The two solutions are mixed, stirred and the highest temperature reached is measured.

For each of the experiments, the energy released can be calculated using: $E_h = cm\Delta T$, where c is the specific heat of water (a reasonable approximation), m is the mass of solution (assumed to be the same as for the same volume of water) and ΔT is the temperature rise.

The enthalpy change for each experiment is then obtained by dividing E_h by the number of moles of potassium hydroxide.

Then: $\Delta H_1 = \Delta H_2 + \Delta H_3$

Calculations using Hess's Law

- Only certain enthalpy changes can be measured directly, as shown in Chapter 2.
- Hess's Law enables the calculation of enthalpy changes which are very difficult or even impossible to measure. The calculation of the enthalpies of formation of carbon compounds from their constituent elements is an example.

Worked Example 14.1

$$2C(s) + H_2(g) \rightarrow C_2H_2(g)$$

Calculate the enthalpy of formation of ethyne gas from carbon and hydrogen using the enthalpies of combustion of carbon, hydrogen and ethyne.

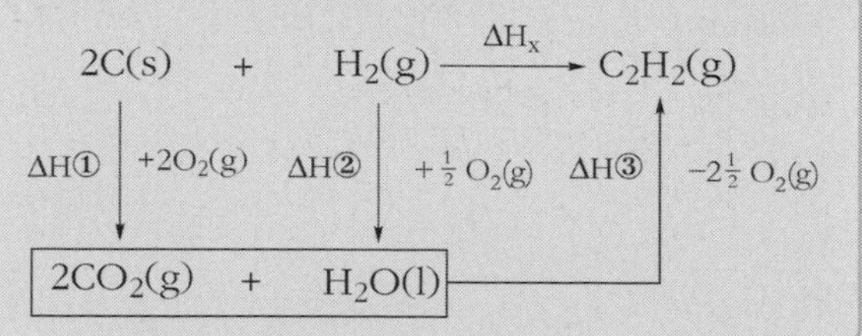

In the alternative route,

Step 1 is the combustion of 2 moles of carbon,

$$2C(s) + 2O_2(g) \rightarrow 2CO_2(g)$$
$$\Delta H_1 = -(2 \times 394)\ kJ = -\ 788\ kJ$$

Step 2 is the combustion of 1 mole of hydrogen,

$$H_2(g) + \tfrac{1}{2}O_2(g) \rightarrow H_2O(l)$$
$$\Delta H_2 = -286\ kJ$$

Step 3 is the **reverse** of combustion of 1 mole of ethyne,

$$2CO_2(g) + H_2O(l) \rightarrow C_2H_2(g) + 2\tfrac{1}{2}O_2(g)$$
$$\Delta H_3 = +1300\ kJ^*$$

Adding these three equations gives the required equation, namely

$$2C(s) + H_2(g) \rightarrow C_2H_2(g)$$

According to Hess's Law,

$$\Delta H_x = \Delta H_1 + \Delta H_2 + \Delta H_3$$
$$= -\ 788 - 286 + 1300$$
$$= +\ 226\ kJ\ mol^{-1}$$

(*The sign has been altered since the reaction has been reversed.)

Worked Example 14.2

The following equation shows the formation of methanol from carbon, hydrogen and oxygen.

$$C(s) + 2H_2(g) + \tfrac{1}{2}O_2(g) \rightarrow CH_3OH(l)$$

Use the enthalpies of combustion of carbon, hydrogen and methanol to calculate the enthalpy change of this reaction.

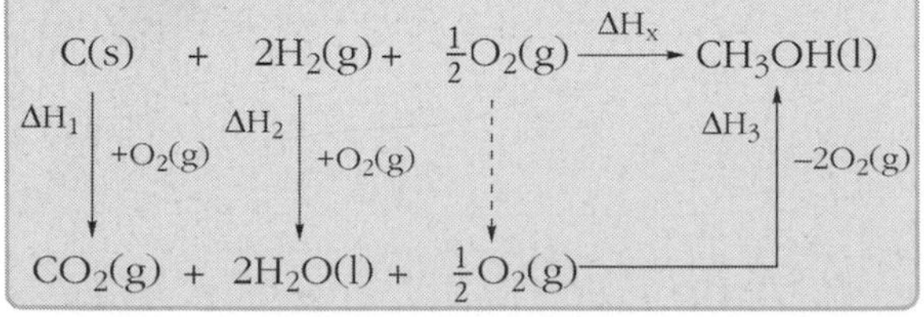

$C(s)$ + $2H_2(g)$ + $\frac{1}{2}O_2(g)$ $\xrightarrow{\Delta H_x}$ $CH_3OH(l)$

ΔH_1 $+O_2(g)$ | ΔH_2 $+O_2(g)$ | ΔH_3 $-2O_2(g)$

$CO_2(g)$ + $2H_2O(l)$ + $\frac{1}{2}O_2(g)$

In the alternative route,

Step 1 is the combustion of 1 mole of carbon,

$$C(s) + O_2(g) \rightarrow CO_2(g) \quad \Delta H_1 = -394\ kJ$$

Step 2 is the combustion of 2 moles of hydrogen,

$$2H_2(g) + O_2(g) \rightarrow 2H_2O(l)$$
$$\Delta H_2 = -(2 \times 286)\ kJ = -572\ kJ$$

Step 3 is the **reverse** of combustion of methanol,

$$CO_2(g) + 2H_2O(l) \rightarrow CH_3OH(l) + \tfrac{3}{2}O_2(g)$$
$$\Delta H_3 = +\ 727\ kJ$$

Adding these three equations gives the required equation, namely

$$C(s) + 2H_2(g) + \tfrac{1}{2}O_2(g) \rightarrow CH_3OH(l)$$

According to Hess's Law,

$$\Delta H_x = \Delta H_1 + \Delta H_2 + \Delta H_3$$
$$= -394 - 572 + 727$$
$$= -239\ kJ\ mol^{-1}$$

Note: Oxygen is one of the elements present in methanol but it is not involved in deriving the required enthalpy change. The calculation is based on enthalpies of combustion. Oxygen gas supports combustion; it does not itself have an enthalpy of combustion. Also note that in all enthalpy calculations, the **states** of the substances involved are important, since changes of state involve enthalpy changes.

Questions

1* Consider the reaction pathway shown below.

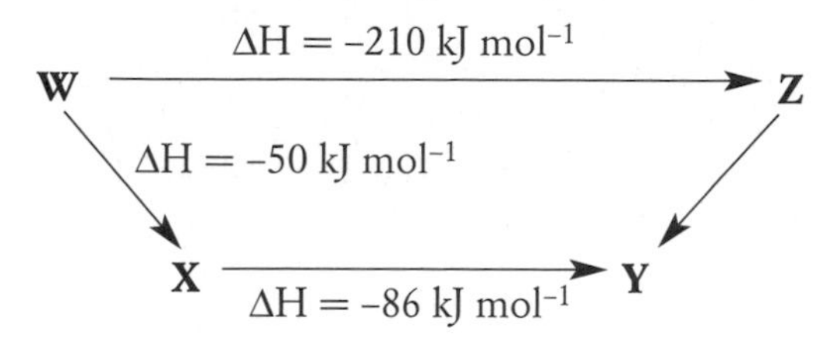

According to Hess' Law, the ΔH value, in $kJ\ mol^{-1}$, for reaction **Z** to **Y** is

A +74   B –74
C +346   D –346

2* $N_2(g) + 2O_2(g) \rightarrow 2NO_2(g) \quad \Delta H = +88\ kJ$
$N_2(g) + 2O_2(g) \rightarrow N_2O_4(g) \quad \Delta H = +10\ kJ$

The enthalpy change for the reaction $2NO_2(g) \rightarrow N_2O_4(g)$ will be

A +98 kJ   B +78 kJ
C –78 kJ   D –98 kJ

3* What is the relationship between a, b, c and d?

$S(s) + H_2(g) \rightarrow H_2S(g) \quad \Delta H = a$

$H_2(g) + \frac{1}{2}O_2(g) \rightarrow H_2O(l) \quad \Delta H = b$

$S(s) + O_2(g) \rightarrow SO_2(g) \quad \Delta H = c$

$H_2S(s) + 1\frac{1}{2}O_2(g) \rightarrow H_2O(l) + SO_2(g) \quad \Delta H = d$

A a = b + c – d
B a = d – b – c
C a = b – c – d
D a = d + c – b

4 Enthalpies of combustion of ethene, hydrogen and ethane can be found in the SQA Data Book.
What is the enthalpy of hydrogenation of ethene to form ethane, in $kJ\ mol^{-1}$?

A +137   B –137
C –149   D +149

5 The enthalpy of combustion of methane is –891 $kJ\ mol^{-1}$. The enthalpy change for the following reaction is –566 kJ.

$$2CO(g) + O_2(g) \rightarrow 2CO_2(g)$$

Using this data, what is the enthalpy of partial combustion of methane, in $kJ\ mol^{-1}$?

$$CH_4(g) + \frac{3}{2}O_2(g) \rightarrow CO(g) + 2H_2O(l)$$

A –325   B –608
C –1174   D –1457

6 Three reactions involving metals and metal oxides are shown below. (All metals and oxides are solids.)

$Mg + FeO \rightarrow Fe + MgO \quad \Delta H = x\ kJ\ mol^{-1}$

$Fe + CuO \rightarrow Cu + FeO \quad \Delta H = y\ kJ\ mol^{-1}$

$Mg + CuO \rightarrow Cu + MgO \quad \Delta H = z\ kJ\ mol^{-1}$

Which of the following is true, according to Hess's Law?

A y + z = –x   B y + z = x
C x + y = –z   D x + y = z

7* A pupil tried to confirm Hess's Law using the reactions shown below.

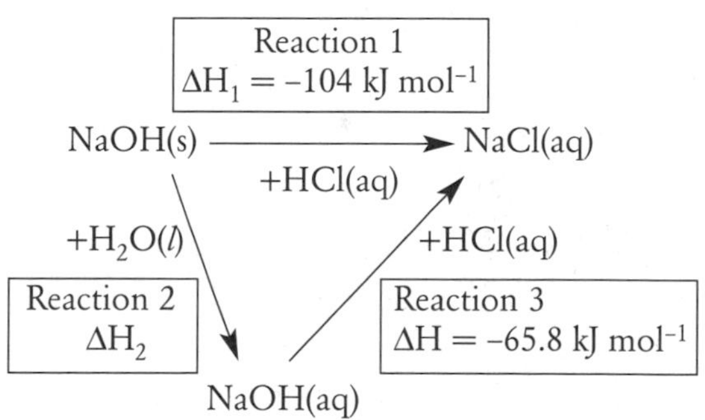

In reaction 1, the pupil measured the mass of NaOH(s) and the temperature change of the reaction mixture.

a) Which further measurement would have been taken?
b) Use the enthalpy changes in the diagram to calculate the enthalpy change for reaction 2.
c) Write, in words, a statement of Hess's Law.

8 Use appropriate enthalpies of combustion from the SQA Data Book to calculate the enthalpy changes of the following reactions.

a) $3C(s) + 4H_2(g) \rightarrow C_3H_8(g)$
(propane)

b) $2C(s) + 2H_2(g) + O_2(g) \rightarrow CH_3COOH(l)$
(ethanoic acid)

9 Methane can be converted into ethyne at very high temperatures (about 1500°C). The equation for the reaction is:

$$2CH_4(g) \rightarrow C_2H_2(g) + 3H_2(g)$$

Use enthalpies of combustion from the SQA Data Book to calculate the enthalpy change for this reaction

a) in kJ mol^{-1} of ethyne
b) in kJ mol^{-1} of methane.

10 The simplest silicon hydride has the formula SiH_4. It is a gas at room temperature. The equation for the complete combustion of this compound is:

$$SiH_4(g) + 2O_2(g) \rightarrow SiO_2(s) + 2H_2O(l)$$

Calculate the enthalpy change for this reaction using the following data:

$Si(s) + 2H_2(g) \rightarrow SiH_4(g)$ $\Delta H = +34$ kJ mol^{-1}

$Si(s) + O_2(g) \rightarrow SiO_2(s)$ $\Delta H = -911$ kJ mol^{-1}

$H_2(g) + \frac{1}{2}O_2(g) \rightarrow H_2O(l)$ $\Delta H = -286$ kJ mol^{-1}

11 Methanal, HCHO, is made industrially by the oxidation of methanol. The boiling point of methanal is –21°C and its enthalpy of combustion is –561 kJ mol^{-1}.

a) Write the equation, including state symbols, for the complete combustion of methanal.
b) Calculate the enthalpy change for the oxidation of methanol to methanal. Refer to the SQA Data Book page 9. The equation for this reaction is:

$$CH_3OH(l) + \frac{1}{2}O_2(g) \rightarrow HCHO(g) + H_2O(l)$$

c) In the industrial process, a catalyst (either copper or silver) is used.
What effect, if any, will the catalyst have on
i) the enthalpy change
ii) the activation energy of the reaction?

d) In the industrial process, methanol vapour is passed over the catalyst at 500°C.
Give one reason why the enthalpy change for the industrial process will be different from your answer in part **b)**.
e) Give **one** important use of methanal.

12* Some rockets have a propellant system which combines dinitrogen tetroxide with methylhydrazine.

$$5N_2O_4 + 4CH_3NHNH_2 \rightarrow \mathbf{x}N_2 + \mathbf{y}H_2O + \mathbf{z}CO_2$$

a) State the values of **x**, **y** and **z** required to balance the above equation.
b) Draw the full structural formula for methylhydrazine.
c) Methylhydrazine burns acccording to the following equation.

$$CH_3NHNH_2(l) + 2\tfrac{1}{2}O_2(g) \rightarrow CO_2(g) + 3H_2O(l) + N_2(g)$$
$$\Delta H = -1305 \text{ kJ mol}^{-1}$$

Use this information, together with information from the SQA Data Book to calculate the enthalpy change for the following reaction.

$$C(s) + N_2(g) + 3H_2(g) \rightarrow CH_3NHNH_2(l)$$

15 Equilibrium

- Reversible reactions attain a state of equilibrium when the rate of the forward reaction is equal to the rate of the reverse reaction.
- The reaction does **not** stop when equilibrium is attained. For this reason, chemical equilibrium is described as being **dynamic.**
- When equilibrium is reached this does not imply that the equilibrium mixture consists of 50% reactants and 50% products. This will only very rarely be the case. The concentrations of reactants and products do however remain constant.
- Equilibrium is reached when the opposing reactions occur at an equal rate. Any condition which changes the rate of one reaction more than the other should change the position of equilibrium, i.e. the relative proportions of reactants and products in the mixture.

Changing the position of equilibrium

This section deals with the influence of changing the concentration, the pressure, the temperature and the catalyst on the equilibrium position.

- The effect of these changes can be summarised by 'Le Chatelier's Principle' which states that:

'If a system at equilibrium is subjected to any change, the system readjusts itself to try and counteract the applied change.'

Note: This statement *only* refers to reversible reactions which have reached equilibrium.

A summary of results of changes is shown in Table 1.

Change applied	Effect on equilibrium position
Concentration	
Addition of reactant or removal of product	Equilibrium shifts to the right
Addition of product or removal of reactant	Equilibrium shifts to the left
(See example 1)	
Temperature	
Increase	Shifts in direction of endothermic reaction
Decrease	Shifts in direction of exothermic reaction
(See example 2)	
Pressure	
Increase	Shifts in direction which reduces the number of molecules in gas phase
Decrease	Shifts in direction which increases the number of molecules in the gas phase
(See example 3)	
Catalyst	No effect on equilibrium position; equilibrium more rapidly attained
(See Figure 1)	

Table 1

Example 1
Bromine water

$$Br_2(l) + H_2O(l) \rightleftharpoons 2H^+(aq) + Br^-(aq) + BrO^-(aq)$$

The addition of NaOH removes H^+ ions and the equilibrium shifts to the right. Adding HCl increases the concentration of H^+ ions moving the equilibrium back to the left.

Example 2

$$\underset{\text{(colourless)}}{N_2O_4(g)} \rightleftharpoons \underset{\text{(dark brown)}}{2NO_2(g)}$$

The forward reaction is endothermic.

Increase in temperature favours the endothermic reaction, the equilibrium moves to the right, the proportion of NO_2 increases and the gas mixture becomes darker in colour.

Decrease in temperature favours the exothermic

reaction, the equilibrium moves to the left, and the gas mixture lightens in colour.

Example 3

$N_2O_4(g) \rightleftharpoons 2NO_2(g)$

1 mole 2 moles

1 volume 2 volumes (at the same temperature and pressure)

Increase in pressure causes the system to counteract this effect, i.e. to reduce the pressure within the system. The equilibrium adjusts to the left, forming more N_2O_4 molecules, reducing the number of molecules per unit volume which reduces the pressure.

Potential energy diagram:

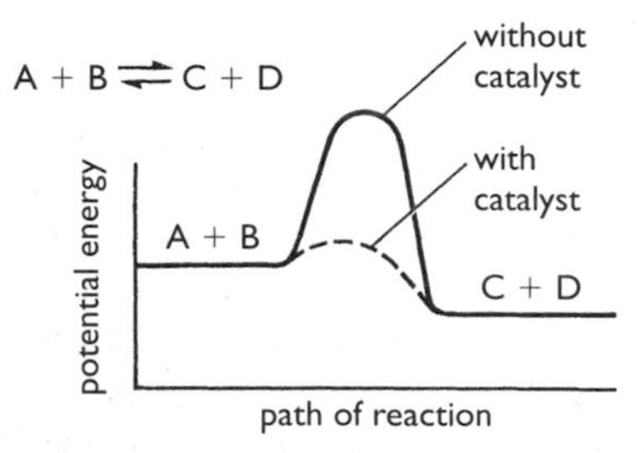

Figure 1 Potential energy: catalysed and uncatalysed reactions

Activation Energy is lowered equally for forward and backward reactions, both are speeded up and the same equilibrium is reached more quickly.

Equilibrium and the Haber process

- The Haber process is used to synthesise ammonia from hydrogen and nitrogen.
- If a closed reaction vessel is used, an equilibrium is set up:

$$N_2(g) + 3H_2(g) \rightleftharpoons 2NH_3(g) \quad \Delta H = -92 \text{ kJ}$$

A summary of the results of changes applied industrially to the Haber Process are shown in Table 2.

New conditions	Equilibrium change	Explanation
Increase pressure	To right, $[NH_3]$ increases	Forward direction involves a decrease in number of moles of gas (4 moles → 2 moles) and hence a decrease in volume. Decrease in volume is assisted by increase in pressure.
Decrease temperature	To right, $[NH_3]$ increases	Decreasing temperature removes energy from system, making reverse endothermic reaction less favourable, less NH_3 splits up, but reaction slows so catalyst needed.
Catalyst	No change	Both forward and reverse reactions are accelerated. The same equilibrium is reached more rapidly.

Table 2

Questions

1* Chemical reactions are in a state of dynamic equilibrium only when

A the rate of the forward reaction equals that of the backward reaction
B the concentrations of reactants and products are equal
C the activation energies of the forward and backward reactions are equal
D the reaction involves no enthalpy change.

2 When a reversible chemical reaction is at equilibrium,

A the concentrations of reactants and products remain equal
B the forward reaction is unable to continue
C the concentrations of reactants and products remain constant
D the forward and reverse reactions proceed at different rates.

Questions 3 and **4** refer to the following equilibrium which exists in bromine water.

$Br_2(aq) + H_2O(l) \rightleftharpoons 2H^+(aq) + Br^-(aq) + BrO^-(aq)$

3 Which of the following substances, when added, would move the equilibrium position to the right?

A Potassium nitrate B Sodium bromide
C Sulphuric acid D Sodium hydroxide

4 Which of the following substances, when added, would increase the pH of the equilibrium mixture?

A Potassium nitrate B Sodium bromide
C Bromine D Sodium chloride

5* The equation refers to the preparation of methanol from synthesis gas.

$CO(g) + 2H_2(g) \rightleftharpoons CH_3OH(g)$
$\Delta H = -91 \text{ kJ mol}^{-1}$

The formation of methanol is favoured by

A high pressure and low temperature
B high pressure and high temperature
C low pressure and low temperature
D low pressure and high temperature.

6* In which of the following systems will the equilibrium be **unaffected** by a change in pressure?

A $2NO(g) + O_2(g) \rightleftharpoons 2NO_2(g)$
B $2NO_2(g) \rightleftharpoons N_2O_4(g)$
C $H_2(g) + I_2(g) \rightleftharpoons 2HI(g)$
D $N_2(g) + 3H_2(g) \rightleftharpoons 2NH_3(g)$

7 $NO_2(g) \rightleftharpoons NO(g) + \frac{1}{2}O_2(g)$
$\Delta H = +56 \text{ kJ mol}^{-1}$

Which two conditions favour the decomposition of NO_2?

A Low temperature, high pressure
B High temperature, low pressure
C Low temperature, low pressure
D High temperature, high pressure

8* Which entry in the table shows the effect of a catalyst on the reaction rates and position of equilibrium in a reversible reaction?

	Rate of forward reaction	**Rate of reverse reaction**	**Position of equilibrium**
A	increased	unchanged	moves right
B	increased	increased	unchanged
C	increased	decreased	moves right
D	unchanged	unchanged	unchanged

9* Which of the following is likely to apply to the use of a catalyst in a chemical reaction?

	Position of equilibrium	**Effect on value of ΔH**
A	Moved to right	Decreased
B	Unaffected	Increased
C	Moved to left	Unaffected
D	Unaffected	Unaffected

10* Consider the following equilibrium:

$N_2(g) + O_2(g) \rightleftharpoons 2NO(g)\ \Delta H_{(forward)} = +180 \text{ kJ}$

How would the equilibrium concentration of nitrogen oxide be affected by:
a) increasing the temperature
b) decreasing the pressure
c) decreasing the concentration of oxygen?

11* Consider the following equilibrium:

$N_2O_4(g) \rightleftharpoons 2NO_2(g)$ $\Delta H_{(forward)}$ is + ve
(pale yellow) (dark brown)

What would be **seen** if the equilibrium mixture was
a) placed in a freezing mixture
b) compressed?

12* Iodine dissolves only slightly in water, the process being endothermic. With excess iodine present, the following equilibrium is set up.

$I_2(s) + aq \rightleftharpoons I_2(aq)$ ΔH positive

The concentration of dissolved iodine was measured over a period of time. The graph below was obtained as the iodine dissolved.

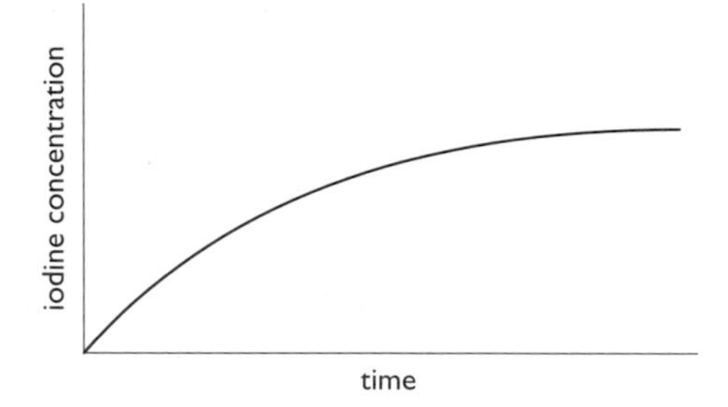

a) Copy the graph and add a curve to show how the iodine concentration would change with time if the measurements were repeated at a higher temperature.

b) The dissolved iodine reacts with water as follows.

$I_2(aq) + H_2O\ (l) \rightleftharpoons 2H^+(aq) + I^-(aq) + IO^-(aq)$

i) Copy and complete the table to show the effect on the equilibrium of adding each of the solids.

Solid	**Effect on equilibrium position**
potassium iodide	
potassium sulphate	

ii) Why does the position of equilibrium move to the right when solid potassium hydroxide is added?

13 An industrial method of producing hydrogen involves the reaction of steam with carbon monoxide. The equation for this reaction is:

$CO(g) + H_2O(g) \rightleftharpoons CO_2(g) + H_2(g)$ $\Delta H_{forward}$ is negativ

a) Explain why a change in pressure does not affect the yield of hydrogen.
b) How will an increase in temperature affect the yield of hydrogen?
c) The reactants are passed over a heated catalyst.
i) Why does the use of a catalyst not affect the yield of hydrogen?
ii) Why then is a catalyst used?
d) How will the removal of carbon dioxide affect the yield of hydrogen?

14 When chlorine is added to water the following reaction occurs:

$Cl_2(g) + H_2O(l) \rightleftharpoons 2H^+(aq) + Cl^-(aq) + ClO^-(aq)$

The efficiency of a chlorine bleach depends on a high concentration of ClO^- ions.

Various substances can be added to achieve this, e.g. sodium carbonate and silver nitrate.
a) Explain why each of these compounds should increase the concentration of ClO^- ions.
b) Why is silver nitrate unlikely to be used in practice?

15 Lead(II) chloride is slightly soluble in cold water. In a saturated solution of lead(II) chloride the following equilibrium is set up.

$PbCl_2(s) \rightleftharpoons Pb^{2+}(aq) + 2Cl^-(aq)$

a) What would be the effect on the equilibrium position of adding
i) hydrochloric acid
ii) lead(II) nitrate solution
iii) silver(I) nitrate solution?
b) Lead(II) chloride is highly soluble in very hot water. What does this indicate about the enthalpy of solution of lead(II) chloride?

16 In the Contact process, sulphur trioxide is formed from sulphur dioxide and oxygen in a reversible reaction. The equation for the reaction is:

$$2SO_2(g) + O_2(g) \rightleftharpoons 2SO_3(g)$$

The forward reaction is exothermic.

a) One litre of sulphur dioxide and one litre of oxygen were mixed under certain conditions and allowed to reach equilibrium. Analysis of the equilibrium mixture showed that 60% conversion of the sulphur dioxide had taken place. Calculate the volume of each gas present in the equilibrum mixture.

b) What effect, if any, will there be on the equilibrium position if

i) the temperature is increased

ii) a catalyst is used

iii) the pressure is increased?

17 Carbon in the form of graphite can be converted into diamond at high temperatures and high pressure. This change can be expressed in the following equation.

$$C(s)\ (graphite) \rightleftharpoons C(s)\ (diamond)$$

a) Calculate the enthalpy change for the forward reaction given the following data.
Enthalpy of combustion of graphite is $-393.5\ kJ\ mol^{-1}$.
Enthalpy of combustion of diamond is $-395.4\ kJ\ mol^{-1}$.

b) Is your answer to **a)** consistent with the use of high temperatures when making diamond from graphite? Explain why.

c) Refer to the densities of graphite and diamond given in the SQA Data Book to suggest why high pressures are also used.

18 Iodine monochloride, a brown liquid, reacts with chlorine to form a yellow solid called iodine trichloride according to the following equation.

$$ICl(l) + Cl_2(g) \rightleftharpoons ICl_3(s)$$

a) What effect, if any, will there be on the mass of yellow solid in an equilibrium mixture if

i) the pressure is increased

ii) some chlorine is removed?

b) Raising the temperature of the equilibrium mixture decreases the mass of yellow solid present. What can you deduce about the forward reaction from this information?

19* Soda water is made by dissolving carbon dioxide in water, under pressure.

$$CO_2(g) + aq \rightleftharpoons CO_2(aq)$$

a) When the stopper is taken off a bottle of soda water, the carbon dioxide gas escapes. Explain why the drink eventually goes **completely** flat.

b) This graph shows the solubility of carbon dioxide in water at different temperatures.

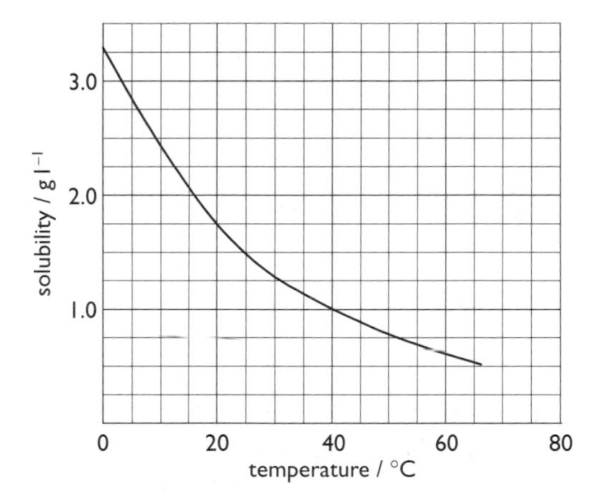

What does the graph indicate about the enthalpy of solution of carbon dioxide in water?

c) When all of the carbon dioxide is removed from one litre of soda water at 0°C, the gas is found to occupy 1.7 litres.
Use information in the graph to calculate the molar volume of carbon dioxide at this temperature.

16 Acids and bases

The pH scale

- The concentration of hydrogen ions in solution is measured in pH units. pH stands for the negative logarithm (to base 10) of the hydrogen ion concentration, i.e. $pH = -\log_{10}[H^+(aq)]$
- The pH scale is a continuous scale with values ranging from less than zero to more than 14. (A concentration of more than 1 mole l^{-1} of H^+ ions will give a positive $\log_{10}[H^+]$ and hence a negative pH value). Integral values of pH can be related to hydrogen ion concentrations $[H^+]$ as in Table 1.

$[H^+]$ mol l^{-1}	$\log_{10}[H^+]$	$pH(-\log_{10}[H^+])$
1	0	0
1/10 or 10^{-1}	–1	1
1/100 or 10^{-2}	–2	2
$1/10^7$ or 10^{-7}	–7	7
$1/10^{14}$ or 10^{-14}	–14	14

Table 1

- For water and neutral solutions in water, the sole source of H^+ and OH^- ions is the very slight ionisation of some of the water molecules:

$$H_2O(l) \rightleftharpoons H^+(aq) + OH^-(aq)$$

- An equilibrium is reached with the concentration of both H^+ and OH^- equal to 10^{-7} moles l^{-1} at 25 °C.

$$\text{i.e. } [H^+] = [OH^-] = 10^{-7} \text{ mol } l^{-1}, \ pH = 7$$

Then:

$$[H^+][OH^-] = 10^{-7} \text{ mol } l^{-1} \times 10^{-7} \text{ mol } l^{-1} = 10^{-14} \text{ mol}^2 \, l^{-2}$$

- This value is called the **ionic product** of water.
- It follows that we can calculate $[H^+]$, pH, $[OH^-]$ for acidic and alkaline solutions using the basic relationship:

$$[H^+][OH^-] = 10^{-14} \text{ mol}^2 \, l^{-2}$$

From which we derive:

$$[H^+] = \frac{10^{-14}}{[OH^-]} \text{ mol } l^{-1}$$

and

$$[OH^-] = \frac{10^{-14}}{[H^+]} \text{ mol } l^{-1}$$

Worked Example 16.1

What is the concentration of OH^- ions in a solution containing 0.01 mol l^{-1} of H^+ ions?

$$[H^+] = 10^{-2} \text{ mol } l^{-1}$$

$$[OH^-] = \frac{10^{-14}}{[H^+]} = \frac{10^{-14}}{10^{-2}} = 10^{-12} \text{ mol } l^{-1}$$

Worked Example 16.2

What is **a)** the concentration of H^+ ions

b) the pH, in a solution containing 0.1 mol l^{-1} of OH^- ions?

Answer:

a) $[OH^-] = 10^{-1}$ mol l^{-1}

$$[H^+] = \frac{10^{-14}}{[OH^-]} = \frac{10^{-14}}{10^{-1}} = 10^{-13} \text{ mol } l^{-1}$$

b) $pH = -\log_{10}[H^+] = 13$

Strong and weak acids

- Strong acids are acids that dissociate completely into ions in solution.

e.g. $HCl(aq) \rightarrow H^+(aq) + Cl^-(aq)$

- Strong acids include HCl, HNO_3 and H_2SO_4.

- Other acids dissociate only partially into ions in solution, i.e. an equilibrium exists between the ions and undissociated molecules.

For ethanoic acid:

$$CH_3COOH(aq) \rightleftharpoons CH_3COO^-(aq) + H^+(aq)$$

For carbon dioxide solution:

$$CO_2(g) + H_2O(l) \rightleftharpoons H_2CO_3(aq)$$

carbonic acid

$$\rightleftharpoons H^+(aq) + HCO_3^-(aq)$$

hydrogencarbonate ion

$$\rightleftharpoons 2H^+(aq) + CO_3^{2-}(aq)$$

carbonate ion

For sulphur dioxide solution:

$$SO_2(g) + H_2O(l) \rightleftharpoons H_2SO_3(aq)$$

sulphurous acid

$$\rightleftharpoons H^+(aq) + HSO_3^-(aq)$$

hydrogensulphite ion

$$\rightleftharpoons 2H^+(aq) + SO_3^{2-}(aq)$$

sulphite ion

- Such incompletely dissociated acids are called weak acids. (Note that '*strong*' and '*weak*' refer to the inherent ability of acids to ionise. These words should not confused with '*concentrated*' and '*dilute*'.)

- Solutions of strong and weak acids differ in pH, conductivity and reaction rates since their hydrogen ion concentrations are different. Solutions should be equimolar for a fair comparison.

For example, 0.1 mol l^{-1} CH_3COOH has a lower hydrogen ion concentration, higher pH, lower conductivity and slower rate of reaction with magnesium or calcium carbonate than 0.1 mol l^{-1} HCl.

- The undissociated molecules of weak acids are in equilibrium with their ions. As a reaction proceeds, consuming H^+ ions, the equilibrium shifts in favour of more dissociation until eventually all the molecules are dissociated. Thus the same number of moles of base is required to neutralise a certain volume of either 0.1 mol l^{-1} HCl or 0.1 mol l^{-1} CH_3COOH. Hence the **stoichiometry** (the mole ratio of reactants) of a neutralisation reaction is the same:

$$NaOH + HCl \rightarrow NaCl + H_2O$$

$$NaOH + CH_3COOH \rightarrow CH_3COONa + H_2O$$

1 mole acid ≡ 1 mole alkali in each case.

Strong and weak bases

- A base is any substance which will neutralise the H^+ ions of an acid to form water.

- Bases which are soluble in water, producing OH^- ions, are alkalis.

- Alkalis which are fully dissociated into ions are strong alkalis, e.g. NaOH and KOH.

$$NaOH(aq) \rightarrow Na^+(aq) + OH^-(aq)$$

- Alkalis showing incomplete dissociation are weak alkalis, e.g. ammonia solution.

$$NH_3(aq) + H_2O(l) \rightleftharpoons NH_4^+(aq) + OH^-(aq)$$

In the example above an equilibrium, normally well to the left, is set up in the ammonia solution. This results in the lower concentration of OH^- ions, lower conductivity and lower pH of a weak alkali compared with an equimolar solution of sodium hydroxide.

- The stoichiometry of neutralisations of strong and weak alkalis is the same.

$$\text{e.g. } NaOH + HCl \rightarrow NaCl + H_2O$$

$$NH_3 + HCl \rightarrow NH_4Cl + H_2O$$

1 mole alkali ≡ 1mole acid in each case

The weak alkali dissociates further as its OH^- ions react with acids. Eventually it will dissociate fully so that a certain volume of either 0.1 mol l^{-1} NaOH or 0.1 mol l^{-1} NH_3 solution will neutralise the same number of moles of acid.

pH of salt solutions

- It is normally assumed that since salts can be made by neutralisation of an acid by an alkali the pH will be neutral, i.e. 7. In fact, the measured pH of salts is often not 7.

In general terms, in solution:

- Salts of **strong acid** and **strong alkali** have **pH 7** (e.g. NaCl and KNO_3)
- Salts of **strong acid** and **weak alkali** have **pH < 7** (e.g. NH_4Cl)
- Salts of **strong alkali** and **weak acid** have **pH > 7** (e.g. CH_3COONa)

Ammonium chloride in water

The pH of ammonium chloride in water can be explained as follows.

Ions present initially, NH_4^+ from the salt and OH^- from the water, combine to form ammonia and water molecules:

$$NH_4^+(aq) + OH^-(aq) \rightarrow NH_3(aq) + H_2O(l)$$

Removal of OH^- ions causes the water equilibrium to move to the right.

$$H_2O(l) \rightleftharpoons H^+(aq) + OH^-(aq)$$

Excess H^+ ions are formed and the solution has pH less than 7.

Sodium ethanoate in water

The pH of sodium ethanoate in water can be explained as follows.

Ions present initially, CH_3COO^- from the salt and H^+ from the water, combine to form ethanoic acid molecules:

$$CH_3COO^-(aq) + H^+(aq) \rightarrow CH_3COOH(aq)$$

Removal of H^+ ions causes the water equilibrium to move to the right.

$$H_2O \rightleftharpoons H^+(aq) + OH^-(aq)$$

Excess OH^- ions are formed, and the solution has pH greater than 7.

Similar considerations apply to sodium and potassium salts of other carboxylic acids and of carbonic and sulphurous acids.

- Soaps are the salts of strong alkalis like NaOH and KOH with weak long chain carboxylic acids such as stearic and oleic acids. Soaps therefore are usually alkaline in solution.

Questions

1 The concentration of $OH^-(aq)$ in a solution is 0.01 mol l^{-1}. What is the pH of the solution?

A 8 B 10 C 12 D 14

2* 0.5 mol of hydrogen chloride is dissolved in water, and the resulting solution is made up to a total of 5 litres. The pH of this solution is

A 0 B 1 C 2 D 3.

3* A trout fishery owner added limestone to his loch to combat the effects of acid rain. He managed to raise the pH of the water from 4 to 6. This caused the concentration of the $H^+(aq)$ to

A increase by a factor of 2
B increase by a factor of 100
C decrease by a factor of 2
D decrease by a factor of 100.

4 The pH of a solution of nitric acid was found to be 1.6. The concentration of $H^+(aq)$ ions in the acid must be

A more than 0.1 mol l^{-1}
B between 0.1 and 0.01 mol l^{-1}
C between 0.01 and 0.001 mol l^{-1}
D less than 0.001 mol l^{-1}.

5 Butanoic acid is described as a weak acid because in water

A its pH is about 4
B its O–H bonds are partially dissociated
C it is only slightly soluble
D it gives only one hydrogen ion per molecule.

6* A fully dissociated acid is progressively diluted by the addition of water. Which of the following would increase with increasing dilution?

A The pH value
B The electrical conductivity
C The rate of its reaction with chalk
D The volume of alkali which it will neutralise

7* Which of the following is the same for equal volumes of equimolar solutions of sodium hydroxide and ammonia?

A pH of solution
B Mass of solute present
C Conductivity of solution
D Moles of acid needed for complete reaction

8* Excess marble chips (calcium carbonate) were added to 100 cm^3 of 1 mol l^{-1} hydrochloric acid. The experiment was repeated using the same mass of the marble chips and 100 cm^3 of 1 mol l^{-1} ethanoic acid. Which would have been the same for both experiments?

A The time taken for the reaction to be completed
B The rate at which the first 10 cm^3 of gas is evolved
C The mass of marble chips left over when reaction has stopped.
D The average rate of the reaction

9 Which of the following correctly describes the concentration of ions in an aqueous solution which has a pH of 10?

	$[H^+]$/mol l^{-1}	$[OH^-]$/mol l^{-1}
A	10^4	10^{-10}
B	10^{-4}	10^{-10}
C	10^{-10}	10^{-4}
D	10^{-10}	10^4

10* On the structure shown, four hydrogen atoms have been replaced by the letters A, B, C and D.

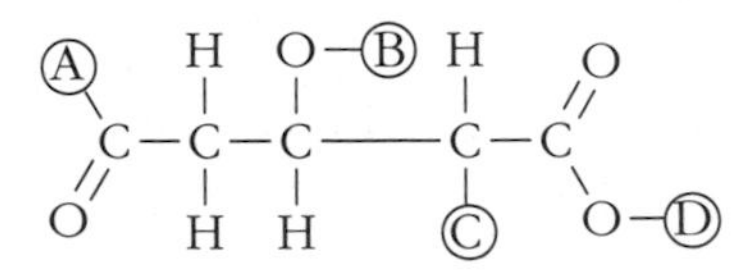

Which letter corresponds to the hydrogen atom which can ionise most easily in aqueous solution?

11 The pH of a solution which has $[OH^-] = 1 \times 10^{-9}$ mol l^{-1} is

A 9 B 7 C 5 D 3

12 Which of the following dissolves in water to give a solution of pH<7 ?
A Potassium ethanoate
B Sodium sulphate
C Calcium hydroxide
D Ammonium chloride

13 Which of the following forms an alkaline solution in water?
A Sodium chloride
B Potassium carbonate
C Ammonium nitrate
D Lithium sulphate

14 Which of the following solutions is most likely to produce hydrogen gas when magnesium is added?
A Ammonium sulphate
B Sodium nitrate
C Potassium chloride
D Sodium ethanoate

15 In which salt solution will $[H^+] = [OH^-]$?
A Potassium nitrate
B Sodium ethanoate
C Ammonium nitrate
D Potassium carbonate

16 Which of the following describes 0.1 mol l^{-1} ammonia solution?
A concentrated solution of a weak alkali
B concentrated solution of a strong alkali
C dilute solution of a weak alkali
D dilute solution of a strong alkali

17 For each of the following solutions, indicate
i) the concentration of H^+ ions
ii) the pH of the solution.

a) 0.001 mol l^{-1} HCl
b) 1.0 mol l^{-1} NaCl
c) 0.5 mol l^{-1} H_2SO_4
d) 0.1 mol l^{-1} NaOH
e) 0.00001 mol l^{-1} KOH

18 Calculate the concentration of OH^- ions in each of the substances listed below.

Solution tested	pH
Trichlorophenol, TCP	2
Handwash	8
Cleaning fluid	11
Lemonade	4

19 Three beakers labelled X, Y and Z contain the following solutions:

X 50 cm^3 0.2 mol l^{-1} ethanoic acid
Y 50 cm^3 0.2 mol l^{-1} hydrochloric acid
Z 50 cm^3 0.2 mol l^{-1} sulphuric acid

a) Explain why:
i) solution **X** has a higher pH than solution **Y**.
ii) solution **Z** has a lower pH than solution **Y**.

b) 50 cm^3 0.2 mol l^{-1} sodium hydroxide solution is added to Z. Show by calculation that the resulting solution has a pH of 1.

20 There are a series of acids based on ethanoic acid in which one or more of the hydrogen atoms of the methyl group are replaced by chlorine atoms. The following table shows a set of results obtained when testing solutions of equal concentration.

Name	Formula	pH
Ethanoic acid	CH_3COOH	3.1
Monochloroethanoic acid	$CH_2ClCOOH$	2.0
Dichloroethanoic acid	$CHCl_2COOH$	1.4
Trichloroethanoic acid	CCl_3COOH	0.9

a) Complete the sentence: 'As more hydrogen atoms in ethanoic acid are replaced by chlorine atoms, the acid strength ____________.'

b) Calculate $[H^+]$ and $[OH^-]$ in monochloroethanoic acid.

c) Write a balanced equation for the reaction between dichloroethanoic acid and potassium hydroxide solution.

d) Ammonium ethanoate solution has a pH of 7. Would you expect the pH of ammonium trichloroethanoate solution to be 7, >7 or <7?

21 A pupil compared two solutions of acids, obtaining the following results.

	0.1 mol l^{-1} Nitric acid	0.1 mol l^{-1} Propanoic acid
Conductivity/ mA	85	15
pH	1.5	3.5

a) What result **should** have been obtained when the pH of nitric acid was tested?

b) The results show that propanoic acid is a weak acid.
Explain what is meant by the term '*weak acid*'.

c) Excess sodium carbonate powder was added to 20 cm^3 of each acid in separate beakers.
i) How would the rate of the two reactions compare?
ii) When both reactions have finished, how would the total volumes of gas produced compare?

22 Two salts can be prepared when sulphurous acid reacts with sodium hydroxide solution. These salts are called sodium sulphite and sodium hydrogensulphite. Sodium sulphite dissolves in water to form an alkaline solution.

a) Write the formulae of **i)** sulphurous acid **ii)** sodium hydrogensulphite.

b) Explain clearly why sodium sulphite solution is alkaline.

c) Sodium hydrogensulphite is also soluble in water. Suggest how the pH of this solution will compare with that of sodium sulphite solution.

17 Redox Reactions

Oxidising and reducing agents

- Reactions in which reduction and oxidation occur are called **redox** reactions. In a redox reaction electron transfer occurs between the reactants. One reactant is reduced while the other is oxidised.

Table 1 summarises important definitions and gives examples of ion-electron equations.

A reducing agent	An oxidising agent
loses electrons and is itself **oxidised**	gains electrons and is itself **reduced**
e.g. $Zn \rightarrow Zn^{2+} + 2e^-$	e.g. $Ag^+ + e^- \rightarrow Ag$
OXIDATION **I**S **L**OSS of electrons	**R**EDUCTION **I**S **G**AIN of electrons

Table 1

Displacement reactions

- In **displacement** reactions one metal displaces another metal from a solution of its salt. The relevant ion-electron equations can be combined to produce a balanced ionic equation for the overall redox reaction.

For example, zinc displaces silver from a solution containing silver(I) ions, e.g. $AgNO_3(aq)$. Each zinc atom loses two electrons whilst each silver ion gains only one electron. The second ion-electron equation must be doubled to balance the number of electrons lost and gained to give the redox equation.

Note that the total charge on each side of the redox equation is the same and that the redox equation does not contain electrons.

Oxidation: $Zn(s) \rightarrow Zn^{2+}(aq) + 2e^-$

Reduction:

$$Ag^+(aq) + e^- \rightarrow Ag(s) \ (\times 2)$$

Redox:

$$2Ag^+(aq) + Zn(s) \rightarrow 2Ag(s) + Zn^{2+}(aq)$$

The negative ions present have not been included since they do not take part in the reaction. It is usual practice to omit **spectator ions** from redox equations.

Redox reactions involving oxyanions

- **Oxyanions** are negative ions which contain oxygen combined with another element. Examples include sulphite ions, SO_3^{2-}, permanganate ions, MnO_4^- and dichromate ions, $Cr_2O_7^{2-}$.

The following examples of redox reactions are given to show how to write:

i) ion-electron equations which involve oxyanions

ii) balanced redox equations for more complex reactions.

(State symbols have been omitted so as not to overload the equations with information.)

Example 1

Iodine solution + sodium sulphite solution

$I_2(aq)$ (**oxidising agent**) **$Na_2SO_3(aq)$** (**reducing agent**)

[spectator ions: Na^+]

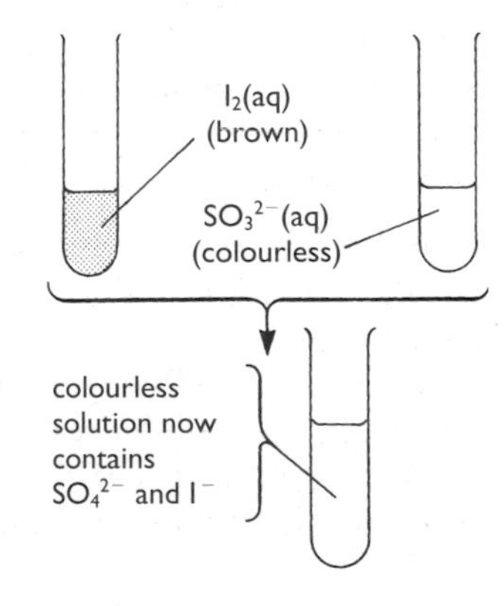

Figure 1

Iodine molecules are reduced to iodide ions.

$$I_2 + 2e^- \rightarrow 2I^-$$

brown **colourless**

Sulphite ions are oxidised to sulphate ions.

$$SO_3^{2-} \rightarrow SO_4^{2-}$$

To complete the ion-electron equation, add H_2O to the left-hand side of the equation to obtain the oxygen atom needed and add $2H^+$ to the right-hand side to give:

$$SO_3^{2-} + H_2O \rightarrow SO_4^{2-} + 2H^+$$

Then add two electrons to the right-hand side so that the charge is the same on each side of the equation giving:

$$SO_3^{2-} + H_2O \rightarrow SO_4^{2-} + 2H^+ + 2e^-$$

The two ion-electron equations can now be combined to give the balanced redox equation:

$$\begin{array}{ll} I_2 + 2e^- & \rightarrow 2I^- \\ SO_3^{2-} + H_2O & \rightarrow SO_4^{2-} + 2H^+ + 2e^- \\ \hline I_2 + SO_3^{2-} + H_2O & \rightarrow 2I^- + SO_4^{2-} + 2H^+ \end{array}$$

Example 2

Acidified potassium dichromate solution $K_2Cr_2O_7(aq) + H^+(aq)$ (oxidising agent)

+ iron(II) sulphate solution, $FeSO_4(aq)$ (reducing agent)

[spectator ions: K^+, SO_4^{2-}]

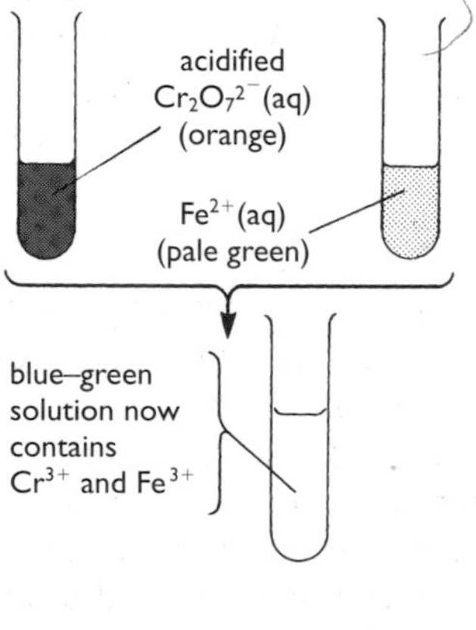

Figure 2

Iron(II) ions are oxidised to iron(III) ions.

$$Fe^{2+} \rightarrow Fe^{3+} + e^-$$

Dichromate ions are reduced to chromium(III) ions.

$$Cr_2O_7^{2-} \rightarrow 2Cr^{3+}$$

orange **blue–green**

The ion-electron equation can be completed in a similar way to the previous example. Only this time oxygen is being removed so that $7H_2O$ is added to the right-hand side and $14H^+$ to the left to give:

$$Cr_2O_7^{2-} + 14H^+ \rightarrow 2Cr^{3+} + 7H_2O$$

Six electrons are then added to the left-hand side to balance the charge giving:

$$Cr_2O_7^{2-} + 14H^+ + 6e^- \rightarrow 2Cr^{3+} + 7H_2O$$

This equation shows why the solution should be acidified.

The two ion-electron equations can now be combined to give the balanced redox equation. To balance the number of electrons lost and gained, the equation involving iron ions has to be multiplied by 6.

$$\begin{array}{c} 6Fe^{2+} \rightarrow 6Fe^{3+} + 6e^- \\ Cr_2O_7^{2-} + 14H^+ + 6e^- \rightarrow 2Cr^{3+} + 7H_2O \\ \hline Cr_2O_7^{2-} + 14H^+ + 6Fe^{2+} \rightarrow 2Cr^{3+} + 6Fe^{3+} + 7H_2O \end{array}$$

Charges:

$$\underbrace{2- \quad 14+ \quad 12+}_{24+} \qquad \underbrace{6+ \quad 18+ \quad 0}_{24+}$$

The total charge is the same on each side of the redox equation.

Summary

To write ion-electron equations involving oxyanions:

1 If a number of oxygen atoms have to be **added**, e.g. $SO_3^{2-} \rightarrow SO_4^{2-}$, add the same number of water molecules to the left-hand side of the equation and twice that number of hydrogen ions to the right-hand side.

OR If a number of oxygen atoms have to be **removed**, e.g. $Cr_2O_7^{2-} \rightarrow 2Cr^{3+}$, add the same number of water molecules to the right-hand side of the equation and twice that number of hydrogen ions to the left-hand side.

2 Complete the equation by adding the number of electrons needed to balance the total charge. Add the electrons to the same side of the equation as the hydrogen ions.

Using the electrochemical series

- Most data books provide a table known as the **electrochemical series**. In this table ion-electron equations are listed as reductions i.e. each equation is written in the form:

oxidising agent + electron(s) → reducing agent

- To use equations from this series to write a redox equation one of the equations must be reversed since an oxidising agent can only react with a reducing agent and vice versa. To decide which equation to reverse, remember that the starting materials, i.e. reactants, **must** appear on the left-hand side of the redox equation.

Example 3

Iron(III) chloride solution oxidises potassium iodide solution to form iron(II) ions and iodine. In this reaction Cl^- and K^+ are spectator ions.

The ion-electron equations obtained from the electrochemical series are:

$$I_2 + 2e^- \rightarrow 2I^-$$
$$Fe^{3+} + e^- \rightarrow Fe^{2+}$$

Since the iodide ions are oxidised the first equation must be reversed. The equation for the reduction of iron(III) ions must be doubled to balance the number of electrons transferred.

$$2I^- \rightarrow I_2 + 2e^-$$
$$2Fe^{3+} + 2e^- \rightarrow 2Fe^{2+}$$

Redox: $$2Fe^{3+} + 2I^- \rightarrow 2Fe^{2+} + I_2$$

Redox titrations

- The concentration of a solution of a reducing agent can be determined using a solution of a suitable oxidising agent of known concentration provided that:
  1. the balanced redox equation is known or can be derived from the relevant ion-electron equations
  2. the volumes of the reactants are accurately measured by pipette and burette, and
  3. some method of indicating the end-point of the titration is available.

Prescribed Practical Activity

To determine the mass of vitamin C in a tablet by redox titration using an iodine solution of known concentration and starch solution as indicator.

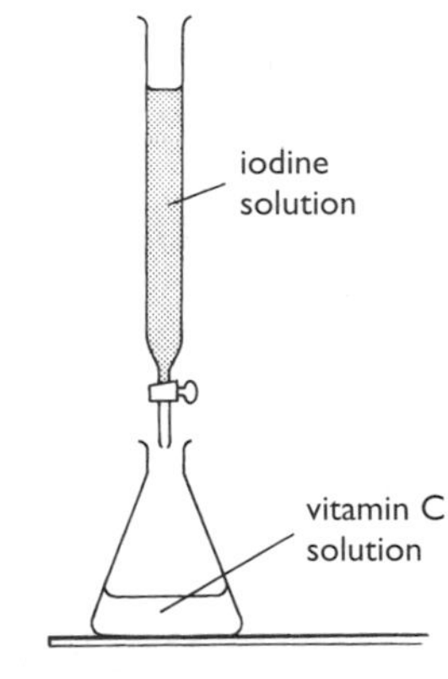

Figure 3

A vitamin C tablet is dissolved in deionised water (about 50 cm^3) in a beaker, transferred with washings to a 250 cm^3 standard flask. The flask is made up to the mark, stoppered and inverted several times to ensure thorough mixing.

25 cm^3 of this solution is transferred by pipette to a conical flask and a few drops of starch indicator are added. Iodine solution of known concentration is added from the burette. The iodine is decolourised at first, but the end-point is the first sign of a permanent blue-black colour. The titration is repeated, with dropwise addition of the iodine solution as the end-point is approached, to obtain concordant titres.

A specimen calculation is given in Worked Example 17.1.

- Redox titration questions can be solved by similar calculations, if the redox equation is known, or, by using the following relationship if the separate ion-electron equations are known:

Concentration × volume × number of electrons gained per mole of oxidising agent = **Concentration × volume × no. of electrons lost per mole of reducing agent**

Electrolysis

- During electrolysis, **reduction occurs at the negative electrode** (or cathode).
- Conversely **at the positive electrode** (or anode) **oxidation takes place**.
- The total number of electrons lost and gained during electrolysis must be the same. In example 3 of Table 2, the cathode reaction occurs twice as often as the anode reaction, i.e. twice as many sodium atoms as bromine molecules will be formed.
- The quantity of electricity, i.e. quantity of electrical charge, required to produce a mole of electrode product can be found by experiment.

Worked Example 17.1

A solution of vitamin C was prepared as described above. 25 cm^3 of this solution was titrated against 0.031 mol l^{-1} iodine solution using starch indicator. The average titre was 17.6 cm^3. Calculate the mass of vitamin C (formula: $C_6H_8O_6$) in the original tablet.

The redox equation is:

$$C_6H_8O_6 + I_2 \rightarrow C_6H_6O_6 + 2H^+ + 2I^-$$

1 mole **1 mole**

Number of moles of iodine used in the titration,

$$n = C \times V = 0.031 \times \frac{17.6}{1000} = 5.456 \times 10^{-4}$$

Hence, number of moles of vitamin C in 25 cm^3 $= 5.456 \times 10^{-4}$

(i.e. in the conical flask)

Number of moles of vitamin C in 250 cm^3 $= 5.456 \times 10^{-3}$

(i.e. in the standard flask)

This is the number of moles of vitamin C in the original tablet.

Gram formula mass of vitamin C, $C_6H_8O_6$ = 176g

Mass of vitamin C present in the tablet $= 176 \times 5.456 \times 10^{-3} = 0.960$g

	Electrolyte	Cathode reaction	Anode reaction
1	Copper(II) chloride solution $CuCl_2$(aq)	$Cu^{2+} + 2e^- \rightarrow Cu$	$2Cl^- \rightarrow Cl_2 + 2e^-$
2	Potassium iodide solution KI(aq)	$2H^+ + 2e^- \rightarrow H_2$	$2I^- \rightarrow I_2 + 2e^-$
3	Molten sodium bromide NaBr(l)	$Na^+ + e^- \rightarrow Na$	$2Br^- \rightarrow Br_2 + 2e^-$

Table 2

Prescribed Practical Activity

Dilute sulphuric acid can be electrolysed using an apparatus such as that shown in Figure 4. To determine the quantity of charge required to produced 1 mole of hydrogen the following measurements need to be made:

1. the volume of hydrogen collected
2. the current used
3. the time during which the solution is electrolysed.

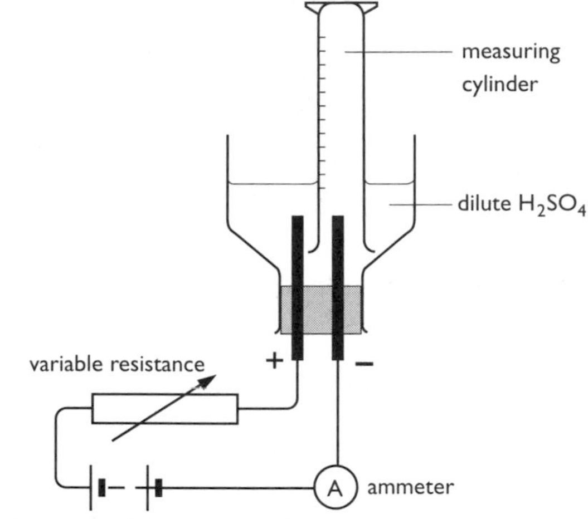

Figure 4

The quantity of electricity, **Q**, used during the experiment is calculated from the following relationship:

Q = I × t,

where **Q** is measured in coulombs (C)
I is the current measured in amps (A), and
t is the time measured in seconds (s).

A specimen calculation is given in worked example 17.2.

Worked Example 17.2

Dilute sulphuric acid was electrolysed using a current of 0.45 A for 6 minutes 50 seconds. The volume of hydrogen collected at room temperature and pressure was 22.8 cm³. Calculate the quantity of electricity required to produce 1 mole of hydrogen.

Quantity of electricity used,

$$Q = I \times t \quad [I = 0.45\ A,\ t = 410\ s]$$
$$= 0.45 \times 410$$
$$= 184.5\ C$$

22.8 cm³ (i.e. 0.0228 litres) of hydrogen was produced by 184.5 C

The molar volume of a gas at room temperature and pressure is approximately 24 litres mol^{-1}.

Hence, 24 litres of hydrogen would require

$$184.5 \times \frac{24}{0.0228} = 194\,210\ C$$
$$= 1.94 \times 10^5\ C$$

An alternative to this experiment is to use an electrolyte which will deposit a metal on the negative electrode. This is cleaned and weighed before electrolysis. The time and current used during electrolysis are measured. At the end of the experiment the negative electrode is removed, rinsed, dried and reweighed to obtain the mass of metal deposited.

Table 3 below summarises the results which would be obtained with various electrolytes.

The quantity of electricity needed to produce one mole of product at the cathode is n × 96 500 C, where n is the number of electrons in the appropriate ion-electron equation. The quantity of electricity, 96 500 C,

Electrolyte	Ion-electron equation for cathode reaction	Quantity of electricity required to produce 1 mole of product
$AgNO_3(aq)$	$Ag^+ + e^- \rightarrow Ag$	96 500 C
$CuCl_2(aq)$	$Cu^{2+} + 2e^- \rightarrow Cu$	193 000 C [or 2 × 96 500 C]
$Al_2O_3(l)$	$Al^{3+} + 3e^- \rightarrow Al$	289 500 C [or 3 × 96 500 C]

Table 3

is a **Faraday.** It is the quantity of electricity which is equivalent to one mole of electrons and, more precisely, has the value 96 500 C *mol*$^{-1}$.

Similar results would be obtained at the positive electrode (anode). The ion-electron equation for the reaction at the anode during the electrolysis of copper(II) chloride solution is:

$$2Cl^- \rightarrow Cl_2 + 2e^-$$

Hence, 193 000 C will be required to produce one mole of chlorine gas.

Calculations involving quantitative electrolysis

The mass of product obtained at an electrode during electrolysis can be calculated if

1 the ion-electron equation for the reaction occurring at the electrode is known, and

2 the current and time during which electrolysis occurs are both given.

❍ In other calculations the mass of product may be given so that the quantity of electricity used may be determined. Then, if the current is given, the time during which electrolysis has taken place can be calculated or *vice versa.* The following relationships should assist you.

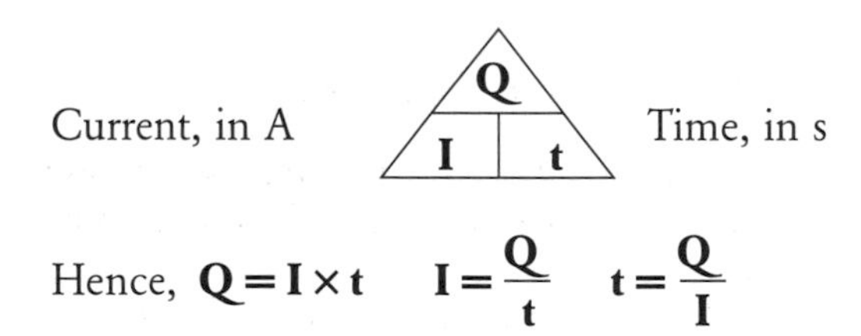

Hence, $\mathbf{Q = I \times t}$ $\quad \mathbf{I = \frac{Q}{t}} \quad \mathbf{t = \frac{Q}{I}}$

Worked Example 17.3

Molten magnesium chloride was electrolysed for 16 minutes 5 seconds using a current of 5 amps. Calculate the mass of product at each electrode.

Quantity of electricity used,

$Q = I \times t \quad [I = 5\ A, t = 965\ s]$
$= 5 \times 965$
$= 4825\ C$

At the **negative** electrode the product is magnesium.

Mg^{2+}	$+ 2e^-$	$\rightarrow$	Mg
1 mol	2 mol		
	$2 \times 96\ 500$ C		24.3 g

Hence, 4825 C will produce

$\frac{24.3}{2} \times \frac{4825}{96\ 500}$ g = 0.61 g of magnesium

At the **positive** electrode the product is chlorine.

$2Cl^- \rightarrow$	Cl_2	$+ 2e^-$
	1 mol	2 mol
	71 g	$2 \times 96\ 500 = 193\ 000$ C

Hence, 4825 C will produce

$71 \times \frac{4825}{193\ 000}$ g = 1.775 g of chlorine

Questions

SECTION A: Redox reactions and titrations

1 Which of the following is a redox reaction?

A $Mg + 2HCl \rightarrow MgCl_2 + H_2$

B $KOH + HCl \rightarrow KCl + H_2O$

C $CuO + 2HCl \rightarrow CuCl_2 + H_2O$

D $ZnCO_3 + 2HCl \rightarrow ZnCl_2 + CO_2 + H_2O$

2 A redox reaction occurs when Zn metal is added to $AgNO_3$ solution. Which statement is **not** true about this reaction?

A Silver metal is displaced.
B Nitrate ions are spectator ions.
C Silver ions are reduced.
D Zinc is the oxidising agent.

3 $MnO_2(s) + xH_2O(l) \rightarrow MnO_4^-(aq) + yH^+(aq) + ze^-$

Which of the following sets of numerical values of x, y and z would balance the ion-electron equation given above?

	x	y	z
A	1	2	1
B	2	4	3
C	3	6	5
D	4	2	1

4 Copper metal can be oxidised by nitric acid. The ion-electron equations are:

$Cu(s) \rightarrow Cu^{2+}(aq) + 2e^-$

$NO_3^-(aq) + 4H^+(aq) + 3e^- \rightarrow NO(g) + H_2O(l)$

The number of moles of nitrate ions reduced by one mole of copper is

A 0.33 B 0.67 C 1.5 D 2.0

5 Which of the following is a redox reaction?

A $Pb^{2+}(aq) + 2I^-(aq) \rightarrow PbI_2(s)$

B $H^+(aq) + OH^-(aq) \rightarrow H_2O(l)$

C $Cl_2(g) + 2Br^-(aq) \rightarrow 2Cl^-(aq) + Br_2(aq)$

D $NH_3(g) + H_2O(l) \rightarrow NH_4^+(aq) + OH^-(aq)$

Questions 6 and **7** relate to the redox reaction between potassium iodide solution and potassium permanganate solution which also contains sulphuric acid.

The redox equation is:

$$10\ I^-(aq) + 2\ MnO_4^-(aq) + 16\ H^+(aq) \rightarrow 2Mn^{2+}(aq) + 5\ I_2(aq) + 8\ H_2O(l)$$

6 The reducing agent in this reaction is:

A $I^-(aq)$
B $MnO_4^-(aq)$
C $H^+(aq)$
D $Mn^{2+}(aq)$

7 The spectator ions in this reaction are

A $H^+(aq)$ and $K^+(aq)$
B $H^+(aq)$ and $MnO_4^-(aq)$
C $I^-(aq)$ and $SO_4^{2-}(aq)$
D $K^+(aq)$ and $SO_4^{2-}(aq)$

8 Complete each of the following ion-electron equations and indicate whether the reactant is a reducing agent or an oxidising agent.

a) $Mn^{2+}(aq) \rightarrow MnO_2(s)$

b) $VO_3^-(aq) \rightarrow V^{2+}(aq)$

c) $H_2O_2(aq) \rightarrow O_2(g)$

d) $Ti(s) \rightarrow TiO^{2+}(aq)$

e) $BrO_3^-(aq) \rightarrow Br_2(aq)$

9 Work out the ion-electron equations for the oxidation steps in the following redox reactions.

a) $Cr + 3Ag^+ \rightarrow Cr^{3+} + 3Ag$
b) $2Fe^{3+} + SO_3^{2-} + H_2O \rightarrow 2Fe^{2+} + SO_4^{2-} + 2H^+$

10 Work out the ion-electron equations for the reduction steps in the following redox reactions.

a) $Br_2 + 2I^- \rightarrow 2Br^- + I_2$
b) $2Fe^{2+} + H_2O_2 + 2H^+ \rightarrow 2Fe^{3+} + 2H_2O$

11 Redox reactions occur between the following pairs of reactants.

a) $Fe^{3+}(aq)$ and $Sn^{2+}(aq)$
b) $I^-(aq)$ and acidified $Cr_2O_7^{2-}(aq)$

Write the redox equation for each reaction (refer to the data book).

12 Acidified potassium dichromate solution can be used to oxidise alcohols.

a) The ion-electron equation for the oxidation of ethanol to ethanal is:

$$C_2H_5OH \rightarrow CH_3CHO + 2H^+ + 2e^-$$

Write the redox equation for the reaction between ethanol and acidified dichromate ions.

b) Ethanal in turn reacts with acidified dichromate ions producing ethanoic acid.

i) Write the ion-electron equation for ethanal changing to ethanoic acid.

ii) How many moles of dichromate ions are needed to oxidise 1 mole of ethanal?

13 Iodine solution is decolourised by sodium thiosulphate solution, $Na_2S_2O_3(aq)$. The equation for this redox reaction is:

$$I_2(aq) + 2S_2O_3^{2-}(aq) \rightarrow 2I^-(aq) + S_4O_6^{2-}(aq)$$

20.0 cm^3 of iodine solution was titrated with 0.1 mol l^{-1} sodium thiosulphate solution using starch indicator. The experiment was repeated and the average titre was 23.4 cm^3.

a) Work out the ion-electron equation for the reducing agent in this redox reaction.
b) What colour change occurs at the end-point of the titration?
c) Calculate the concentration of the iodine solution in mol l^{-1}.

14* Hydrogen peroxide solution, $H_2O_2(aq)$, has several uses, e.g. as hair bleach and as a disinfectant. The concentration of a hydrogen peroxide solution can be found by a redox titration. The hydrogen peroxide is acidified and titrated against potassium permanganate solution of known concentration.

The equation for the reaction which takes place is:

$$5H_2O_2(aq) + 2MnO_4^-(aq) + 6H^+(aq) \rightarrow 2Mn^{2+}(aq) + 5O_2(g) + 8H_2O(l)$$

a) Write an ion-electron equation for the reduction of permanganate ions.
b) Why is it **not** necessary to add an indicator to detect the end-point of the titration?
c) i) In one experiment, 20.0 cm^3 of hair bleach was diluted accurately to 500 cm^3. Explain, with full experimental detail, how you would carry out this dilution.
ii) It was found that 20.0 cm^3 of the **diluted** hair bleach required 24.6 cm^3 of 0.02 mol l^{-1} potassium permanganate when titrated. Use these results to calculate the concentration of hydrogen peroxide, in mol l^{-1}, in the **original** hair bleach.

15* For people who suffer from bronchitis, even low concentrations of ozone, O_3, irritate the lining of the throat and can cause headaches. NO_2 gas from car exhausts reacts with oxygen to form ozone as follows.

$$NO_2(g) + O_2(g) \rightleftharpoons NO(g) + O_3(g)$$

Car exhaust gases also contain volatile organic compounds (VOCs), which can combine with NO gas.

a) Explain how a rise in VOC concentration will change the ozone concentration.

b) In an experiment to measure the ozone concentration of air in a Scottish city, 100,000 litres of air were bubbled through a solution of potassium iodide solution. Ozone reacts with KI solution, releasing iodine.

$$2KI(aq) + O_3(g) + H_2O(l) \rightarrow I_2(aq) + O_2(g) + 2KOH(aq)$$

The iodine formed was titrated with sodium thiosulphate solution, $Na_2S_2O_3(aq)$.

$$I_2(aq) + 2S_2O_3^{2-}(aq) \rightarrow 2I^-(aq) + S_4O_6^{2-}(aq)$$

22.45 cm^3 of 0.01 mol l^{-1} sodium thiosulphate solution was required. Calculate

i) the number of moles of iodine produced, and hence
ii) the volume of ozone in **one litre** of air. (Take V_{mol} of ozone to be 24 litres mol^{-1}.)

SECTION B: Quantitative electrolysis

1* The reduction of zinc ions during electroplating can be represented as:

$$Zn^{2+}(aq) + 2e \rightarrow Zn(s)$$

What is the quantity of electricity needed to produce 0.25 mol of zinc?

A 24 125 C B 48 250 C
C 96 500 C D 193 000 C

2* If a steady current of 0.4 A is passed through silver nitrate solution, concentration 1 mol l^{-1} for 40 minutes, what amount of silver will be liberated?

A 0.001 mol B 0.01 mol
C 0.1 mol D 1.0 mol

3* If 96 500 C of electricity are passed through separate solutions of copper(II) chloride and nickel(II) chloride, then

A equal masses of copper and nickel will be deposited
B the same number of atoms of each metal will be deposited
C the metals will be plated on the positive electrode
D different numbers of moles of each metal will be deposited.

4 Which of the following combinations of current and time would produce 900 coulombs?

A 4 amps for 450 seconds
B 0.2 amps for 1 hour
C 0.5 amps for 30 minutes
D 2 amps for 10 minutes

5 What volume of hydrogen, in litres, would be released at room temperature and pressure if 20 000 C of electrical charge is passed through dilute hydrochloric acid?
(V_{mol} = 24 litres mol^{-1})

A 231.6 B 25.0 C 5.0 D 2.5

6 A solution of chromium(III) sulphate is electrolysed using chromium electrodes. The passage of 96 500 C results in

A the negative electrode gaining 17.3 g
B the positive electrode gaining 17.3 g
C the negative electrode gaining 34.7 g
D the positive electrode gaining 34.7 g

7 One mole of an element was produced by electrolysis when 386 000 C was used. Which equation shows the production of this element?

A $2Br^- \rightarrow Br_2 + 2e^-$
B $Al^{3+} + 3e^- \rightarrow Al$
C $4OH^- \rightarrow O_2 + 2H^+ + 4e^-$
D $Ag^+ + e^- \rightarrow Ag$

8 Which is the correct quantity of product when 9650 C of electrical charge is passed through excess copper(II) chloride solution?

A 0.05 mol Cl_2
B 0.10 mol Cl_2
C 0.15 mol Cu
D 0.20 mol Cu

9 Use relevant ion-electron equations from the SQA Data Book when answering the following questions.

a) Calculate the length of time required to plate a metal spoon with 0.2 g of silver when a current of 1.5 A is passed through silver(I) nitrate.

b) From the data given below calculate the current used when nickel-plating a sheet of metal. The plating solution contains nickel(II) ions.

Mass of metal sheet before plating = 198.76 g
Mass of metal sheet after plating = 200.30 g
Time taken for plating = 20 minutes

c) Calculate the mass of aluminium, in kg, produced per hour when a current of 175 000 A is passed through molten aluminium oxide.

10* Gases are produced by the electrolysis of $Na_2SO_4(aq)$.

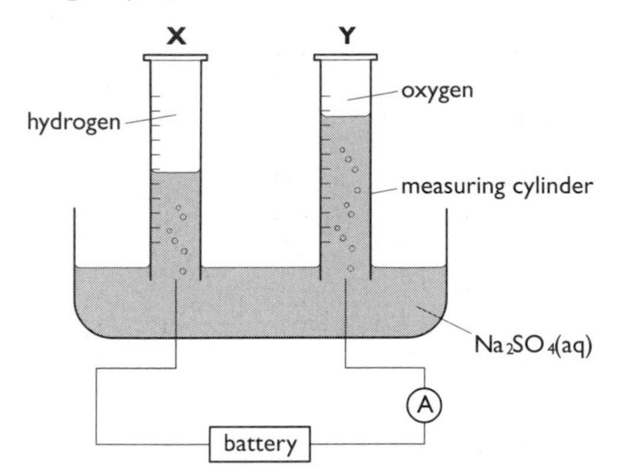

The ion-electron equations are shown.

Electrode **X**:

$2H_2O(l) + 2e^- \rightarrow H_2(g) + 2OH^-(aq)$

Electrode **Y**:

$H_2O(l) \rightarrow \frac{1}{2}O_2(g) + 2H^+(aq) + 2e^-$

a) Explain what happens to the pH at each electrode.

b) A current of 2 A was passed through the apparatus for 5 min and 20 s. Calculate the volume of hydrogen gas produced.
(Take the molar volume of hydrogen gas to be 24 litres mol^{-1}.)

11 In an experiment to determine the Avogadro Constant (L) by quantitative electrolysis, a steady current of 0.26 A was passed through dilute sulphuric acid, using platinum electrodes, for 15 minutes. The volume of hydrogen produced was 29.1 cm^3. The equation for the production of hydrogen is:

$$2H^+(aq) + 2e^- \rightarrow H_2(g)$$

a) Use the information given above to calculate the quantity of electric charge
i) needed to produce 1 mole of hydrogen gas (V_{mol} = 24 litres mol^{-1})
ii) equivalent to 1 mole of electrons.

b) To obtain a value for L, the charge on 1 electron is required. This is obtained by a different experiment and its value is 1.6×10^{-19} C.
Calculate L using this value and your answer to **a) ii)** above.

12

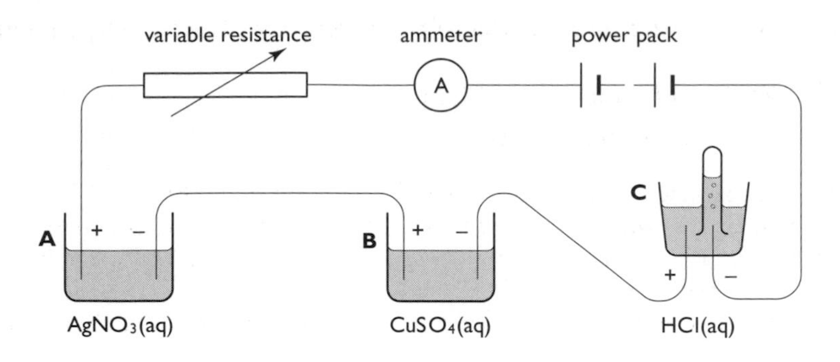

In an experiment using the apparatus shown above, 0.125 g of silver was produced in cell **A**. In answering the following questions about this experiment you should assume that

i) the same quantity of electric charge is passed through all three cells, and
ii) the electrodes are inert, i.e. they do not react with the electrode products.

a) With reference to the SQA Data Book, calculate
i) the quantity of electric charge passed through each cell
ii) the mass of copper produced in cell **B**
iii) the volume of hydrogen produced in cell **C** (V_{mol} = 24 litres mol^{-1}).

b) Both hydrogen and chlorine are produced in cell **C**.
i) Explain why the volume of each gas should be the same.
ii) In the experiment, much less chlorine is obtained than hydrogen. Suggest a reason for this.

c) In cells **A** and **B**, the product at the positive electrode is oxygen. The relevant ion-electron equation is:

$$2H_2O(l) \rightarrow O_2(g) + 4H^+(aq) + 4e^-$$

i) Calculate the volume of oxygen which should be produced in each cell in this experiment.
ii) What effect, if any, does electrolysis have on the pH of the solutions in cells **A** and **B**?

13* The concentration of a solution of sodium thiosulphate can be found by reaction with iodine.

The iodine is produced by electrolysis of an iodide solution using the apparatus shown.

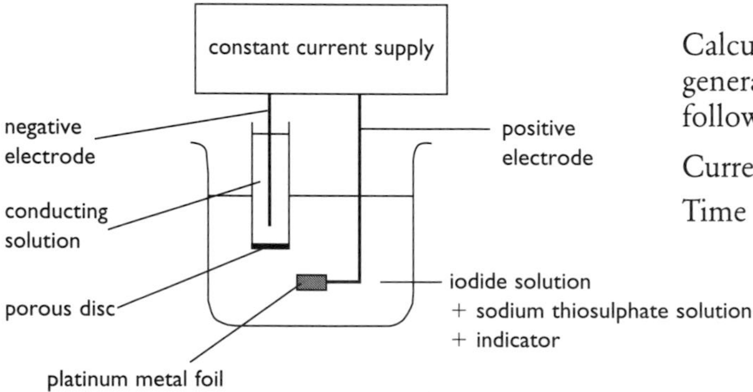

The current is noted and the time when the indicator detects the end-point of the reaction is recorded.

a) Iodine is produced from the iodide solution according to the following equation:

$$2I^-(aq) \rightarrow I_2(aq) + 2e^-$$

Calculate the number of moles of iodine generated during the electrolysis given the following results.

Current = 0.010 A
Time = 1 min 37 s

b) The iodine produced reacts with the thiosulphate ions according to the equation:

$I_2(aq) + 2S_2O_3^{2-}(aq) \rightarrow 2I^-(aq) + S_4O_6^{2-}(aq)$
iodine thiosulphate ions

At the end-point of the reaction, excess iodine is detected by the indicator.

i) Name the indicator which could be used to detect the excess iodine present at the end-point.

ii) In a second experiment it was found that 1.2×10^{-5} mol of iodine reacted with 3.0 cm^3 of the sodium thiosulphate solution.
Use this information to calculate the concentration of the sodium thiosulphate solution, in mol l^{-1}.

iii) The production of iodine takes place at the surface of the platinum foil at the tip of the positive electrode.
Suggest what could be done to the solution during the reaction to increase the accuracy of the results.

18 Nuclear chemistry

- Radioactive elements continue to display activity in their compounds, therefore radioactivity is unaffected by changes in electron arrangement and must originate in the nucleus of atoms.
- The stability of nuclei depends on the ratio of neutrons to protons (Figure 1).
- Radioactive emissions change the neutron to proton ratio and at the same time release energy.
- Radioactive emissions can be shown to be of three types by their behaviour in an electrical field:

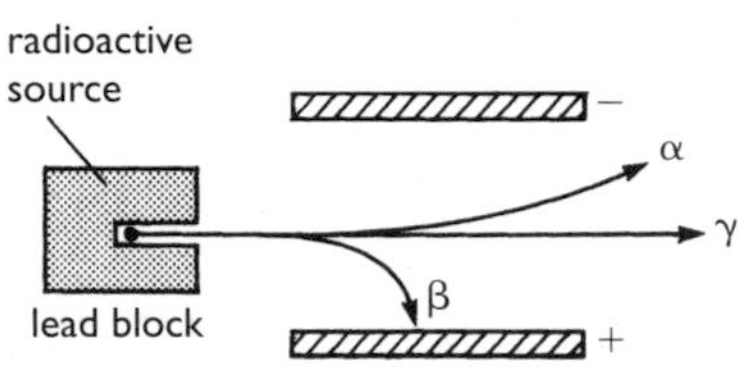

Figure 2 Radiation deflected by an electric field between charged plates

Other experiments provide the information given in Table 1.

- Loss of an α particle causes a loss of 2 units of charge and 4 units of mass.

  e.g. ${}^{238}_{92}U \rightarrow {}^{234}_{90}Th + {}^{4}_{2}He$

- A β particle is an electron. It is believed to be formed by:

  $${}^{1}_{0}n \rightarrow {}^{1}_{1}p + {}^{0}_{-1}e$$

  neutron proton electron

- Loss of a β particle causes a **gain** of 1 unit of charge and no change in mass.

  e.g. ${}^{234}_{90}Th \rightarrow {}^{234}_{91}Pa + {}^{0}_{-1}e$

- Radioactive isotopes can be created by bombarding stable isotopes with fast moving particles, usually neutrons, which are not repelled by the positively charged nucleus.

  e.g. ${}^{27}_{13}Al + {}^{1}_{0}n \rightarrow {}^{24}_{11}Na + {}^{4}_{2}He$

Name	Penetration	Nature	Symbol	Charge	Mass
α (alpha)	few cm in air	He nucleus	${}^{4}_{2}He$	2^+	4
β (beta)	thin metal foil	electron	${}^{0}_{-1}e$	1^-	$\frac{1}{2000}$
γ (gamma)	great thickness of concrete	EMR	none	none	none

Table 1

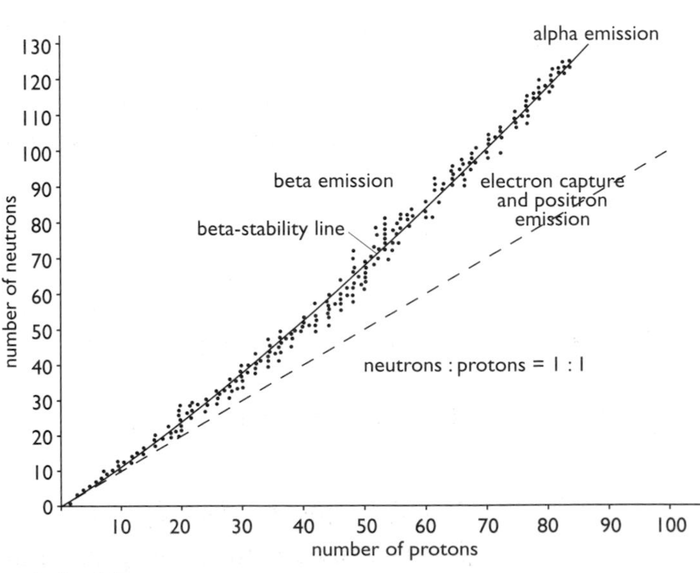

 Figure 1 Band of stability

The sodium isotope produced then decays by β emission:

$^{24}_{11}Na \rightarrow ^{24}_{12}Mg + ^{0}_{-1}e$

- In all nuclear equations, note that on each side of the equations, the sum of the mass numbers (at the top of the symbols) must be the same, and similarly the sum of the atomic numbers (below the symbols) must be the same.
- The half-life of a radioisotope is the time taken for the mass, or activity, of the isotope to halve by radioactive decay. The half-life, $t_{1/2}$, is characteristic of the isotope.

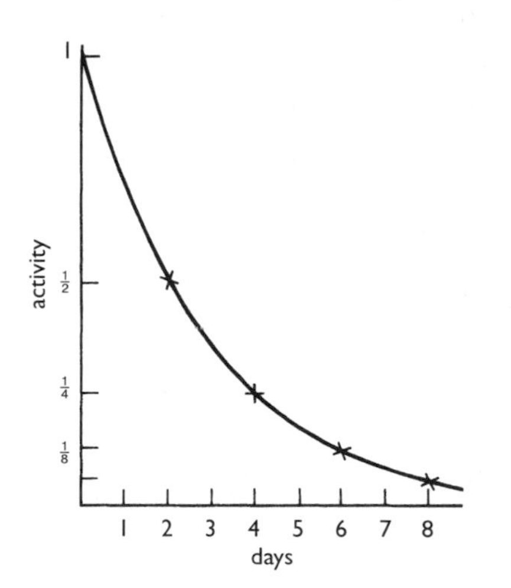

Figure 3 Change in activity with time for an isotope of half-life 2 days

- The half-life of any isotope is independent of the mass of the sample being investigated. It is independent of temperature, pressure, concentration, presence of catalysts or chemical state of the isotope. In the half-life, half of the atoms of the isotopes decay but the process is completely random and it is not possible to predict the time of decay of any individual atom.
- After 'n' half-lives, the fraction of the original activity which remains is given by $(1/2)^n$.
- The quantity of radioisotope, the half-life or time elapsed can be calculated, given the other two variables.

Worked Example 18.1

Rhodium-106 has a half life of 30 seconds. What fraction of a sample's activity will remain after 3 minutes?

$t_{1/2} = 30$ seconds, so 3 minutes $= 180$ seconds $= 6\ t_{1/2}$

Fraction of activity remaining $= (\frac{1}{2})^6 = \frac{1}{64}$

Worked Example 18.2

An archaeological sample shows a $^{14}_{6}C:^{12}_{6}C$ ratio only $\frac{1}{8}$ of that expected in a modern sample. If the half-life of $^{14}_{6}C$ is 5600 years, calculate an approximate age for the archaeological sample.

Fraction of activity remaining $= \frac{1}{8} = (\frac{1}{2})^3$

Therefore the sample has decayed through 3 half-lives i.e. 3 × 5600 years.

The archaeological sample is approximately 16 800 years old.

Uses of radioisotopes

Radioisotopes in medicine

The γ emitter $^{60}_{27}Co$ is used to treat deep-seated tumours.

The β emitter $^{32}_{15}P$ is used to treat skin cancer.

The short half-life β emitter $^{131}_{53}I$ is used to assess the condition of the thyroid gland which absorbs iodine. A scan shows the concentration of iodine throughout the gland.

Radioisotopes in industry

The γ emitter $^{60}_{27}Co$ is used to check the condition of welds in steel and to irradiate food to kill bacteria and fungi, so increasing 'shelf-life'.

The α emitter $^{241}_{95}Am$ is used in domestic smoke alarms.

Radioisotopes in scientific research

$^{32}_{15}P$ can be used to trace the uptake of phosphate fertilisers by plants.

$^{14}_{6}C$ with a half-life of 5600 years can be used to date archaeological specimens with an origin in living material. In living material the ratio $^{14}_{6}C$:$^{12}_{6}C$ is constant. After death, the ratio decreases at a rate determined by the $t_{1/2}$ of $^{14}_{6}C$. If the present ratio is measured the approximate age of the specimen can be determined.

Radioisotopes in energy production by nuclear fission

When impact by relatively slowly moving neutrons occurs on the nucleus of $^{235}_{92}U$, the nucleus splits, or **fissions**, producing smaller nuclei and more neutrons. These neutrons can, when there is more than a critical mass of uranium, go on to split more nuclei and a self-sustaining chain reaction results. The total mass of the products of fission is slightly less than the mass of the starting materials. This mass defect is converted into a large amount of energy. An example of a fission reaction is:

$$^{235}_{92}U + ^{1}_{0}n \rightarrow ^{236}_{92}U \rightarrow ^{140}_{54}Xe + ^{94}_{38}Sr + 2^{1}_{0}n$$

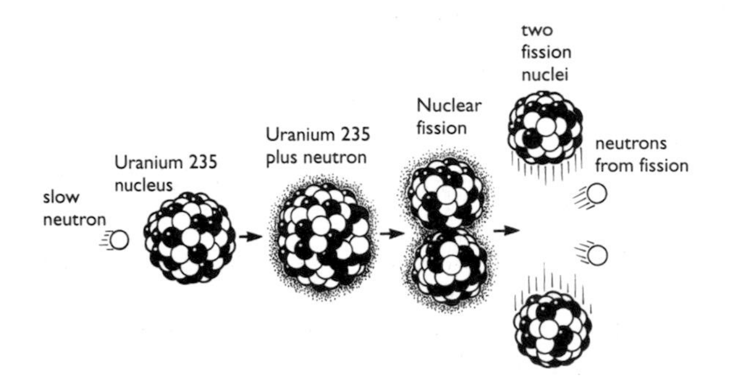

Figure 4 Uranium fission

Radioisotopes in energy production by nuclear fusion

If small nuclei can be induced to **fuse** together,

e.g.

$$^{2}_{1}H + ^{3}_{1}H \rightarrow ^{4}_{2}He + ^{1}_{0}n$$

then mass is 'lost' and converted into energy. So far the process has only been carried out on a large scale in a hydrogen bomb, but several major research programmes are attempting to generate energy by peaceful means by fusion.

The origin of the elements

- In space there are large quantities of hydrogen and in the interstellar clouds this is concentrated so that fusion occurs.

e.g. $^{1}_{1}H + ^{2}_{1}H \rightarrow ^{3}_{2}He$

$^{2}_{1}H + ^{3}_{1}H \rightarrow ^{4}_{2}He + ^{1}_{0}n$

- The matter coalesces into stars, the energy produced by fusion initiates more fusion.

e.g. $3^{4}_{2}He \rightarrow ^{12}_{6}C$

$^{12}_{6}C + ^{4}_{2}He \rightarrow ^{16}_{8}O$

- By processes such as these, all the elements have been produced, essentially from hydrogen.

Advantages of nuclear power
No 'greenhouse' gases emitted
No SO_2 emitted to increase 'acid rain'
Fewer deaths and injuries in uranium mining than in coal mining
Fuel reserves will last longer than fossil fuel reserves
Power stations have less visual impact than coal or oil-fired stations or wind farms
Alternative source of energy for countries with no fossil or renewable energy sources

Disadvantages of nuclear power
Finite (but low) probability of a disastrous accident
Contribution to 'background' radiation
Difficulty of disposing of spent fuel which remains radioactive for many years
Capital cost of plant
Cost of decommissioning obsolete stations
Stations slow to respond to rapid changes in demand for power, conventional stations faster
Plutonium can be produced, possibly leading to proliferation of nuclear weapons

Questions

1* β-particles emitted by certain radioactive atoms are

A electrons from the outer shell
B electrons from the nucleus
C particles consisting of 2 protons and 2 neutrons
D electromagnetic radiations of very short wavelength.

2 A radioactive isotope of a halogen emits a β-particle. The product will be an isotope of an element in

A Group 6 B Group 7
C Group 0 D Group 1

3* Which particle will be formed when an atom of $^{211}_{83}Bi$ loses an α-particle and the decay product then loses a β-particle?

A $^{210}_{79}Au$ B $^{209}_{80}Hg$
C $^{208}_{81}Tl$ D $^{207}_{82}Pb$

4* The stability of the nucleus of an ion depends on the ratio of

A mass : charge
B neutrons : protons
C neutrons : electrons
D protons : electrons.

Questions 5 and **6** refer to the following section of a natural radioactive series.

$$^{w}X \xrightarrow{\alpha} {}^{223}Fr \xrightarrow{\beta} {}^{y}Z$$

5 Which of the following isotopes is ^{w}X?

A ^{225}Ac B ^{223}Rn
C ^{227}Pa D ^{227}Ac

6 Which of the following isotopes is ^{y}Z?

A ^{223}Ra B ^{223}Rn
C ^{219}At D ^{224}Ra

7 $^{238}_{92}U$ can absorb an α-particle with the emission of a neutron. What is the product of this reaction?

A $^{241}_{93}Np$ B $^{239}_{94}Pu$
C $^{241}_{94}Pu$ D $^{242}_{95}Am$

8* When some zinc pellets containing radioactive zinc are placed in a solution of zinc chloride, radioactivity soon appears in the solution. Compared with that of the pellets, the half-life of the radioactive solution will be

A shorter
B the same
C longer
D dependent upon how long the zinc is in contact with the solution.

9* Which of the following has an electrical charge?

A α-particles B X-rays
C Neutrons D γ-rays

10 Several isotopes of polonium appear in the two radioactive decay series on page 8 of the SQA Data Book. The mass number of the isotope of polonium with the shortest half-life is

A 218 B 216
C 214 D 212

11 ^{220}Rn has a half-life of 55 s. What fraction of the original number of ^{220}Rn atoms will remain after 5.5 minutes?

A $\frac{1}{4}$ B $\frac{1}{6}$ C $\frac{1}{16}$ D $\frac{1}{64}$

12 Which sentence describes what happens when an isotope emits gamma rays?

A The atomic number increases and the mass number does not change.
B Both the atomic number and the mass number decrease.
C The atomic number decreases and the mass number does not change.
D Neither the atomic number nor the mass number change.

Questions 13 and **14** refer to the following processes which cause nuclear changes to occur.

A Neutron capture.
B Nuclear fusion.
C Proton capture.
D Nuclear fission.

13 Which process occurs in the following nuclear change?

$$^{23}_{11}\mathbf{Na} + x \rightarrow {}^{24}_{11}\mathbf{Na}$$

14 Which process occurs in the following nuclear change?

$$^{2}_{1}\text{H} + ^{3}_{1}\text{H} \rightarrow ^{4}_{2}\text{He} + y$$

15 The following sequence shows part of a radioactive series.

$$^{235}_{92}\text{U} \rightarrow {}^{\beta}\text{Q} \rightarrow {}^{\beta}\text{R} \rightarrow {}^{\alpha}\text{X} \rightarrow {}^{\alpha}\text{Z}$$

Identify the true statement.

A Q and Z are isotopes of the same element.
B Atoms of Q have 91 protons.
C Atoms of R have 139 neutrons.
D X is an isotope of uranium.

16 The following diagram shows how mass number and atomic number alters during a radioactive series.

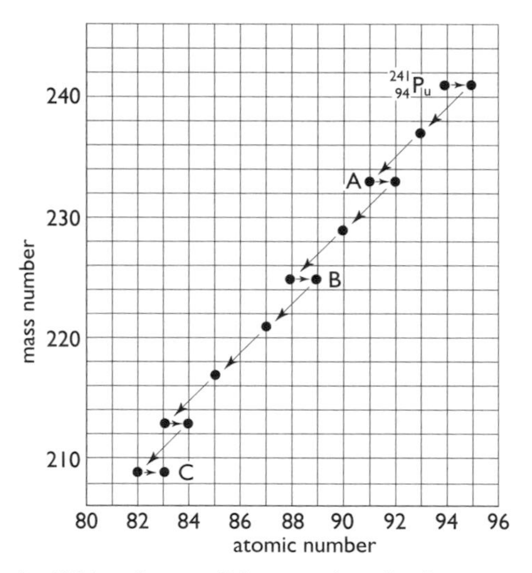

a) Write the nuclide notation for isotopes A, B and C.

b) How many **i)** α particles **ii)** β particles, are released **in total** in the series?

c) Draw a similar diagram to represent the first 6 stages of the Thorium Series given in Table 2 on page 8 of the SQA Data Book.

d) The series shown above is sometimes described as the '4n + 1 Series', while the Thorium Series is denoted the '4n Series' where 'n' is an integer. These designations are related to the mass numbers of the isotopes present in the series.
What designation should be given to the Plutonium-Uranium Series shown in Table 1 on page 8 of the SQA Data Book?

17 **a)** Write the appropriate symbol of the particle produced when

i) ^{210}At loses an α particle

ii) an astatine atom gains an electron in its outer shell.

b) Suggest why the mass number is specified in **a) i)** but not in **a) ii)**.

18 Nuclear changes which are used to produce radioactive isotopes can be summarised as follows:

T(x,y)P

when: T = target nucleus,
P = product nucleus,
x = bombarding particle
y = ejected particle.

For example, the nuclear equation:

$^{14}_{7}\text{N} + ^{4}_{2}\text{He} \rightarrow ^{17}_{8}\text{O} + ^{1}_{1}\text{p}$

can be summarised: $^{14}_{7}\text{N}(\alpha,\text{p})^{17}_{8}\text{O}$.

a) Write the nuclear equation for the change: $^{27}_{13}\text{Al}(\text{n},\alpha)^{24}_{11}\text{Na}$.

b) Write the summarised form of the following nuclear equation.

$$^{32}_{16}\text{S} + ^{1}_{0}\text{n} \rightarrow ^{32}_{15}\text{P} + ^{1}_{1}\text{p}$$

19 For each of the following pairs, indicate whether or not the quantities would have

i) the same half-life
ii) the same intensity of radiation.

a) 0.1 g of ^{214}Pb and 1 g of ^{214}Pb

b) 0.1 g of ^{214}Pb and 0.1 g of $^{214}PbCl_2$

c) 0.1 mol of ^{210}Pb and 0.1 mol of ^{214}Pb

d) 0.1 mol of ^{210}Pb and 0.1 mol of $^{210}PbCl_2$

20 In 1995 the New Scientist reported that research scientists in Germany had discovered a new element, atomic number 111. They had bombarded bismuth-209 with billions of nickel atoms.
The new atoms produced had a mass number of 272 and decayed into two previously unknown isotopes of elements 109 and 107.

a) What type of nuclear process is described in the first paragraph?

b) What type of radiation is suggested by the information in the second paragraph?

c) Predict the mass numbers of the previously unknown isotopes of elements 109 and 107.

21* Many granite rocks contain radioactive elements which decay to release radon gas. The gas is an alpha-emitter with a half-life of 55 s and contributes to background radiation.

a) Give another source of background radiation.
b) Write a balanced nuclear equation for the alpha-decay of radon-220.
c) A sample of radon had a count rate of 80 counts min^{-1}. How long would it take for the count rate to fall to 5 counts min^{-1}?
d) What effect would a temperature rise of 20°C have on the half-life of radon-220?

22* Complete the following nuclear equations and identify **X** and **Y**.

a) ${}^{238}_{92}U + {}^{4}_{2}He \rightarrow {}^{239}_{94}Pu + 3\mathbf{X}$

b) ${}^{6}_{3}Li + {}^{1}_{0}n \rightarrow {}^{4}_{2}He + \mathbf{Y}$

23* The Avogadro constant (L) may be estimated in a number of ways, one of which is described below.
When a radioactive substance emits α particles, helium gas is formed by the reaction:

$${}^{4}_{2}He^{2+} + 2e^{-} \rightarrow {}^{4}_{2}He(g)$$

A Geiger counter is used to count the number of α particles emitted and, over a period of time, the volume of helium collected is measured. In an actual experiment, a sample of radium-226 emitted 4.4×10^{10} α particles **per second** over a period of 24 hours. The total volume of helium collected (at room temperature and pressure) was 1.50×10^{-4} cm^3. ($V_{mol} = 24$ litres mol^{-1}.)

a) Write a nuclear equation for the radioactive decay of radium-226.
b) Use the SQA Data Book to find the value for the half-life of the radioactive **product** obtained in **a)**.
c) How many α particles were emitted during the 24 hour period?
d) How many moles of helium are present in 1.50×10^{-4} cm^3 of the gas?
e) Use your answers to **c)** and **d)** to calculate L.

24 Torness Nuclear Power Station

Refer to the sketch map on page 87.

a) Suggest a reason why a relatively remote site was chosen for the power station.
b) Why are nuclear stations in the UK usually sited on the coast?
c) What other features of the surrounding area might have been useful during construction of the power station?

Questions 7, 8 and 9 on pages 122–3 are based on the Unit 3 Prescribed Practical Activities.

1 The effect of changing concentration on the rate of a reaction can be investigated using a 'clock reaction' involving aqueous solutions of hydrogen peroxide and potassium iodide.

25 cm^3 of KI(aq) is mixed with sodium thiosulphate solution, dilute sulphuric acid and starch solution. A few cm^3 of $H_2O_2(aq)$ are added and the time taken for the reaction to occur is measured. The rate of reaction is obtained by calculating the reciprocal of time. The experiment is repeated four times using smaller volumes of KI(aq) but with water added, e.g. 20 cm^3 KI(aq) + 5 cm^3 H_2O; 15 cm^3 KI(aq) + 10 cm^3 H_2O etc.

The graph shows typical results from this experiment.

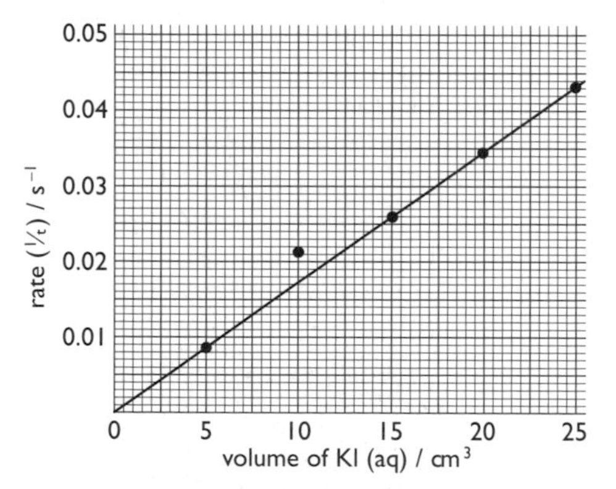

a) What variables should be kept constant for a fair comparison?
b) The method of dilution of KI(aq) described above ensures that the total volume is kept constant. Why is this important in this experiment?
c) Use information from the graph to calculate the reaction time when 20 cm^3 of KI(aq) was used.
d) One of the results shown on the graph appears to be wrong.
   i) What should have been the rate for this volume of KI(aq)?
   ii) Apart from errors in timing suggest how this wrong result may have occurred.
e) What conclusion can be drawn regarding the rate of this reaction from the results shown on the graph?

2 The effect of temperature change on the rate of a reaction can be studied in the following experiment.

Potassium permanganate solution, which has been acidified with dilute sulphuric acid, is heated to about 40°C. A few cm^3 of oxalic acid solution is added and the time for the reaction to reach completion is measured. The temperature is also noted. The experiment is repeated three times at higher temperatures. The table below shows typical results.

Temperature/ °C	Reaction time (t)/s	Rate ($^1/t$)/s^{-1}
42	62.5	0.016
49	35.7	**X**
58	**Y**	0.052
66	11.0	0.091

a) What colour change occurs at the end-point of the reaction?
b) What variables should be kept constant in this experiment?
c) Calculate the values of **X** and **Y** to complete the table of results.
d) Draw a line graph of rate against temperature using the above data.
e) Use the graph to find
   i) the rate of reaction at 55°C
   ii) the temperature at which the rate is twice this value.

3* A page of a pupil's notebook shows instructions on how to measure the enthalpy of combustion of an alcohol.

Experimental procedure

1. Measure out 100 cm^3 of water into a beaker.
2. Take steps to insulate the apparatus.
3. Read the water temperature before and after using the alcohol burner to heat it.
4. Weigh the alcohol burner before and after the experiment.

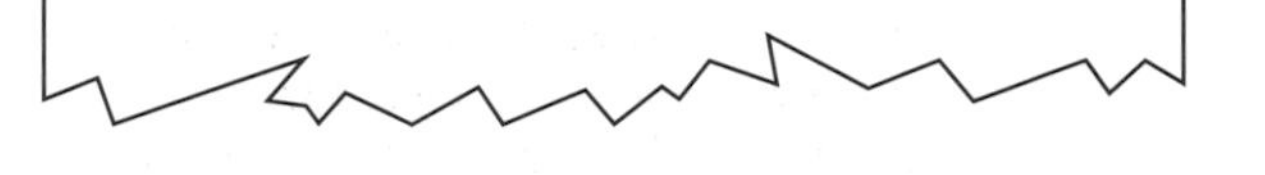

a) Draw a neat labelled diagram of the apparatus which the pupil could use to carry out this experiment.
b) Write the equation corresponding to the enthalpy of combustion of methanol.
c) The pupil found that when 0.23 g of methanol burned, the heat produced raised the temperature of 100 g of water by 9.2°C.
Using information in the SQA Data Book, calculate the enthalpy of combustion of methanol.
d) The pupil's result is well below the value in the SQA Data Book. Even with insulation, much heat is lost to the surroundings, including the apparatus. Suggest one **other** reason why the experimental result is low.

4 **P** and **Q** are carbonyl compounds which can be distinguished by using various reagents.

Test tube **1** contains a few drops of **P** added to acidified $K_2Cr_2O_7(aq)$.
Test tube **2** contains a few drops of **Q** added to $AgNO_3(aq)$ containing $NH_3(aq)$.
Water is heated as shown in the diagram until the temperature reaches about 60 °C. Test tubes **1** and **2** are placed in the hot water for several minutes.

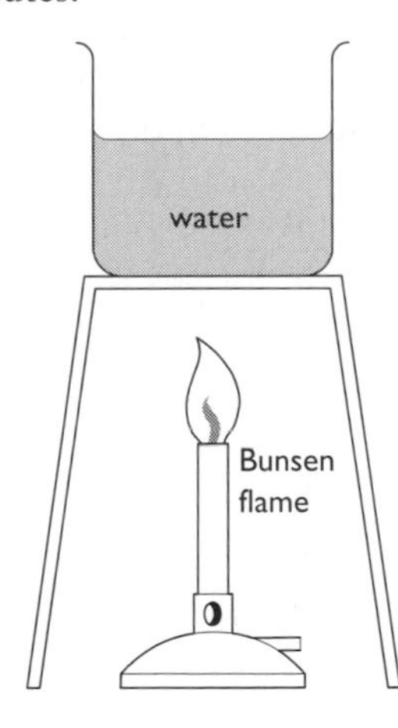

a) What precaution should be taken **before** putting the test tubes in the water?
b) In test tube **1** no colour change was observed.
What type of carbonyl compound is **P**?
c) In test tube **2** a silver 'mirror' is formed.
i) Write an ion-electron equation to explain the formation of the 'mirror'.
ii) What type of carbonyl compound is **Q**?
d) Supply the words which are missing from the following paragraph.

When $CH_3CH_2CH_2CHO$ reacts with acidified $K_2Cr_2O_7(aq)$, the solution changes colour from __________ to __________. The reaction produces a carbon compound called ____________ .

e) Another reagent can be used to distinguish **P** and **Q**. This reagent is a blue solution which gives an orange–red precipitate with **Q**, but does not react with **P**.
i) Name this reagent.
ii) Name the orange–red precipitate and give its formula.

5 The diagram illustrates one way of preparing an ester.

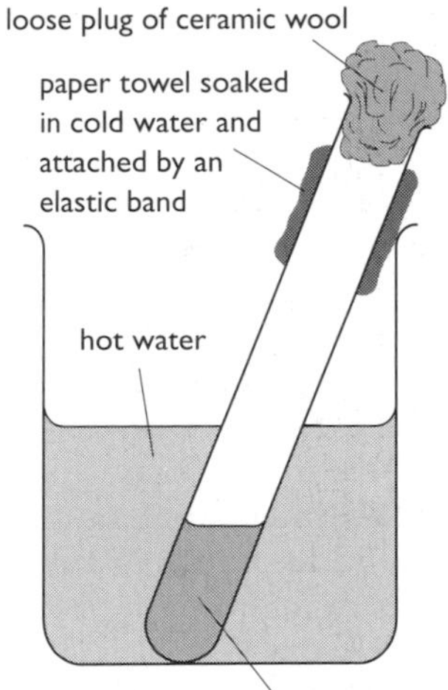

a) What other chemical should be present in the test tube?
b) What is the purpose of the 'paper towel soaked in cold water'?
c) After heating as shown above for several minutes, the contents of the test tube are poured into a beaker containing $NaHCO_3(aq)$.
i) Apart from its smell, how can you tell that an ester has been formed?
ii) Bubbles are also seen in the solution. Why does this happen?
d) An ester was prepared from methanoic acid and butan-1-ol. Write an equation showing full structural formulae for this reaction.

6

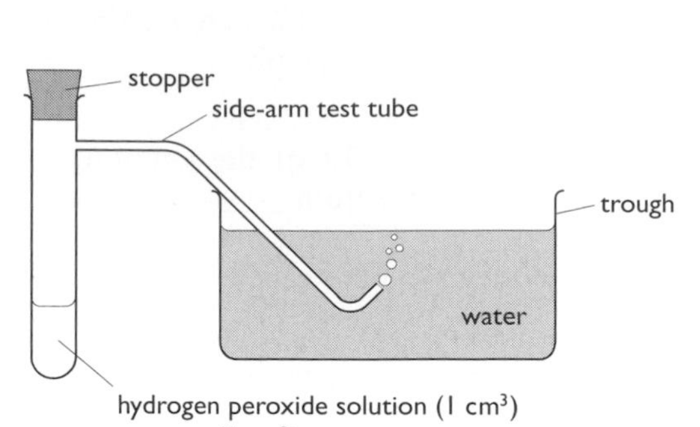

The diagram shows part of an apparatus which can be used when investigating the effect of temperature and pH on catalase, an enzyme which catalyses the decomposition of hydrogen peroxide solution.

a) What can be added to the test tube as the source of catalase?
b) What other pieces of equipment are needed when investigating the effect of temperature on catalase?
c) What observation enables you to judge how changing the temperature affects the catalase in this experiment?
d) It is said that an enzyme operates most effectively at an optimum temperature. How do the results of this experiment illustrate the concept of 'optimum temperature'?
e) How would the experiment be adapted to investigate the effect of pH on the ability of catalase to decompose hydrogen peroxide?
f) How would you ensure that a fair comparison was being made when altering the pH in this investigation?

7 A group of S5 pupils carried out a series of experiments to illustrate Hess's Law.

a) One pupil added 0.02 mol of KOH(s) to 50 cm^3 of water in a polystyrene cup. She measured the temperature change and calculated the enthalpy change to be –52.25 kJ mol^{-1} of KOH.
   i) Name the enthalpy change which has been obtained.
   ii) Why was a polystyrene cup used in the experiment?
   iii) Calculate the temperature rise in this experiment.

b) Another pupil carried out an experiment to determine the enthalpy of neutralisation of KOH(aq) by HCl(aq) by mixing 20 cm^3 of 1.0 mol l^{-1} KOH and 20 cm^3 of 1.0 mol l^{-1} HCl.
He obtained the following data.

Initial temperature of the alkali = 21.2 °C
Initial temperature of the acid = 20.8 °C
Highest temperature of the mixture = 27.4 °C

Calculate the enthalpy of neutralisation of KOH(aq) by HCl(aq).

c) i) Outline the experiment which a third pupil should carry out if the group are to verify Hess's Law.
   ii) Write the equation, including physical states, for this reaction.
   iii) What would be the enthalpy of this reaction if the group's results verify Hess's Law?

8

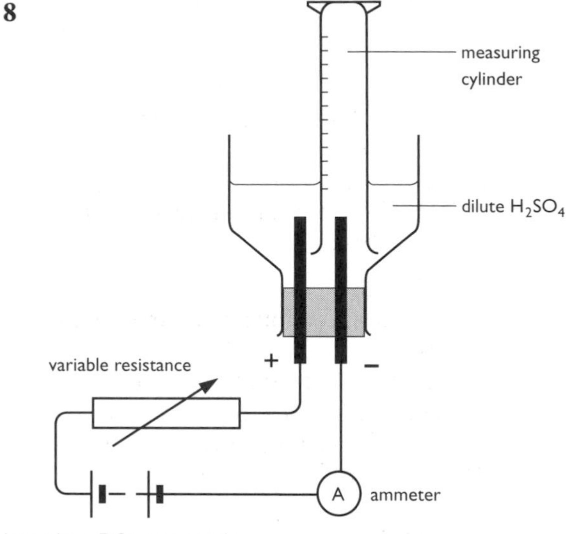

A prescribed practical activity was carried out using the apparatus shown above.

a) Supply the missing words in the following sentence.

   The aim of this experiment is to determine the quantity of electric charge required to produce ________________ by electrolysis of dilute sulphuric acid.

b) What measurements, with units, are made in the experiment to enable the quantity of electric charge to be calculated?

c) Write the ion-electron equation for the reaction occurring at the negative electrode.

d) The volume of hydrogen that is collected is measured.
What other information about hydrogen is needed to complete the calculation?

e) Some pupils obtained a result of 188 000 C from this experiment.
What is the expected result?

9 To find the mass of vitamin C in a tablet, a solution (volume: 250 cm^3) is firstly prepared and then titrated with an iodine solution of known concentration.

a) Describe in detail how the vitamin C solution is prepared.

b) Supply the words missing from the following paragraph which describes the titration.

25.0 cm^3 of a vitamin C solution was measured by __________ and transferred to a ____________________ . A few drops of __________ indicator were added. Iodine solution of known concentration was added from a ___________ until the indicator had just turned _________ after shaking. The titration was repeated at least twice to obtain ______________________.

c) Calculate the mass of vitamin C present in a tablet from the following experimental results.
25.0 cm^3 of a vitamin C solution was titrated against 0.024 mol l^{-1} iodine solution. The average titre was 11.8 cm^3.
The equation for the reaction is:

$$C_6H_8O_6 + I_2 \rightarrow C_6H_6O_6 + 2H^+ + 2I^-$$

Practice Prelim: Units 1 and 2

SECTION A [40 marks]

1 Which of the following substances is a non-conductor when solid, but becomes a good conductor on melting?
A Argon
B Potassium
C Potassium fluoride
D Tetrachloromethane

2 Which gas would dissolve in water to form an alkali?
A HBr
B NH_3
C CO_2
D CH_4

3 An iron nail is covered with water

Which of the following procedures would **not** increase the rate at which the iron nail corrodes?
A Adding some sodium sulphate to the water
B Adding some glucose to the water
C Attaching a copper wire to the nail
D Passing carbon dioxide through the water

4 When copper is added to a solution containing zinc nitrate and silver nitrate
A deposits of both zinc and silver form
B a deposit of zinc forms
C a deposit of silver forms
D no new deposit forms.

5 A mixture of calcium chloride and calcium sulphate is known to contain 0.6 mol of chloride ions and 0.2 mol of sulphate ions.

How many moles of calcium ions are present?
A 0.4
B 0.5
C 0.8
D 1.0

6 What volume of 0.4 mol l^{-1} sodium hydroxide solution is needed to neutralise 50 cm^3 of 0.1 mol l^{-1} sulphuric acid?
A 25 cm^3
B 50 cm^3
C 100 cm^3
D 200 cm^3

7 In which of the following compounds do **both** ions have the same electron arrangement as argon?
A Calcium sulphide
B Magnesium oxide
C Sodium sulphide
D Calcium bromide

8 Isotopes of an element must have
A the same mass number
B different numbers of protons
C the same number of protons and neutrons
D different numbers of neutrons

9 Excess marble chips (calcium carbonate) were added to 25 cm^3 of hydrochloric acid, concentration 2 mol l^{-1}.

Which measurement, taken at regular intervals and plotted against time, would give the graph shown below?

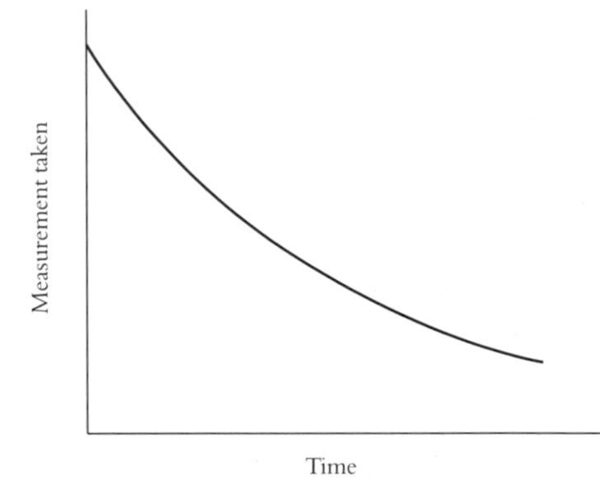

A Temperature
B Volume of gas produced
C pH of solution
D Mass of the beaker and contents

10 The following potential energy diagram is for an uncatalysed reaction.

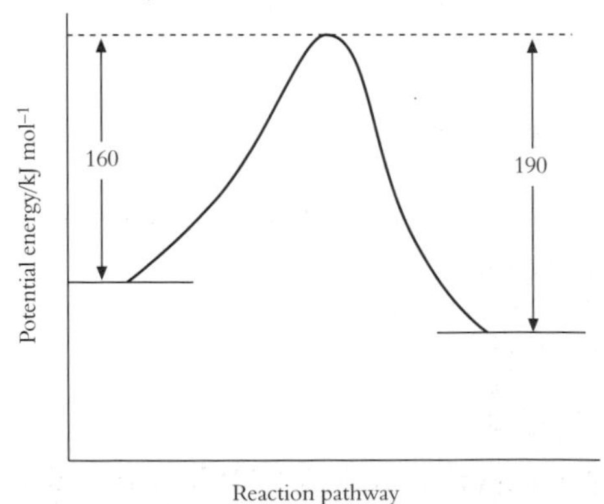

When a catalyst is used the activation energy of the forward reaction is reduced to 35 kJ mol^{-1}.

What is the activation energy of the catalysed reverse reaction, in kJ mol^{-1}?
A 35
B 65
C 125
D 155

11 The enthalpy of neutralisation in an acid/alkali reaction is **always** the energy released in
A the neutralisation of one mole of acid
B the neutralisation of one mole of alkali
C the formation of one mole of water
D the formation of one mole of salt.

12 Which of the following elements would require the most energy to convert one mole of gaseous atoms into gaseous ions each carrying two positive charges?

(You may wish to use the data booklet.)
A Scandium
B Titanium
C Vanadium
D Chromium

13 As the relative atomic mass in the halogens increases
A the boiling point increases
B the density decreases
C the first ionisation energy increases
D the atomic size decreases.

14 Which type of structure is found in a fullerene?
A Ionic lattice
B Metallic lattice
C Covalent network
D Covalent molecular

15 Which element exists as a discrete molecular solid at room temperature and pressure?
A Sulphur
B Sodium
C Chlorine
D Boron

16 Which of the following compounds has polar molecules?
A CH_4
B CO_2
C NH_3
D CCl_4

17 A compound boils at $^{-}33$°C. It also dissolves in water to give an alkaline solution.

Which type of bonding is present within the compound?
A Metallic
B Covalent (polar)
C Ionic
D Covalent (non-polar)

18 In which molecule will the chlorine atom carry a partial positive charge ($\delta+$)?
A Cl–Br
B Cl–Cl
C Cl–F
D Cl–I

19 Which of the following chlorides is most likely to be soluble in tetrachloromethane, CCl_4?
A Barium chloride
B Caesium chloride
C Calcium chloride
D Phosphorus chloride

20 Which of the following compounds is a liquid at 25°C? (Refer to page 6 of the Data Book.)
A Fluorine oxide
B Silicon chloride
C Phosphorus oxide
D Magnesium chloride

21 Which substance has hydrogen bonding between molecules?
A Ethane
B Hydrogen fluoride
C Hydrogen
D Lithium hydroxide

22 The Avogadro constant is the same as the number of
A molecules in 32 g of oxygen
B ions in 500 cm^3 of 2 mol l^{-1} sodium chloride solution
C atoms in 6 g of carbon
D molecules in 14 g of nitrogen

23 Which of the following contains the **largest** number of molecules?
A 0.10 g of hydrogen gas
B 0.17 g of ammonia gas
C 0.32 g of methane gas
D 0.35 g of chlorine gas

24 What volume of oxygen, in litres, is required for the complete combustion of 1 litre of ethane?
A 1
B 2
C 3.5
D 7

25 Which of the following equations represents a reaction which takes place during reforming?

A $C_6H_{14} \rightarrow C_6H_6 + 4H_2$

B $C_4H_8 + H_2 \rightarrow C_4H_{10}$

C $C_2H_5OH \rightarrow C_2H_4 + H_2O$

D $C_8H_{18} \rightarrow C_4H_{10} + C_4H_8$

26 Which pollutant, produced during internal combustion in a car engine, is **not** the result of incomplete combustion?

A Nitrogen dioxide
B Hydrocarbons
C Carbon
D Carbon monoxide

27 What is the product when one mole of ethyne reacts with one mole of chlorine?

A 1,1-dichloroethene
B 1,1-dichloroethane
C 1,2-dichloroethene
D 1,2-dichloroethane

28 The extensive use of which type of compound is thought to contribute significantly to the depletion of the ozone layer?

A Oxides of carbon
B Hydrocarbons
C Oxides of sulphur
D Chlorofluorocarbons

29 Which statement about benzene is correct?

A Benzene is an isomer of cyclohexane.
B Benzene reacts with bromine solution as if it is unsaturated.
C The ratio of carbon to hydrogen atoms in benzene is the same as in ethyne.
D Benzene undergoes addition reactions more readily than hexene.

30 Which of the following is **not** a correct statement about methanol?

A It is a primary alkanol.
B It can be oxidised to methanal.
C It can be made from synthesis gas.
D It can be dehydrated to an alkene.

31 Which of the following is an isomer of hexanal?

A 2-methylbutanal
B 3-methylpentan-2-one
C 2,2-dimethylbutan-1-ol
D 3-ethylpentanal

32 Which consumer product is least likely to contain esters?

A Flavourings
B Medicines
C Solvents
D Dyes

33 Which mixture of gases is known as synthesis gas?

A Methane and oxygen
B Carbon monoxide and oxygen
C Carbon dioxide and hydrogen
D Carbon monoxide and hydrogen

34 Ethene is used in the manufacture of addition polymers.

What type of reaction is used to produce ethene from ethane?

A Addition
B Cracking
C Hydrogenation
D Oxidation

35 A part of the formula for nylon is shown.

```
 H          H O           O
 |          | ||          ||
-N-(CH2)6-N-C-(CH2)4-C-
```

$$-\overset{H}{\underset{|}{N}}-(CH_2)_6-\overset{H}{N}-\overset{O}{\overset{\|}{C}}-(CH_2)_4-\overset{O}{\overset{\|}{C}}-$$

This polymer is classed as a

A synthetic addition polymer
B synthetic condensation polymer
C natural condensation polymer
D natural addition polymer.

36 Which of the following polymers is used in making bullet-proof vests?

A Kevlar
B Biopol
C Poly(ethenol)
D Poly(ethyne)

37 Fats have higher melting points than oils because comparing fats and oils

A fats have more hydrogen bonds
B fats have more cross-links between molecules
C fat molecules are more loosely packed
D fat molecules are more saturated.

38 Which of the following **must** contain nitrogen?

A A polyester
B An oil

C A protein
D A carbohydrate

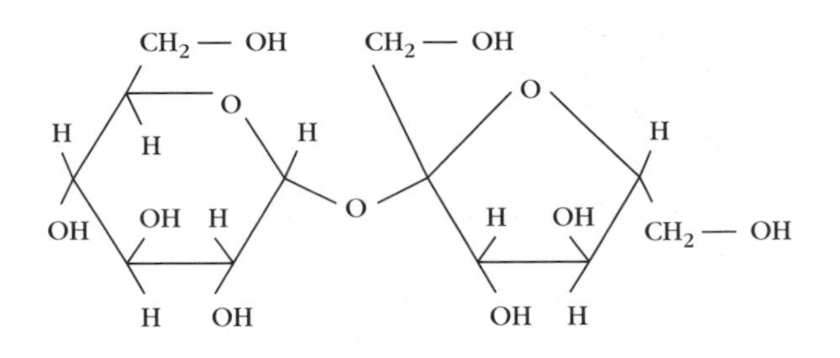

39 Olestra is a calorie free fat made by reacting fatty acids with sucrose. The structure of a sucrose molecule can be represented as shown.

How many fatty acid molecules can react with one molecule of sucrose?

A 3
B 5
C 8
D 11

40 Which statement is true?

A Proteins are a more concentrated source of energy than carbohydrates.
B Denaturing of proteins involves changes in the structure of the molecules.
C Proteins can be made in animals but not in plants.
D Globular proteins are the major structural materials of animal tissue.

SECTION B [60 marks]

Show clear working in all calculations.

1 The three statements below are taken from a note made by a student who is studying trends in the Periodic Table.

> 1 First Ionisation Energy
>
> The energy required to remove one mole of electrons from one mole of atoms in the gaseous state.
>
> 2 Second Ionisation Energy
>
> The energy required to remove a second mole of electrons.
>
> 3 ____________________
>
> The measure of the attraction an atom has for the shared electrons in a bond.

a) Complete the note above to give the heading for the third statement **1**

b) What is the trend in the first ionisation energy across a period from left to right? **1**

c) Why is the second ionisation energy of sodium so much greater than its first ionisation energy? **1**

(3)

2 Diamond and graphite are forms of carbon with very different properties.

Graphite can mark paper, is a lubricant and is a conductor of electricity.

Diamond has none of these properties.

a) Draw a diagram to show the structure of diamond. **1**

b) Why is graphite an effective lubricant? **1**

c) A pupil uses a graphite pencil to write her signature 100 times on a piece of weighed paper.

Results		
Number of signatures	=	*100*
Mass of blank paper	=	*4.895 g*
Mass of paper + 100 signatures	=	*4.905 g*

Use her results to calculate the number of carbon atoms present in one signature. **2**

(4)

3 The structure of a molecule found in olive oil can be represented as shown.

$$\begin{array}{l} \qquad\qquad\qquad\qquad CH_2-O-\overset{O}{\overset{\|}{C}}-\sim\sim\sim \\ \sim\sim\sim-\overset{O}{\overset{\|}{C}}-O-CH \\ \qquad\qquad\qquad\qquad CH_2-O-\overset{O}{\overset{\|}{C}}-\sim\sim\sim \end{array}$$

a) Olive oil can be hardened using a nickel catalyst to produce a fat.

i) What type of catalyst is nickel in this reaction? **1**

ii) In what way does the structure of a fat molecule differ from that of an oil molecule? **1**

b) Olive oil can be hydrolysed using sodium hydroxide solution to produce sodium salts of fatty acids.

i) Name the other product of this reaction. **1**

ii) Give a commercial use for sodium salts of fatty acids. **1**

(4)

4 The structures of two antiseptics are shown. Both are aromatic.

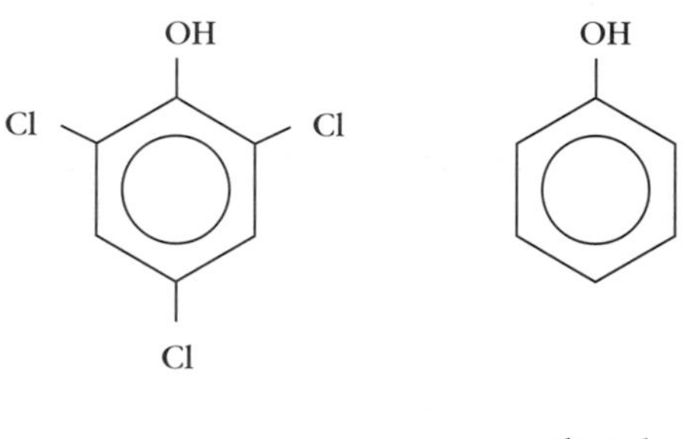

TCP phenol

a) i) What gives the aromatic ring its stability? **1**

ii) Write the molecular formula for TCP. **1**

iii) The systematic name for TCP is 2,4,6-trichlorophenol.
The systematic name for Dettol, another antiseptic, is 4-chloro-3,5-dimethylphenol.
Draw a structural formula for Dettol. **1**

b) The feedstocks for the production of antiseptics are made by reforming the naphtha fraction of crude oil.
Give another use for reformed naphtha. **1**

(4)

5 a) Ethanol and propanoic acid can react to form an ester.

i) Draw a structural formula for this ester. **1**

ii) Draw a labelled diagram of the assembled apparatus that could be used to prepare this ester in the laboratory. **2**

b) Pyrolysis (thermal decomposition) of esters can produce two compounds, an alkene and an alkanoic acid, according to the following equation.

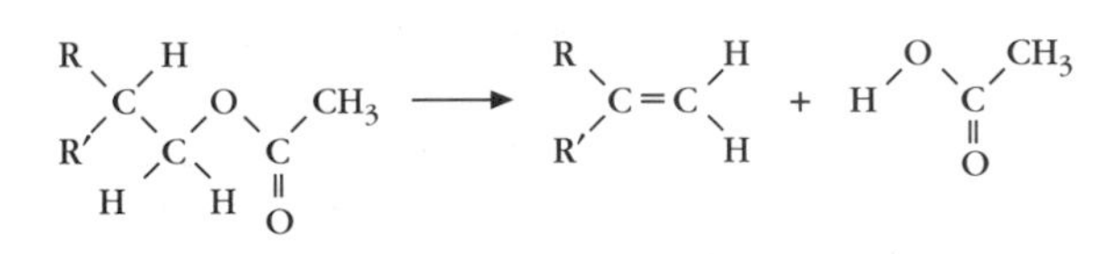

(R and R′ represent alkyl groups)

Draw a structural formula for the ester that would produce 2-methylbut-1-ene and methanoic acid on pyrolysis. **1**

(4)

6 A calorimeter, like the one shown, can be used to measure the enthalpy of combustion of ethanol.

The ethanol is ignited and burns completely in the oxygen gas. The heat energy released in the reaction is taken in by the water as the hot product gases are drawn through the coiled copper pipe by the pump.

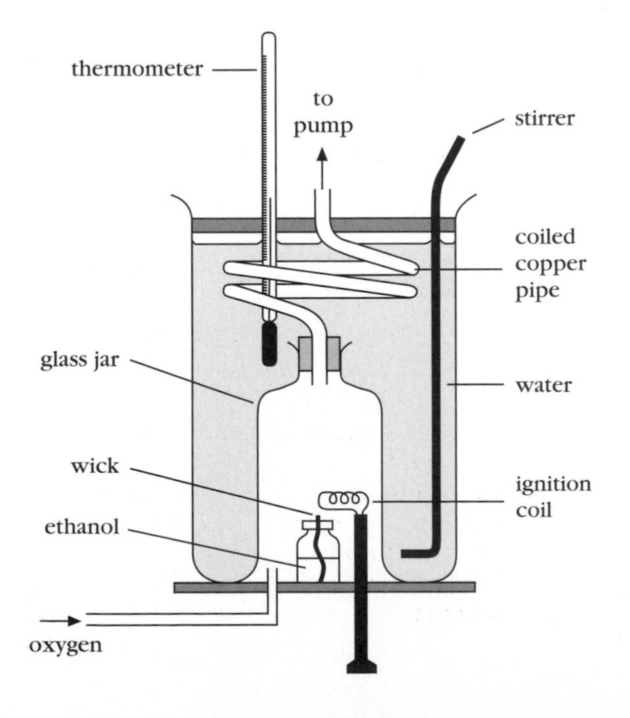

a) Why is the copper pipe coiled as shown in the diagram? **1**

b) The value for the enthalpy of combustion of ethanol obtained by the calorimeter method is higher than the value obtained by the typical school laboratory method. One reason for this is that more heat is lost to the surroundings in the typical school laboratory method.
Give **one** other reason for the value being higher with the calorimeter method. **1**

c) In one experiment the burning of 0.980 g of ethanol resulted in the temperature of 400 cm^3 of water rising from 14.2°C to 31.6°C.
Use this information to calculate the enthalpy of combuston of ethanol. **3**

(5)

7 Diphosphine, P_2H_4, is a hydride of phosphorus. All of the covalent bonds in diphosphine molecules are non-polar.

a) Refer to your data booklet to show that the bonds in diphosphine are non-polar. **1**

b) The balanced equation for the complete combustion of diphosphine is:

$$2P_2H_4(g) + 7O_2(g) \rightarrow P_4O_{10}(s) + 4H_2O(l)$$

What volume of oxygen would be required for the complete combustion of 10 cm^3 of diphosphine? **1**

c) Calculate the volume occupied by 0.330 g of diphosphine. (V_{mol} = 24.0 litres mol^{-1}) **1**

(3)

8 Although aldehydes and ketones have different structures, they both contain the carbonyl functional group.

a) i) In what way is the structure of an aldehyde different from that of a ketone? **1**

ii) As a result of the difference in structure, aldehydes react with Fehling's (or Benedict's) solution and Tollens' reagent but ketones do not. What colour change would be observed when propanal is heated with Fehling's (or Benedict's) solution? **1**

iii) When propanal reacts with Tollens' reagent a silver 'mirror' is formed and the following change occurs:
$C_3H_6O \rightarrow C_2H_5COOH$.
What happens to the oxygen-hydrogen ratio in ths change and what does this show about the type of change? **1**

iv) Name the compound with the formula C_2H_5COOH. **1**

b) As a result of both containing the carbonyl group, aldehydes and ketones react in a similar way with hydrogen cyanide.
The equation for the reaction of propanal and hydrogen cyanide is shown.

```
   H  H  H                          H  H  H
   |  |  |                          |  |  |
H– C– C– C= O  +  H– CN  ——→   H– C– C– C– OH
   |  |                             |  |  |
   H  H                             H  H  CN
```

i) Suggest a name for this type of reaction. **1**

ii) Draw a structure for the product of the reaction between propanone and hydrogen cyanide. **1**

(6)

9 a) Kevlar and Nomex are examples of recently manufactured polymers. Their properties are different because they are made from different monomers.

The diamine monomer used to make Nomex is 1,3-diaminobenzene.

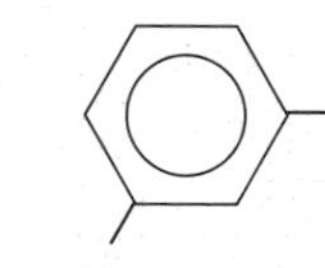

1,3 diaminobenzene

This reacts with the other monomer to form the repeating unit shown.

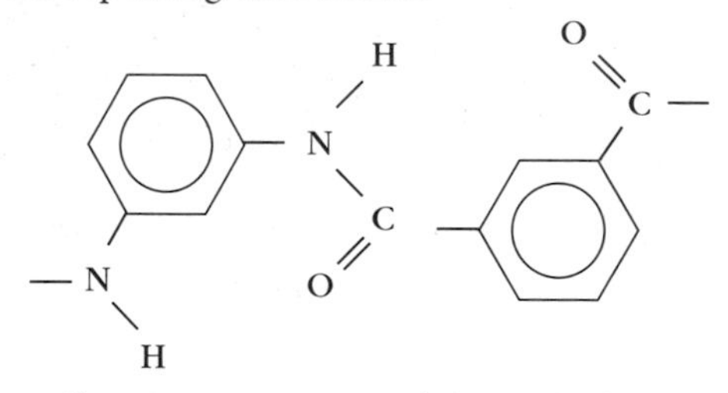

i) Draw a structural formula for the other monomer. **1**

ii) The repeating unit in Kevlar is:

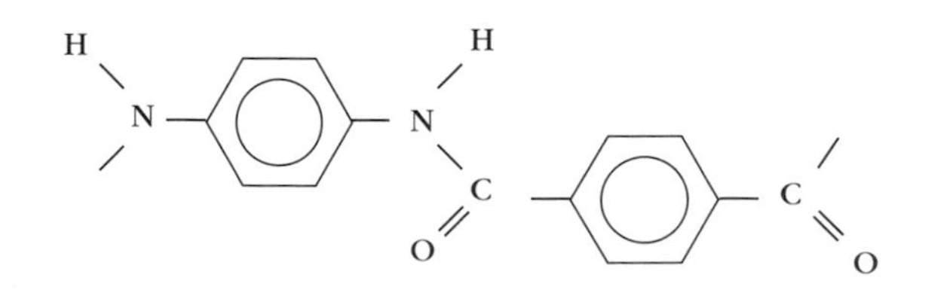

Name the diamine used to make Kevlar. **1**

b) Another recently manufactured polymer is polyvinylcarbazole.
Give the unusual property of polyvinylcarbazole which makes it suitable for use in photocopiers. **1**

(3)

10 In an experiment to produce hydrogen gas, a pupil added 200 cm^3 4.0 mol l^{-1} hydrochloric acid to 32.7 g of zinc.

Once the reaction is complete the pupil decides to produce more hydrogen. Show by calculation whether he will need to add more acid or more zinc to do this. The equation for the reaction is

$$Zn + 2HCl \rightarrow ZnCl_2 + H_2$$ **(2)**

11 a) Mordenite is a porous, crystalline material with a surface area of over 500 m^2g^{-1}.

It is used in an isomerisation reaction, part of a sequence which converts pentane into 2-methylbutane for blending into petrol.

pentane
↓
pent-2-ene —isomerisation→ 2-methylbut-2-ene
↓
2-methylbutane

i) Name an isomer of pent-2-ene belonging to a different homologous series. **1**

ii) What role does mordenite play in the isomerisation reaction? **1**

iii) Why is 2-methylbutane a more suitable component than pentane when used in unleaded petrol? **1**

b) Mordenite consists mainly of silicon dioxide.
Name the structure and type of bonding in silicon dioxide. **1**

(4)

12 Hydrogen peroxide, H_2O_2, decomposes very slowly to produce water and oxygen.

a) The activation energy (E_A) for the reaction is **75 kJ mol^{-1}** and the enthalpy change (ΔH) is **−26 kJ mol^{-1}**.

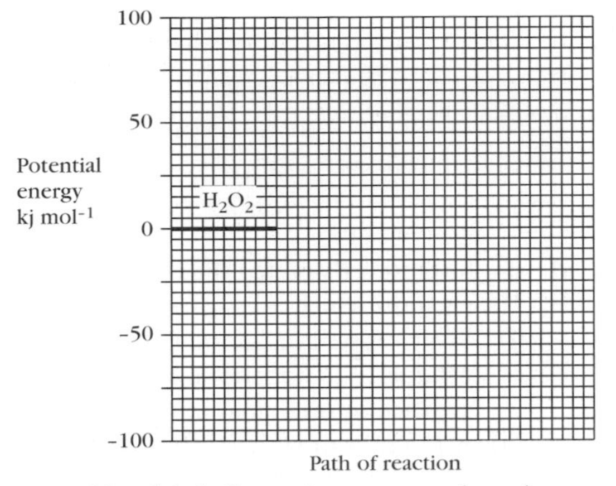

Use this information to complete the potential energy diagram for the reaction. **1**

b) Powdered manganese dioxide catalyses the decomposition of hydrogen peroxide solution.
Add a dotted line to the above diagram to show the path of the reaction when the catalyst is used. **1**

c) The balanced equation for the reaction is:

$$2H_2O_2(aq) \rightarrow 2H_2O(l) + O_2(g)$$

i) The following graph is obtained for the volume of oxygen released over time.
Calculate the average rate of reaction between 10 and 20s. **1**

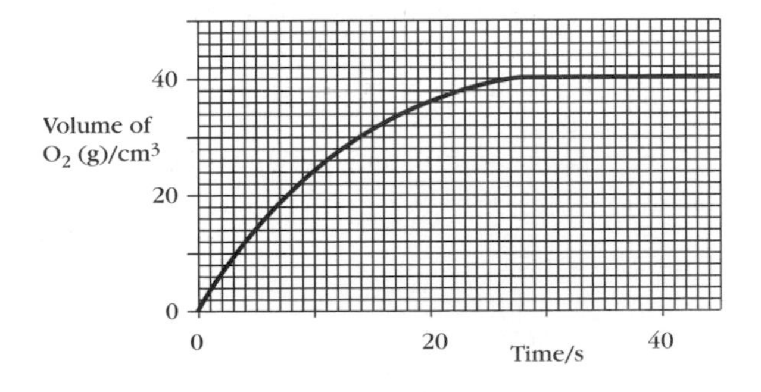

ii) Using information from the above graph, calculate the mass of hydrogen peroxide used in the reaction, assuming all the hydrogen peroxide decomposed.
(Take the molar volume of oxygen to be 24 litres mol^{-1}.) **2**

(5)

13 The effect of temperature on reaction rate can be studied using the reaction between oxalic acid and acidified potassium permanganate solutions.

a) What colour change would indicate that the reaction was complete? **1**

b) A student's results are shown on the graph.

i) Use the graph to calculate the reaction time, in s, at 40°C. **1**

ii) Why is it difficult to obtain an accurate reaction time when the reaction is carried out at room temperature? **1**

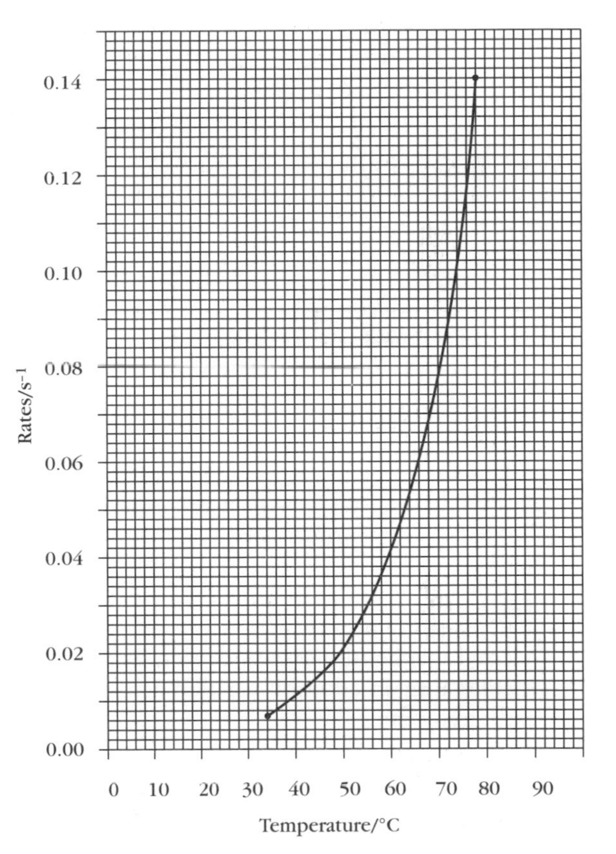

c) The diagram below shows the energy distribution of molecules in a gas at a particular temperature.
Copy the diagram and add a second curve to show the energy distribution of the molecules in the gas at a higher temperature.
Label the diagram to indicate why an increase in temperature has such a significant effect on reaction rate. **1**

(4)

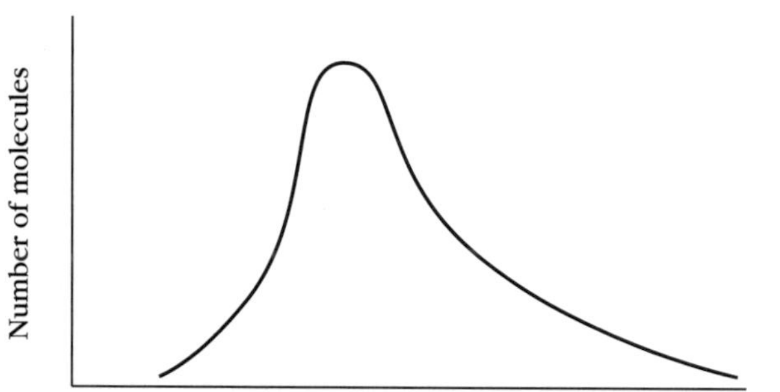

14 Alkanols can be prepared by the reaction of carbonyl compounds with methyl magnesium bromide. The reaction takes place in two stages.

Stage 1

Methyl magnesium bromide reacts with methanal in an addition reaction across the carbonyl group.

```
H                 H
 \                |
  C=O   +   H-C-MgBr   ——→
 /                |
H                 H

methanal    methyl magnesium
                 bromide

                         H  H
                         |  |
                       H-C-C-O-MgBr
                         |  |
                         H  H
```

Stage 2

Reaction of the product with water produces ethanol.

```
  H  H
  |  |
H-C-C-O-MgBr  +  H2O  ——→
  |  |
  H  H

                     H  H
                     |  |
                   H-C-C-OH  +  MgBrOH
                     |  |
                     H  H

                    ethanol
```

a) i) Suggest a name for the type of reaction which takes place in Stage 2. **1**

ii) Which **type** of alkanol would be produced if propanone had been used in place of methanal in this reaction. **1**

iii) A reaction in which 5.01 g of methanal was used yielded 5.75 g of ethanol. Calculate the percentage yield. **2**

b) State an important industrial use for methanal. **1**

(5)

15 Perfumes normally contain three groups of components called the **top note**, the **middle note** and the **end note**.

a) The **top note** components of a perfume form vapours most easily. Two compounds found in **top note** components are:

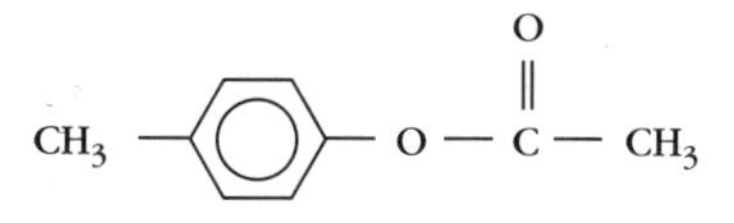

p-cresyl acetate

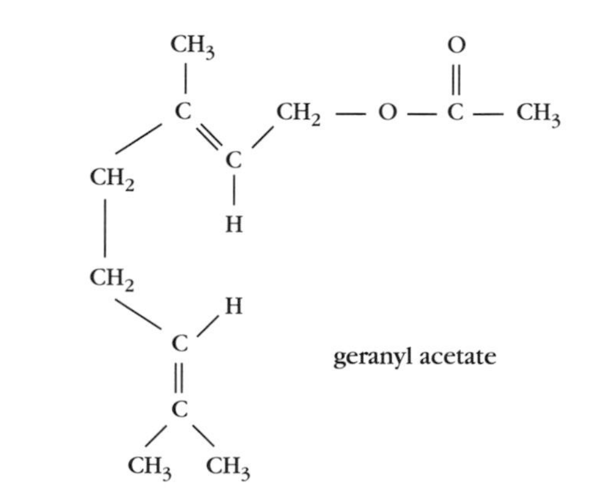

geranyl acetate

i) With reference to the structure of these compounds, why are they likely to have pleasant smells? **1**

ii) Describe a chemical test which would distinguish between these two compounds and give the result of the test. **1**

b) The **middle note** compounds form vapours less readily than the **top note** compounds. A typical compound of the **middle note** is:

2-phenylethanol $C_6H_5 — CH_2 — CH_2 — O — H$

Due to hydrogen bonding 2-phenylethanol forms a vapour less readily than *p*-cresyl acetate.

Draw two molecules of 2-phenylethanol and use a dotted line to show where a hydrogen bond exists between them. **1**

c) The **end note** of a perfume has a long lasting odour which stays with the user. An example of an **end note** compound is:

civetone

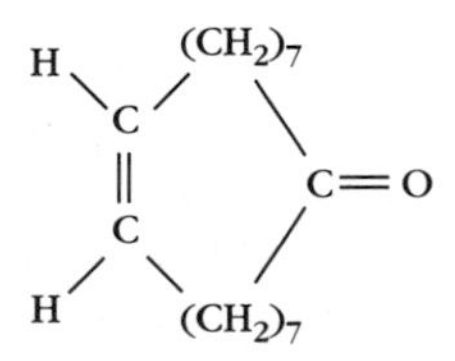

Draw the structure of the alcohol which would be formed by the reduction of civetone. **1**

(4)

END OF QUESTION PAPER

Unit 3 Test

SECTION A [16 MARKS]

1 Which compound is **not** a raw material in the chemical industry?
A Benzene
B Water
C Iron oxide
D Sodium chloride

2 Which of the following is produced by a batch process?
A Sulphuric acid from sulphur and oxygen
B Aspirin from salicylic acid
C Iron from iron ore
D Ammonia from nitrogen and hydrogen

3 Consider the reaction pathway shown.

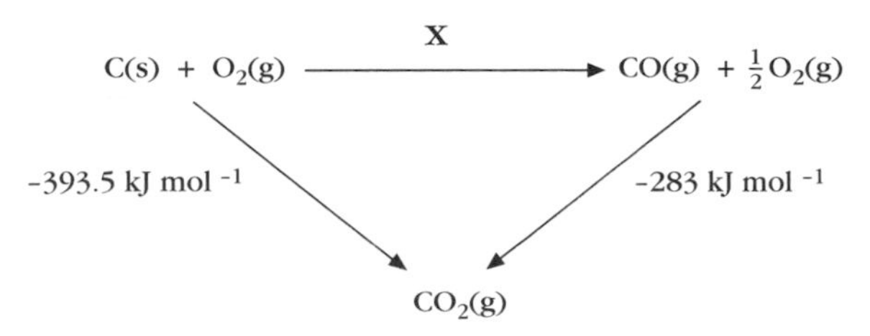

According to Hess's Law, what is the enthalpy change for reaction **X**?
A $+110.5$ kJ mol^{-1}
B -110.5 kJ mol^{-1}
C -676.5 kJ mol^{-1}
D $+676.5$ kJ mol^{-1}

4 $Cl_2(g) + H_2O(l) \rightleftharpoons Cl^-(aq) + ClO^-(aq) + 2H^+(aq)$

Which substance would move the above equilibrium to the right?
A Hydrogen chloride
B Sodium bromide
C Chlorine
D Hydrogen

5 $ICl(l) + Cl_2(g) \rightleftharpoons ICl_3(s)\ \Delta H = -106$ kJ mol^{-1}

Which line in the table identifies correctly the changes that will cause the greatest increase in the proportion of solid in the above equilibrium?

	Temperature	Pressure
A	decrease	decrease
B	decrease	increase
C	increase	decrease
D	increase	increase

6 A catalyst is added to a reaction at equilibrium.

Which of the following does **not** apply?
A The rate of the forward reaction increases.
B The rate of the reverse reaction increases.
C The position of equilibrium remains unchanged.
D The position of equilibrium shifts to the right.

7 The concentration of $OH^-(aq)$ ions in a solution is 1×10^{-2} mol l^{-1}

What is the concentration of $H^+(aq)$ ions, in mol l^{-1}?
A 1×10^{-2}
B 1×10^{-5}
C 1×10^{-9}
D 1×10^{-12}

8 When a certain aqueous solution is diluted, its conductivity decreases but its pH remains constant.

The solution could be
A ethanoic acid
B sodium chloride
C sodium hydroxide
D nitric acid.

9 Which line in the table is correct for 0.1 mol l^{-1} sodium hydroxide compared with 0.1 mol l^{-1} aqueous ammonia?

	pH	Conductivity
A	higher	lower
B	higher	higher
C	lower	higher
D	lower	lower

10 A solution has a pH of 3. Its concentration of hydrogen ions, in mol l^{-1}, is

A 1×10^{-3}
B 3×10^{-1}
C 1×10^{-11}
D 3×10^{-11}

11 Which of the following is the best description of a 0.1 mol l^{-1} solution of hydrochloric acid?

A Dilute solution of a strong acid
B Dilute solution of a weak acid
C Concentrated solution of a weak acid
D Concentrated solution of a strong acid

12 The ion-electron equations for a redox reaction are:

$2I^-(aq) \rightarrow I_2(aq) + 2e^-$

$MnO_4^-(aq) + 8H^+(aq) + 5e^- \rightarrow Mn^{2+}(aq) + 4H_2O(l)$

How many moles of iodide ions are oxidised by one mole of permanganate ions?

A 0.2
B 0.4
C 2
D 5

13 Ammonia reacts with magnesium as shown.

$3Mg(s) + 2NH_3(g) \rightarrow (Mg^{2+})_3(N^{3-})_2(s) + 3H_2(g)$

In this reaction, ammonia is acting as

A an acid
B a base
C an oxidising agent
D a reducing agent.

14 Which of the following has an electrical charge?

A β-particles
B X-rays
C Neutrons
D γ-rays

15 A radioactive isotope of a noble gas emits an α-particle.

The decay product will be an isotope of an element in

A Group 0
B Group 2
C Group 4
D Group 6

16 Which of the following equations represents a nuclear fission process?

A $^{40}_{19}K + ^{0}_{-1}e \rightarrow ^{40}_{18}Ar$

B $^{2}_{1}H + ^{3}_{1}H \rightarrow ^{4}_{2}He + ^{1}_{0}n$

C $^{235}_{92}U + ^{1}_{0}n \rightarrow ^{90}_{38}Sr + ^{144}_{54}Xe + 2^{1}_{0}n$

D $^{14}_{7}N + ^{1}_{0}n \rightarrow ^{14}_{6}C + ^{1}_{1}p$

SECTION B [30 marks]

Show clear working in all calculations.

1 Vitamin C, $C_6H_8O_6$, is a powerful reducing agent. The concentration of vitamin C in a solution can be found by titrating it with a standard solution of iodine, using starch as an indicator. The equation for the reaction is:

$$C_6H_8O_6(aq) + I_2(aq) \rightarrow C_6H_6O_6(aq) + 2H^+(aq) + 2I^-(aq)$$

a) Write an ion-electron equation for the oxidation half-reaction. **1**

b) A work card gave the following instructions as part of an investigation into the vitamin C content of a tablet. Some instructions have been omitted.

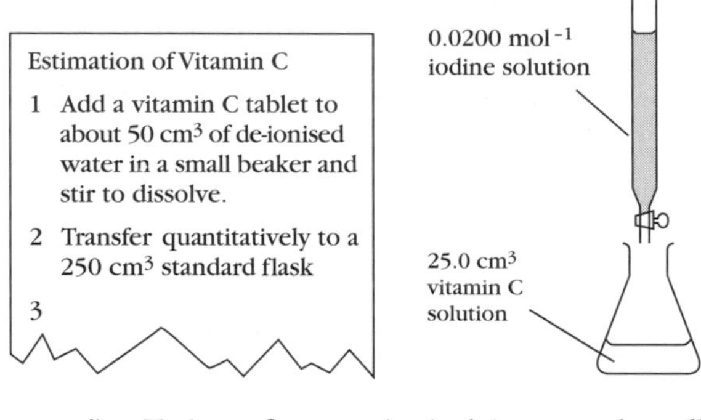

i) To 'transfer quantitatively' means that **all** of the vitamin C must be transferred into the standard flask. Describe how this is carried out in practice. **1**

ii) What colour change indicates that the end-point of the titration has been reached? **1**

c) In one investigation, it was found that an average of 29.5 cm^3 of 0.02 mol l^{-1} iodine solution was required to react completely with 25.0 cm^3 of vitamin C solution.
Use this result to calculate the mass, in grams, of vitamin C present in each tablet. **3**

(6)

2 An 8 g sample of ^{24}Na undergoes β-decay as shown in the graph opposite. A stable isotope is produced.

a) Write a nuclear equation for the the decay of ^{24}Na. **1**

b) What happens to the proton-neutron ratio during β-decay? **1**

c) From the graph, what is the half-life of ^{24}Na? **1**

d) What mass of **product** would be formed from the sample after 45 hours? **1**

(4)

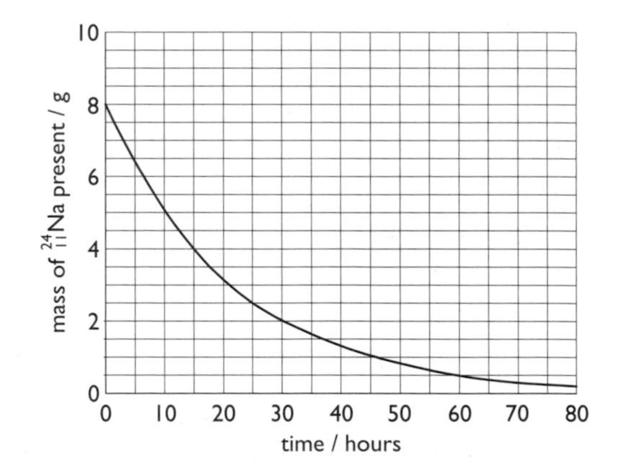

3 Vinegar is a dilute solution of ethanoic acid.

a) Hess's Law can be used to calculate the enthalpy change for the formation of ethanoic acid from its elements.

$$\underset{\text{(graphite)}}{2C(s)} + 2H_2(g) + O_2(g) \rightarrow CH_3COOH(l)$$

Calculate the enthalpy change for the above reaction, in kJ mol^{-1}, using information from the data booklet and the following data.

$CH_3COOH(l) + 2O_2(g) \rightarrow 2CO_2(g) + 2H_2O(l) \quad \Delta H = -876 \text{ kJ mol}^{-1}$ **2**

b) Ethanoic acid can be used to prepare the salt, sodium ethanoate, CH_3COONa.
Explain why sodium ethanoate solution has a pH greater than 7.
In your answer you should mention the **two** equilibria involved. **3**

(5)

4 a) Some carbon monoxide detectors contain crystals of hydrated palladium(II) chloride. These form palladium in a redox reaction if exposed to carbon monoxide.

$$CO(g) + PdCl_2.2H_2O(s) \rightarrow CO_2(g) + Pd(s) + 2HCl(g) + H_2O(l)$$

Write the ion-electron equation for the reduction step in this reaction. **1**

b) Another type of detector uses an electrochemical method to detect carbon monoxide.
At the positive electrode:

$$CO(g) + H_2O(l) \rightarrow CO_2(g) + 2H^+(aq) + 2e^-$$

At the negative electrode:

$$O_2(g) + 4H^+(aq) + 4e^- \rightarrow 2H_2O(l)$$

Combine the two ion-electron equations to give the overall redox equation. **1**

(2)

5 If the conditions are kept constant, reversible reactions will attain a state of equilibrium.

a) Choose the correct words in the table to show what is true for reactions at equilibrium.

Rate of forward reaction compared to rate of reverse reaction	faster / same / slower
Concentrations of reactants compared to concentrations of products	usually different / always the same

1

b) The following equilibrium involves two compounds of phosphorus.

$$PCl_3(g) + 3NH_3(g) \rightleftharpoons P(NH_2)_3(g) + 3HCl(g)$$

i) An increase in temperature moves the above equilibrium to the left.
What does this indicate about the enthalpy change for the forward reaction? **1**

ii) What effect, if any, will an increase in pressure have on the above equilibrium? **1**

(3)

6 Potassium hydroxide can be used in experiments to verify Hess's Law. The reactions concerned can be summarised as follows.

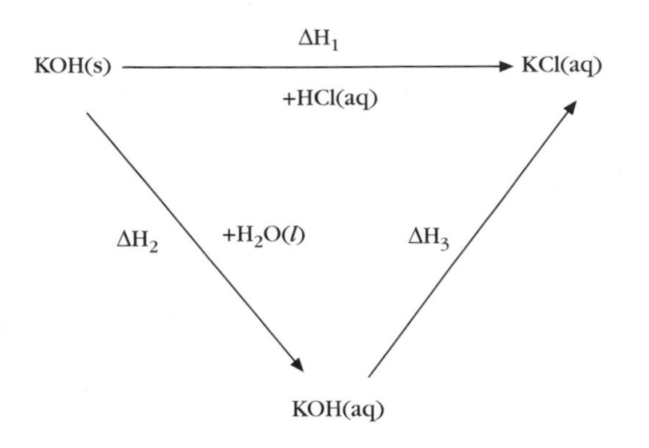

a) State Hess's Law. **1**

b) Four measurements have to be made in order to calculate ΔH_2. One of these is the mass of potassium hydroxide.
What are the other 3 measurements needed? **1**

c) What solution must be added to the potassium hydroxide solution in order to calculate ΔH_3? **1**

(3)

7 Aluminium is extracted from bauxite. This ore contains aluminium oxide along with iron(III) oxide and other impurities.

The process is shown in the flow diagram.

a) Add an arrow to the flow diagram to show how the process could be made more economical. **1**

b) In **Stage 1** of the process, aluminium oxide reacts with sodium hydroxide solution. State whether aluminium oxide is behaving as an acidic oxide or as a basic oxide in this reaction. **1**

c) What type of reaction takes place during **Stage 3**? **1**

d) During **Stage 4**, aluminium is manufactured in cells by the electrolysis of aluminium oxide dissolved in molten cryolite.
What mass of aluminium is produced each hour, if the current passing through the liquid is 180 000 A? **3**

e) In **Stage 4**, the carbon blocks that are used as positive electrodes must be regularly replaced.
Suggest a reason for this. **1**

(7)

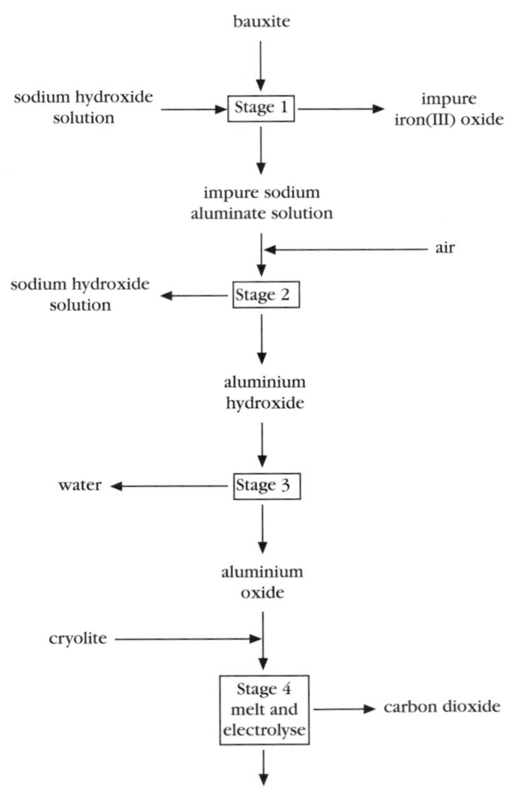

END OF TEST

Summary of Bonding and Structure

Intermolecular Bonds (bonds between molecules)

Hydrogen, Dipole-dipole and Van der Waals'

Shown by: ------------------------------

and

Intramolecular Bonds (bonds inside molecules)

Covalent and Polar Covalent

Shown by: ________________

Hydrogen bonds

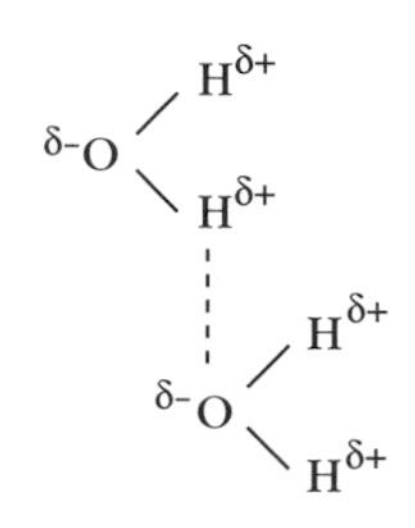

(Van der Waals' also present)

Dipole-dipole bonds

$^{\delta+}I$ — $Cl^{\delta-}$

$^{\delta+}I$ — $Cl^{\delta-}$

(Van der Waals' also present)

Van der Waals' bonds

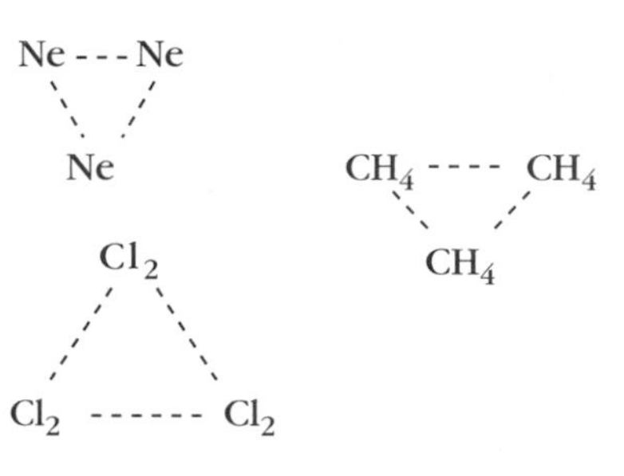

Comparing bond strengths:

Covalent and Polar Covalent bonds >> Hydrogen bonds > Dipole-dipole bonds > Van der Waals' bonds

Elements may be:

metallic, e.g. Li, Be, Na, Fe, Cu

covalent molecular, e.g. H_2, O_2, P_4, S_8, C_{60}

covalent network, e.g. B, C (diamond, graphite)

monatomic, e.g. noble gases

Compounds may be:

ionic, e.g. NaCl, MgO, CaF_2, $CuSO_4$

covalent molecular, e.g. CH_4, H_2O, NH_3, CO_2, $C_6H_{12}O_6$, C_2H_5OH

covalent network, e.g. SiO_2, SiC

Summary of Bonding and Structure (continued)

Substance Type	Bonding	Structure	Properties
Metallic Substances	Outer electrons of metal atoms are delocalised and move randomly from one atom to another. This type of bond occurs in mixtures of metals (alloys), as well as in metallic elements.	Consist of giant lattices of positively-charged ions held together by the delocalised electrons. See Fig. 8, Page 26.	Can conduct electricity when solid or liquid (molten) by movement of these delocalised electrons, and do not change in the process. Most metals are solid with high Melting points. (Data Book p. 4) – exceptions include Group 1 metals which have only one outside electron giving weaker bonding. Metals vary in hardness, but can be distorted by impact or pressure, i.e. they are malleable and ductile.
Ionic Compounds	Atoms of elements with low electro-negativity, i.e. metals, lose their outside electrons to atoms of elements with high electronegativity, i.e. non-metals. Thus metals form positive ions and non-metals form negative ions. For example: Na(atom) + Cl(atom) $\rightarrow$ Na^+ + Cl^- 2.8.1 2.8.7 2.8 2.8.8	Consist of giant lattices of oppositely-charged ions. See Fig. 1, Page 29.	Can conduct electricity when molten or in aqueous solution but not when solid as the ions are trapped in the lattice. When direct current is used, electrolysis occurs. Are solids with high melting points due to the strong attraction between ions of opposite charge. Form crystals which are hard but brittle. Most are soluble in water and insoluble in non-polar solvents.

Summary of Bonding and Structure (continued)

Substance Type	Bonding	Structure	Properties
Covalent Substances	Atoms in a covalent bond are held together by electrostatic forces between the nuclei and shared electrons. The polarity of the bond depends on the electronegativity difference between the bonded atoms. **Pure Covalent** (non-polar) Occurs in non-metallic elements and compounds where elements of similar electronegativities are combined. The bonding electrons are equally shared, e.g. Cl_2, O_2, P_4, S_8, PH_3, CH_4, NCl_3.	A **Covalent Molecular** structure consists of discrete molecules held together by intermolecular forces. **All** the substances listed to the left are examples, as is the fullerene form of carbon	**Covalent Molecular** Low melting and boiling points due to weak intermolecular forces. Often gases or liquids; but if solids they are easily melted. Usually insoluble (or immiscible) in water and soluble (or miscible) in non-polar solvents. Do not conduct electricity. *Polar liquids* contain polar bonds arranged unsymmetrically and are attracted to a charged rod. *Non-polar liquids* contain only non-polar bonds or polar bonds arranged symmetrically and are not attracted to a charged rod.
	Polar Covalent Occurs in compounds formed by elements with an electronegativity difference greater than about 0.4. The bonding electrons are unequally shared. Atoms of the element with greater electronegativity have a slight negative charge, $\delta-$, for example: $^{\delta+}I-Cl^{\delta-}$ $^{\delta-}O(-H^{\delta+})_2$	A **Covalent Network** structure consists of a giant lattice of covalently-bonded atoms. Elements: boron, diamond and graphite – Figs 5 and 6, pages 25, 26. Compounds: silicon dioxide, SiO_2 – Fig. 3, p 30, and silicon carbide, SiC.	**Covalent Network** Solids have very high melting points due to strong covalent bonds between atoms. Insoluble in water or other solvents. Diamond, SiO_2, and SiC are very hard. Graphite is very soft due to sliding of layers and is the **only** covalent substance to conduct electricity.

Summary of the Relationships of the Main Organic Homologous Series

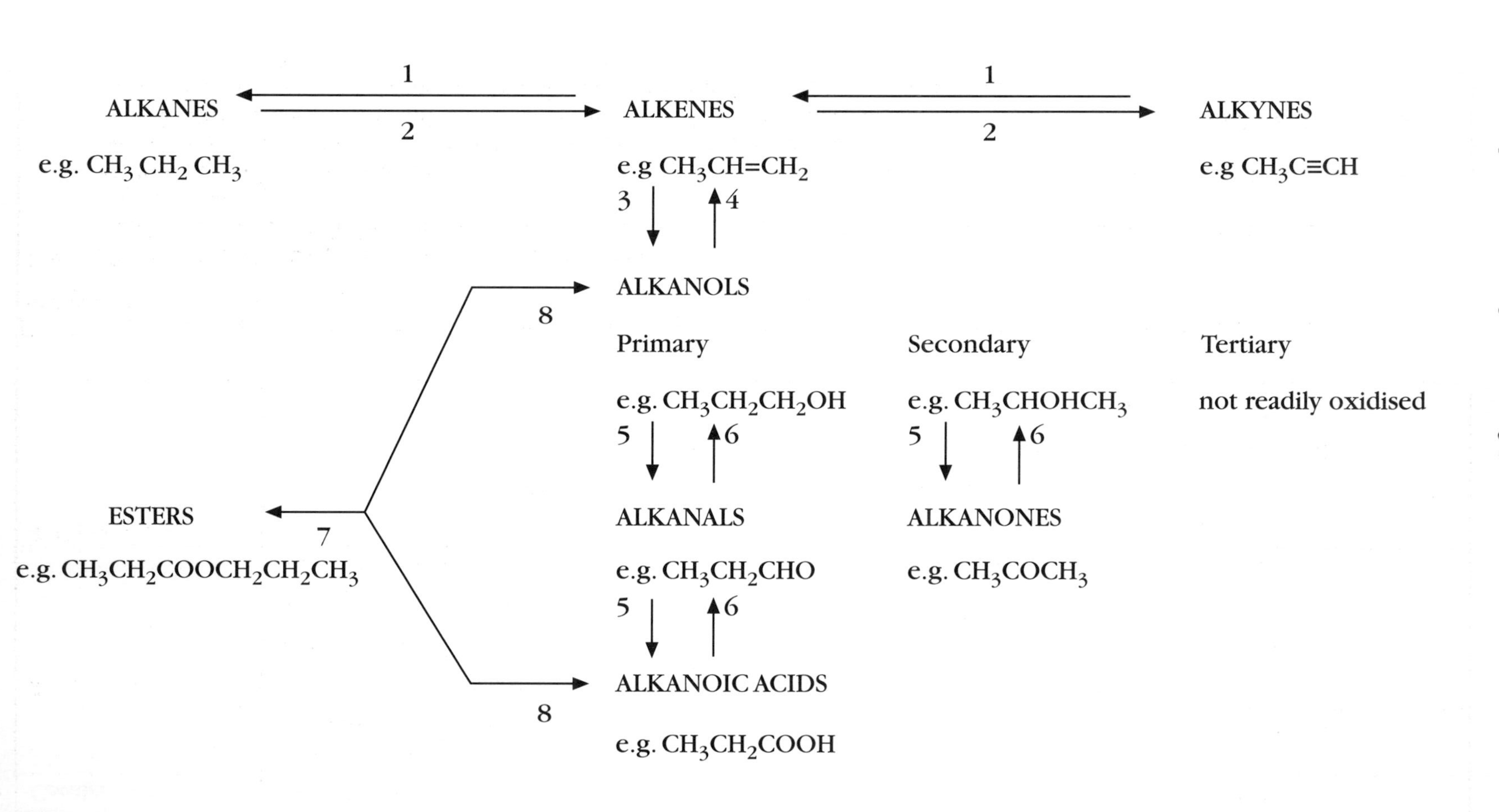

The numbers refer to the chart following: Named Organic Reactions and Reaction Conditions.

Named Organic Reactions and Reaction Conditions

Type of Reaction	Reactants and Reaction Conditions	Reverse Reaction	Reactants and Reaction Conditions
1 Hydrogenation	Alkene or alkyne + hydrogen → alkane (Ni catalyst + heat)	**2** Dehydrogenation (or cracking)	Alkane → alkene (Aluminium oxide catalyst + heat)
3 Hydration	Alkene + steam → alkanol (Phosphoric acid at 300°C)	**4** Dehydration	Alkanol → alkene (Aluminium oxide catalyst + heat)
5 Oxidation	Primary alkanol → alkanal Secondary alkanol → alkanone Alkanal → alkanoic acid (Acidified $K_2Cr_2O_7$ + heat in water bath)	**6** Reduction	Alkanal → primary alkanol Alkanone → sec. alkanol Alkanoic acid → alkanal (Conditions not required at Higher)
7 Condensation	Alkanol + alkanoic acid → ester (Conc. H_2SO_4 + heat in water bath)	**8** Hydrolysis	Ester → alkanol + alkanoic acid (Dil. acid + heat under reflux) Ester → alkanol + salt of alkanoic acid (Alkali soln. + heat under reflux)

The above chart has been drawn up specifically for homologous series, but holds true for more general sets of compounds, i. e. alcohols, aldehydes, ketones, carboxylic acids.

Note: **Hydration** and **hydrogenation** are examples of **addition** reactions.
Condensation and **hydrolysis** also occur in other situations, e.g. in the chemistry of oils and fats and of proteins, but are always opposite reactions.
Hydration involves *one* organic molecule gaining *one* molecule of water to form *one* molecule of product. **Hydrolysis** involves *one* organic molecule reacting with *one* molecule of water and splitting up into *two* products.
Dehydration involves *one* organic molecule losing *one* molecule of water to form *one* molecule of product. **Condensation** involves *two* organic molecules jointly losing *one* molecule of water to form *one* molecule of product.

Answers

CHAPTER 1:

1 D 2 B 3 B 4 C 5 B
6 A 7 D 8 C 9 B

10 a) Purple to colourless
b) i) Concentrations and volumes of all solutions should be constant.
ii) Graph similar to fig 8 page 4.
c) Diagram similar to fig 10 or 11 page 5. Increasing the temperature increases the number of particles with energy greater than the activation energy.
d) Homogeneous since catalyst and reactants are in the same physical state.

11 D

12 a) Homogeneous b) Homogeneous
c) Heterogeneous

13 a) i) 0.011 mol l^{-1} s^{-1}
ii) 0.007 mol l^{-1} s^{-1}
iii) 0.33 mol l^{-1} min^{-1}
b) Acid

14 a) i) $Mg + H_2SO_4 \rightarrow MgSO_4 + H_2$
ii) Sulphuric acid
b)

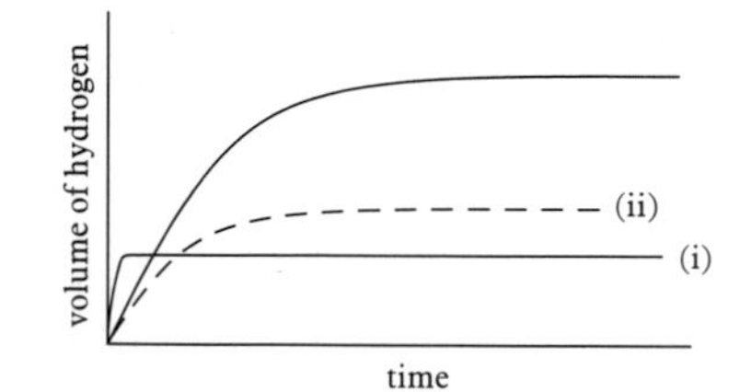

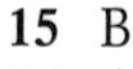

15 B

16 A

17 a) A gas is released.
b) and c) – see graph.

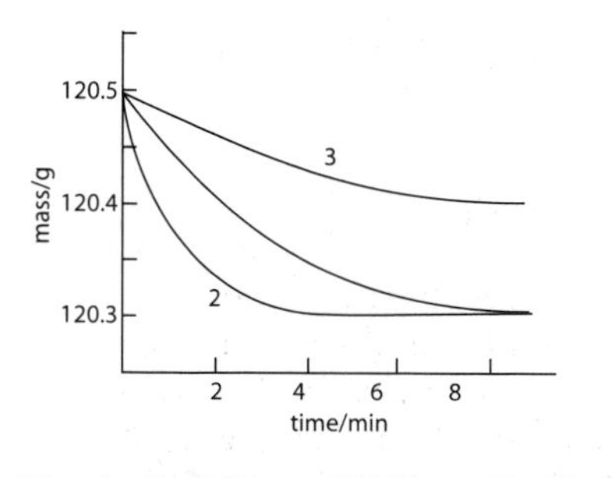

18 a) $CaCO_3 + 2HCl \rightarrow CaCl_2 + CO_2 + H_2O$
b) 0.33 g $minute^{-1}$
c) The concentration of the acid is decreasing.
d) 3.0 minutes
e)

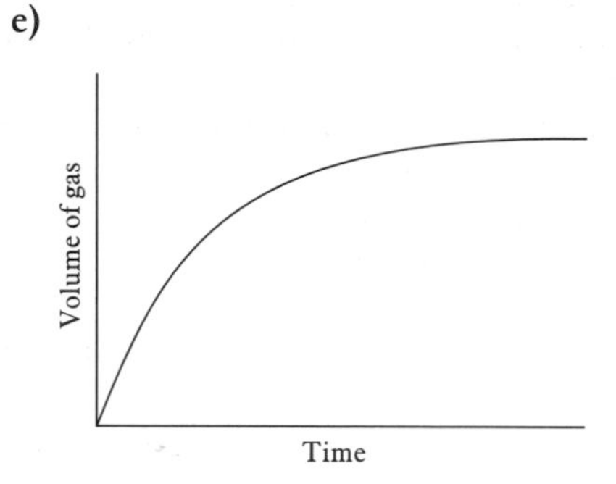

19 a) i) Hydrochloric acid ii) 1.43 g
b) i) Silver nitrate ii) 0.85 g

20 a) To react with the sodium produced when sodium azide decomposes.
b) 130 g.
c) 4.2

CHAPTER 2:

1 D 2 C 3 D 4 A 5 C
6 B 7 A 8 A 9 A 10 C

11 a) i) 40 kJ mol^{-1} and 70 kJ mol^{-1}.
ii) −40 kJ mol^{-1} and 30 kJ mol^{-1}.
b) Activation energy values.

12 a) Heterogeneous
b)

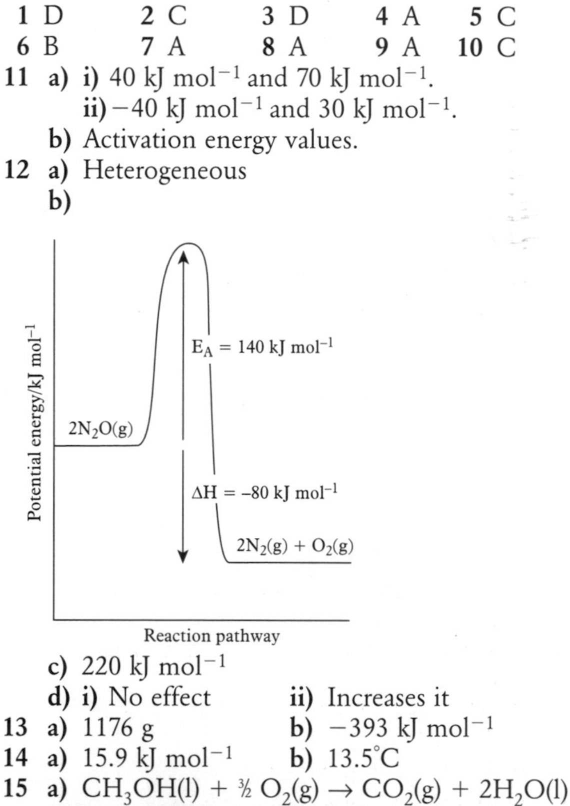

c) 220 kJ mol^{-1}
d) i) No effect ii) Increases it

13 a) 1176 g b) −393 kJ mol^{-1}

14 a) 15.9 kJ mol^{-1} b) 13.5°C

15 a) $CH_3OH(l) + \frac{3}{2} O_2(g) \rightarrow CO_2(g) + 2H_2O(l)$
$\Delta H = -727$ kJ mol^{-1} b) 0.31 g

CHAPTER 3:

1 B 2 A 3 C 4 D 5 B
6 C 7 B 8 B 9 A 10 D

11 a) $K(g) \rightarrow K^+(g) + e^-$
b) i) K has one proton less than Ca or K has a larger covalent radius.
ii) K has one more energy level of electrons than Na.
iii) The second electron is removed from an inner energy level.

12 a) Noble gases.
b) Smaller covalent radius or stronger nuclear charge.
c) The second electron comes from the outer energy level of L, but an inner level of K.

13 a) Larger covalent radius or smaller nuclear charge.
b) In an alkali metal the second electron is removed from an inner energy level.
c) 7500 kJ mol^{-1}

14 a) Covalent radius of Tc should be
i) greater than that of Mn,
ii) less than that of Cd.
b) Tc has one more energy level of electrons than Mn.

15 a) Increasing number of electron energy levels.
b) Increasing number of protons.

CHAPTER 4

1 B 2 C 3 B 4 C 5 D
6 B 7 A 8 D 9 A

10 a) i) Chlorine ii) Argon
b) Sulphur has discrete molecules or it has van der Waals' forces between molecules. Silicon has a network structure.
c) Stronger metallic bonding as Al has more delocalised electrons than Na.

11 a) Covalent b) van der Waals'
c) Metallic d) Covalent
e) van der Waals'

12 a) Phosphorus or sulphur.
b) i) Network
ii) Covalent bonds are very strong.
c) Magnesium has more delocalised electrons than sodium.

13 a) The outer energy level of electrons has been removed.
b) Si^{4-} has one more electron energy level than Al^{3+}.
c) The same number of electrons.
d) Increasing number of protons.

CHAPTER 5:

1 C 2 C 3 D 4 A 5 D
6 B 7 A 8 A 9 B 10 D
11 A 12 B 13 C

14 a) $I_2(s) + Cl_2(g) \rightarrow 2\ ICl(l)$
b) ICl has a molecular mass of 162.4; it has polar covalent bonds within its molecules and permanent dipole: permanent dipole attractions between its molecules. Bromine has a boiling point of 59°C and has van der Waals' forces between molecules.
c) Stronger bonding between molecules.

15 a) C–S b) H–N

16 a) i) $Si + 2Cl_2 \rightarrow SiCl_4$
ii) $SiCl_4 + 2H_2 \rightarrow Si + 4HCl$
b) $SiCl_4$
c) $SiCl_4$: discrete covalent molecules; Si: covalent network.
d) Diagram should show tetrahedral shape.
e) i) Chlorine and hydrogen.
ii) Produces chlorine for step 1 and hydrogen for step 3.

17 a) Stronger van der Waals' forces between molecules or greater molecular mass.
b)

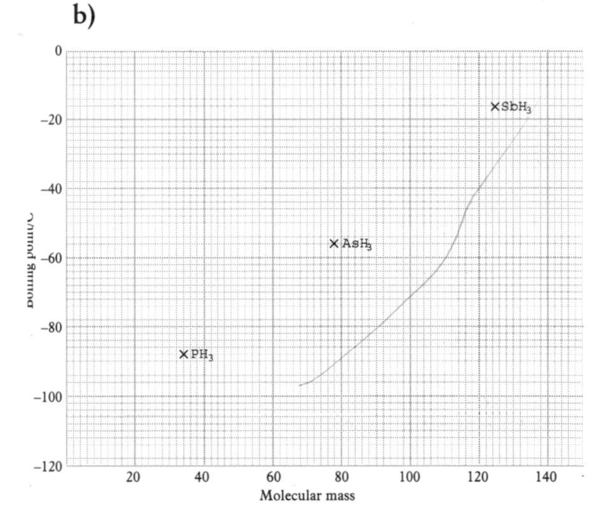

c) −100°C
d) Hydrogen bonding

18 Both compounds have weak intermolecular van der Waals' forces, due to temporary dipoles caused by uneven distribution of electrons moving within atoms. Ethanol also has stronger hydrogen bonds arising from highly polar O–H bonds inside its molecules. Slightly positive H atoms in one molecule are then attracted to slightly negative O atoms in another one, see Fig 10 (a) page 32.

CHAPTER 6:

SECTION A

1 A	2 D	3 D	4 B
5 D	6 B	7 C	8 C
9 C	10 D	11 A	12 C
13 D	14 C	15 B	16 B

17 a) Hydrazine: 32 g Hydrogen sulphide: 34 g
Benzene: 78 g
b) Hydrazine: 8 g Hydrogen sulphide: 17 g
Benzene: 13 g

18 a) i) 1.5 L **ii)** 4 L **iii)** 0.5 L
b) i) 250 cm^3 **ii)** 500 cm^3

SECTION B

1 C	2 B	3 D	4 A
5 B	6 A	7 C	8 B
9 B	10 C	11 B	12 A

13 a) Carry out reaction in a fume cupboard.
b) 2.4 litres **c)** 9.75 g **d)** 1.0 mol l^{-1}

14 a) 2.16 g **b)** 0.36 litres
c) Decreases to 0.12 litres since CO_2 dissolves in the alkali.

15 a) i) Oxygen, 25 cm^3 **ii)** 50 cm^3 CO_2
b) i) Isomers have the same molecular formula but different structural formulae.
ii) No hydrogen bonding between molecules.

16 a) 0.02 mol l^{-1} **b)** 0.24 litres

17 a) i) 450 cm^3 **ii)** 300 cm^3
b) $4NH_3 + 5O_2 \rightarrow 4NO + 6H_2O$

18 a) 7.62 g **b)** 2 cm^3 **c)** 0.96 litres
d) i) pH increases.
ii) No effect since ammonia is very soluble in water.

CHAPTER 7

1 C	2 B	3 D	4 B	5 B
6 A	7 D	8 A	9 A	10 C

11 a) To reduce 'knocking' or auto-ignition
b) Lead compounds, which are poisonous, were present in the exhaust gases.
c) 64.1%

12 a) Reaction II: reforming, isomerisation
Reaction III: reforming, dehydrogenation
Reaction IV: addition, polymerisation
b) Reaction III
c) It has a branched-chain structure.
d) i) Benzene **ii)** Poly(propene)
e) Carbon, it contaminates the catalyst.

13 a) Hydrogen, methane, ethanol
b) Ethanol (29.7 kJ g^{-1}), methane (55.7 kJ g^{-1}), hydrogen (143 kJ g^{-1})
c) Hydrogen's very low mass means that it has very high energy-mass ratio.

CHAPTER 8

1 A	2 D	3 C	4 A	5 B

6 a)
```
  H  CH3 H   H  H
  |   |  |   |  |
H-C - C -C - C -C-H
  |   |  |   |  |
  H   H C2H5 H  H
```

b)
```
      H  H
      |  |
    H-C -C-CH3
      |  |
  CH3-C -C-H
      |  |
      H  H
```

c)
```
  H        H  H
  |        |  |
H-C-C=C -C -C-H
  | |  |  |  |
  H H  H  H  H
```

d)
```
H CH3 H  H
|  |  |  |
C= C -C -C-H
|     |  |
H     H  H
```

e)
```
  H  H        H  H
  |  |        |  |
H-C -C -C≡C -C -C-H
  |  |        |  |
  H  H        H  H
```

f)
```
         H   H
         |   |
H-C≡C -  C - C-H
         |   |
        CH3  H
```

g)
```
 C2H5
  |
(benzene ring)
```

h)
```
       CH3
        |
  (benzene ring)
   /         \
CH3           CH3
```

7 a)
```
        C2H5
         |
CH3CH2CCH2CH3
         |
        C2H5
```

b)
```
   CH3             CH3
    |               |
CH3CCH2CH2CH2 CHCH3
    |
   CH3
```

c)
```
         CH3
          |
CH3CH2C=CHCH2CH3
```

d)
```
               C2H5
                |
CH2 = CHCHCH2CH3
```

e)
```
           CH3
            |
CH3C≡CCCH2CH3
            |
           CH3
```

f) $CH_2ClCH_2CH_2Cl$

8 a) 2,2,4-trimethylhexane
b) 4-ethylhex-2-ene
c) 1,2-dimethylcyclopentane
d) 2,2-dibromobutane
e) 4-methylpent-2-yne
f) 1,3-dimethylbenzene

9 a)–**f)**

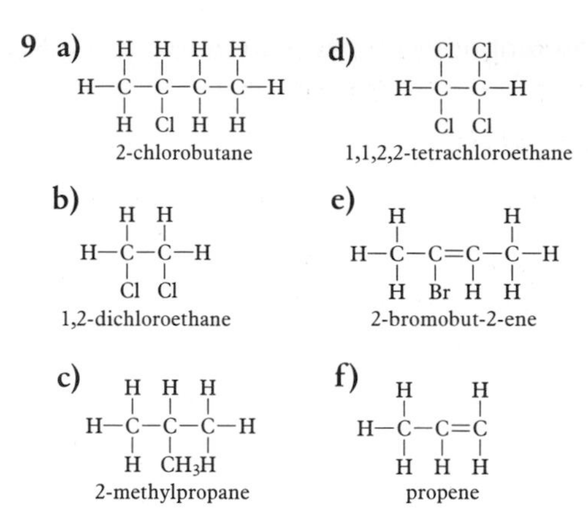

10 a) Addition

b) CH_3CHICH_2Cl (H–C–C–C–H with H, I, Cl below) and $CH_3CHClCH_2I$ (H–C–C–C–H with H, Cl, I below)

c) The double bond is in the middle of the carbon chain and the halogen atoms add on to the middle pair of carbon atoms thus forming only one product.

11 a) Iodobutane

b) $CH_3CH(CH_3)–CH(CH_3)CH_3$

c) Butane, hexane

12 a) Reactions 3 and 4

b) Reactions 2 and 7

c) i) Dehydration

ii) Propanol vapour is passed over heated aluminium oxide; diagram – Fig 2 page 50.

d) $C_3H_8 + Br_2 \rightarrow C_3H_7Br + HBr$

13 a) C_6H_5OH or C_6H_6O; $CH_3C_6H_4OH$ or C_7H_8O; $C_6H_2Cl_3OH$ or $C_6H_3Cl_3O$

b) H^+, $C_6H_5O^-$

c) 2,4,6-trichlorophenol

14 a) $CH_3C \equiv CCH_2CH_3$

b) i) Pent-1-yne

ii) 3-methylbut-1-yne

iii) Cyclopentene

15 a) I: $CH_2=CHCl$ (H, Cl / C=C / H, H); II: $CH_2Cl–CH_2Cl$ (H–C–C–H with H, Cl above and H, Cl below); III: $CH_2=CH_2$ (H, H / C=C / H, H)

b) A is hydrogen, B is chlorine, C is hydrogen chloride.

CHAPTER 9

1 C	2 D	3 B	4 C
5 C	6 B	7 A	

8 a) $CH_3CH_2CH(OH)CH_2CH_2CH_3$ secondary

b) $CH_3C(CH_3)(OH)CH_2CH_3$ tertiary

c) $CH_3C(CH_3)_2CH_2CH_2OH$ primary

d) $CH_3CH_2CH(C_2H_5)CH(OH)CH_3$ secondary

9 a) Pentan-3-ol, secondary

b) 3-ethylpentan-3-ol, tertiary

c) 3-methylbutan-1-ol, primary

d) 4,4-dimethylpentan-1-ol, primary

10 a) H–C–C–C–C–H (H, H, –, H above; H, H, O (double bond), H below), i.e. $CH_3CH_2COCH_3$

b) Butan-2-ol

c) Hydration or addition

d) Acidified potassium dichromate solution

11 a) Y is propan-2-ol, Z is propene.

b) i) Y changes acidified potassium dichromate solution from orange to blue-green.

ii) Z decolourises bromine water.

12 The O:H ratio for benzoic acid is 2:6 or 0.33:1; for the product of reaction (1) it is 1:8 or 0.13:1 and for the product of reaction (2) it is 1:6 or 0.17:1.
Therefore, reaction (1) is reduction and reaction (2) is oxidation.

13 a) 3-ethylpent-2-ene

b) Pentan-3-one

c) Ethanal changes acidified potassium dichromate solution from orange to blue-green, while pentanone does not react (or appropriate effect on Fehling's solution or Tollen's reagent.)

d) The double bond is in the middle of the carbon chain, so that each hex-3-ene molecule splits to form 2 molecules of propanal.

e) $(CH_3)_2C=C(CH_3)_2$

14 C
15 B
16 a) i) 2-Methylpropan-2-ol
ii) Propan-2-ol
iii) Propan-2-ol
b) i) Propanone and butanal
ii) Pentan-3-ol and propan-2-ol
iii) Butan-1-ol and 2-methylpropan-2-ol
c) i) Butanal
ii) $CH_3CH_2CH_2COOH$ or C_3H_7COOH
iii) Orange to blue-green

CHAPTER 10

1 D 2 B 3 C 4 C 5 A

6 a) $CH_3CH_2CH_2CH_2CH_2COOH$
b) $CH_3CH(CH_3)CH(CH_3)COOH$ (drawn with CH_3 branches above and below)
c) $CH_3CH(CH_3)CH(C_2H_5)CH_2COOH$ (drawn with CH_3 branch above and C_2H_5 below)
d) $HCOOCH_2CH_2CH_2CH_3$
e) $CH_3CH_2COOCH_2CH_2CH_3$

7 a) 4-ethylhexanoic acid
b) Methyl ethanoate
c) 3,3-dimethylbutanoic acid
d) Propyl methanoate

8 a) Concentrated sulphuric acid
b) A heated water bath
c) Condensation or esterification
d) 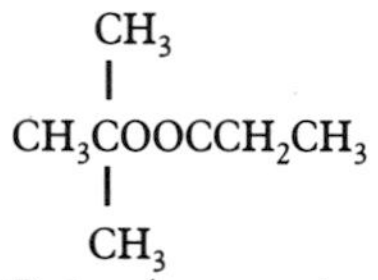 $CH_3COOC(CH_3)_2CH_2CH_3$ (drawn with CH_3 above and below the C)
e) Pour the reaction mixture into $NaHCO_3(aq)$. The ester forms an immiscible layer on the surface.
f) 70.6%

9 a) Hydrolysis
b) Diagram similar to fig 5 page 65.
c) 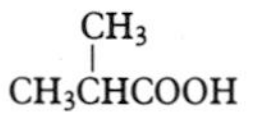 $CH_3CH(CH_3)COOH$ — 2-methylpropanoic acid 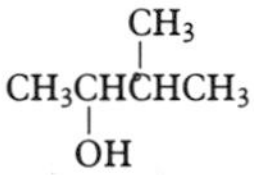
$CH_3CH(CH_3)CH(OH)CH_3$ — 3-methylbutan-2-ol
d) 3.62 g

10 a) H–C(H)(H)–C(H)(H)–O–C(=O)–H
b) Concentrated sulphuric acid
c) Pour the reaction mixture into $NaHCO_3(aq)$.

11 a) See fig 1 page 63.
b) Careful addition of concentrated sulphuric acid to prevent boiling or extinguish flame as reactants and products are inflammable.
c) Solvent or flavouring
d) $CH_3C(=O)H–O–CH_3$ (printed as CH_3H—O—CH_3 with O double-bonded above)

12 A
13 D
14 C
15 D

16 a) Ethene
b) H–C(H)(H)–C(H)(H)–O–C(H)(H)–C(H)(H)–H No hydrogen bonding between molecules
c) 28.0%

17 a) $C_6H_3OCl_3$
b) i) 0.298 ii) 21.45 litres iii) 48.85 g

18 a) $C_9H_8O_4$
b) Ester, hydroxyl
c) i) Methyl salicylate ii) 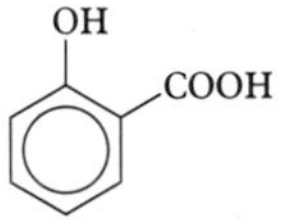 benzene ring with OH and COOH on adjacent carbons
d) 73.5 %

19 a) H–C(H)(H)–C(H)(H)–C≡N
b) Hydration
c) $C_2H_5CN + 2H_2O + HCl \rightarrow C_2H_5COOH + NH_4Cl$
d) i) Ethyl propanoate, H–C(H)(H)–C(H)(H)–O–C(=O)–C(H)(H)–C(H)(H)–H
ii) Add a few drops of concentrated sulphuric acid to a mixture of ethanol and propanoic acid in a test tube and heat for several minutes in a water bath.

20 a) i) B ii) E
b) A: aldehyde C: ketone D: ester
F: primary alcohol
c) i) B ii) F iii) C iv) D
d) $C_8H_8O_2$
e) B, neutralisation; D, hydrolysis
f) Add each to acidified potassium dichromate and heat. A turns it from orange to blue-green; C does not react (or appropriate effect with Fehling's solution or Tollen's reagent).

21 a) 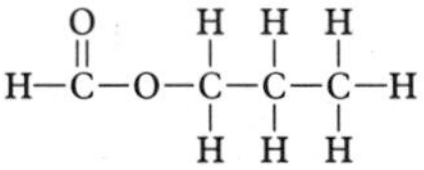 H–C(=O)–O–C(H)(H)–C(H)(H)–C(H)(H)–H
b) Propanal
c) Hydrogen bonds are not present between molecules of X or Y, only van der Waals' forces.

d) i) It is reduced to copper.
ii) Dehydration, propene

CHAPTER 11

1 B **2** D **3** D **4** D **5** B **6** A

7 a) i) Compounds A and C.
ii)

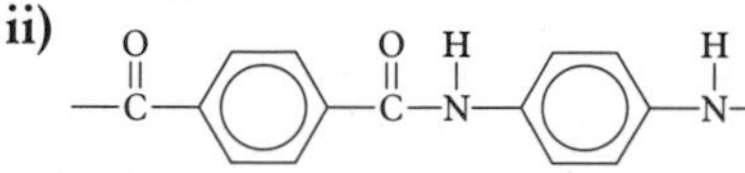

iii) Bullet-proof clothing, crash helmets, sails, tyre strengthening

b) i) Compounds A and D.
ii)

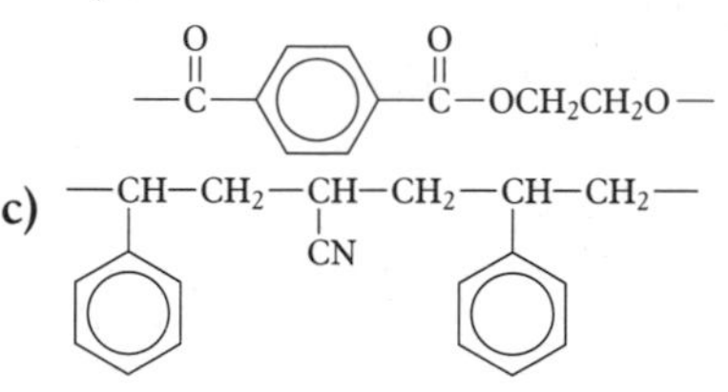

c)

d) Condensation - molecules combine with the loss of water or some other small molecule. Thermosetting – harden when heated and cannot be remoulded.

8 a) HCl is the by-product instead of water.
b) —O—C(=O)—O—
c) It neutralises the HCl produced.
d) They are immiscible.

9 a) Alkynes
b) $H_2C=CH-C\equiv C-H$
c) Addition
d) —CH(Cl)—CH(H)—CH(Cl)—CH(H)—CH(Cl)—CH(H)— (H Cl H Cl H Cl above; H H H H H H below: $-C-C-C-C-C-C-$)

10 a) —O—[X]—O—C(=O)—N(H)—[Y]—N(H)—C(=O)—O—[X]— (repeat unit boxed: O—C(=O)—N(H) or N(H)—C(=O)—O)

b) i) Monomers do not contain C=C bonds.
ii) No by-product, e.g. water.
c) Propan-1,2- diol, $CH_2(OH)\,CH(OH)\,CH_3$
or
propan-1,3-diol, $CH_2(OH)\,CH_2\,CH_2(OH)$
d) This allows cross-linking to occur.
e) To produce polyurethane foam.
f) CFCs damage the ozone layer.

CHAPTER 12:

SECTION A

1 D **2** A **3** A **4** D
5 B **6** D **7** C **8** C

9 a) To provide energy.
b) It has more van der Waals' bonds between molecules due to greater molecular size.
c) Glyceryl trierucate is unsaturated and has a distorted 'tuning-fork' shape which restricts close-packing of the molecules (see fig 2 page 76). Hence the intermolecular bonding is weaker.

10 a) Ester
b) There are 3 ester groups per molecule.
c) $C_{17}H_{33}COOH$ or $C_{18}H_{34}O_2$

11 a) i) Hydrogenation **ii)** Hydrolysis
b) It is a catalyst.
c) Margarine
d) Glycerol, $CH_2(OH)-CH(OH)-CH_2(OH)$
e) $CH_3(CH_2)_7CH=CH(CH_2)_7COONa$ or $C_{17}H_{33}COONa$
f) The hydrocarbon 'tail' is attracted to grease forming negatively-charged droplets, which repel each other.

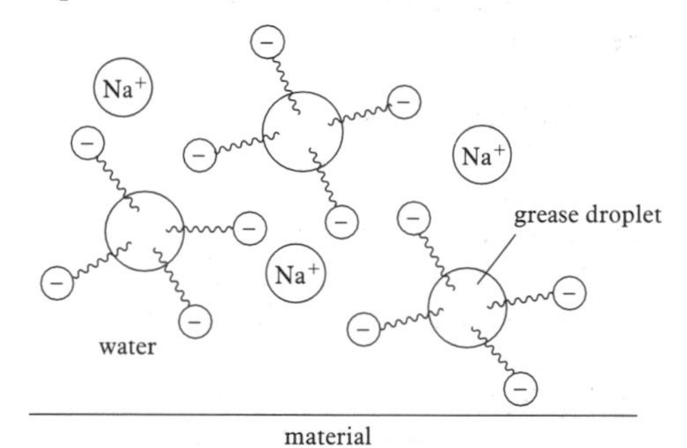

SECTION B

1 B **2** D **3** B **4** C
5 C **6** A **7** B

8 a) Carbon, hydrogen, oxygen and nitrogen.
b) It is denatured.
c) An enzyme has a characteristic shape so that it fits the substrate which it is hydrolysing.

9 a) –NH–CO–
b) >CHOH and $-CH_2OH$
c) –COOH and $-NH_2$
d) –CHO and –COOH

10 a) i) Any 2 of the following 3 amino acids:

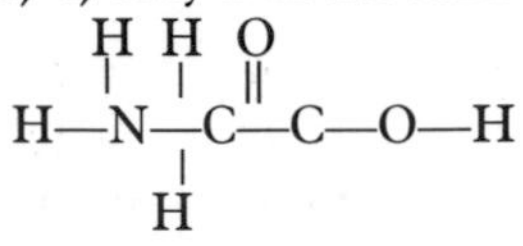

```
H   CH3 O
|   |   ||
H—N—C—C—O—H
    |
    H
```

```
CH3 CH3
  \ /
H  CH O
|  |  ||
H—N—C— C—O—H
    |
    H
```

ii) Condensation

b) Essential amino acids

CHAPTER 13

1 B   2 D   3 C

4 a) Increased price of feedstock or increased cost of transport.

b) Increased cost of energy required for production or increased cost of transport.

c) Increased cost of transport or increased cost of production of plastic or glass bottles.

5 a)

Capital Costs	Fixed Costs	Variable Costs
Construction of new plant	Staff training	Fuel bills
Construction of rail siding	Advertising	Landfill tax
	Land rental	Wages
	Development of new product	*[Some of the above answers may depend on the accounting period e.g. salaries may be fixed costs due to longer-term contracts]*

6 a) Long term damage – all aquatic animals killed including insects.

b) Main source of water – unsafe for a long time.

c) Neighbouring countries likely to blame Rumania for harmful effects.

7 a) Motor fuels

b) More suitable for large-scale process; requires a smaller workforce.

c) Raw material is a solid – harder to work with in a continuous process.

8 Favourable factors: Deep water harbour; readily-available labour force.
Unfavourable factors: Cost of National Grid electricity; remote site – poor transport links.

9 a) i) Limestone, coal **ii)** Road and rail links

b) i) Reduces fuel costs.

ii) Disposes of hazardous waste.

c) Use as a fuel, reducing costs.

10 a) Ethane, naphtha

b) $2C_2H_4 + O_2 \rightarrow 2(CH_2)_2O$

c) Products of combustion, i.e. CO_2 and H_2O.

d) Process (2): Advantage – neutral solution.
Disadvantage – high pressure or high temperature.
Process (4): Advantage – atmospheric pressure or low temperature.
Disadvantage – acidic conditions (corrosive).

e) Antifreeze, manufacture of polyester.

f) Construction of high pressure plant or initial purchase of silver catalyst.

g) Feedstock costs, fuel costs or wages.

11 a) To lower the melting point of the electrolyte.

b) Continuous process – diagram shows addition of salt and removal of products.

c) i) Liquid **ii)** Its melting point is lower than that of the electrolyte.

CHAPTER 14

1 A   2 C   3 A   4 B   5 B   6 D

7 a) Volume of acid.

b) -38.2 kJ mol^{-1}

8 a) -106 kJ mol^{-1} **b)** -484 kJ mol^{-1}

9 a) 376 kJ mol^{-1} of ethyne

b) 188 kJ mol^{-1} of methane

10 -1517 kJ mol^{-1}

11 a) $HCHO(g) + O_2(g) \rightarrow CO_2(g) + H_2O(l)$

b) -166 kJ mol^{-1}

c) i) No effect   **ii)** Decreases it.

d) Methanol is a gas (not a liquid) in the industrial process.

e) Manufacture of bakelite or thermosetting polymers.

12 a) x = 9, y = 12, z = 4

b)

```
  H  H    H
  |  |   /
H—C—N—N
  |      \
  H       H
```

c) 53 kJ mol^{-1}

CHAPTER 15

1 A 2 C 3 D 4 B 5 A
6 C 7 B 8 B 9 D

10 **a)** It is increased. **b)** It is unaltered.
c) It is decreased.

11 **a)** Mixture turns pale yellow.
b) Mixture turns pale yellow.

12 **a)**

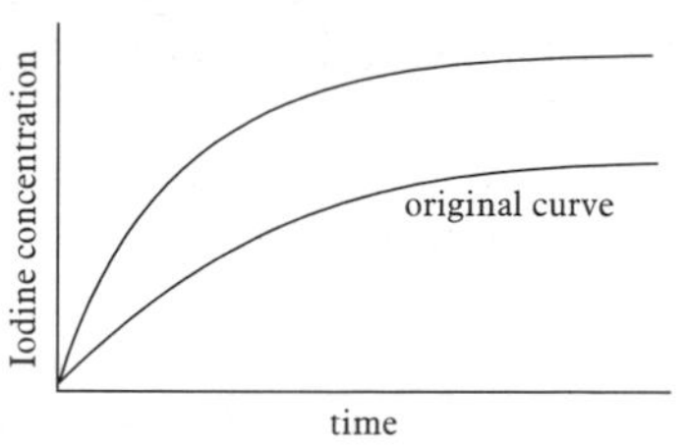

b) i) KI: Goes to the left. K_2SO_4: No effect.
ii) KOH lowers the concentration of hydrogen ions.

13 **a)** There are the same number of moles of gas on each side of the equation.
b) The yield of hydrogen decreases.
c) i) A catalyst speeds up both reactions to the same extent.
ii) Equilibrium is reached sooner.
d) It will increase.

14 **a)** Na_2CO_3 reacts with H^+ ions (to form CO_2 and H_2O) so equilibrium shifts to the right.
Ag^+ combine with Cl^- (to form insoluble AgCl) so equilibrium shifts to the right.
b) It is much more expensive.

15 **a)** Equilibrium moves to the left in (i) and (ii), and to the right in (iii).
b) Enthalpy of solution is positive (or reaction is endothermic).

16 **a)** 0.4 litre SO_2, 0.7 litre O_2, 0.6 litre SO_3.
b) i) It moves to the left.
ii) No effect.
iii) It moves to the right.

17 **a)** 1.9 kJ mol^{-1}.
b) High temperature favours the forward reaction since it is endothermic.
c) Diamond is more dense, i.e. its atoms are more closely-packed.

18 **a) i)** It increases. **ii)** It decreases.
b) It is exothermic.

19 **a)** The equilibrium moves to the left until no CO_2 is left in solution.
b) It is negative (or reaction is exothermic).
c) 23.0 litres

CHAPTER 16

1 C 2 B 3 D 4 B 5 B
6 A 7 D 8 C 9 C 10 D
11 C 12 D 13 B 14 A 15 A
16 C

17 **a) i)** 10^{-3} mol l^{-1} **ii)** 3
b) i) 10^{-7} mol l^{-1} **ii)** 7
c) i) 1.0 mol l^{-1} **ii)** 0
d) i) 10^{-13} mol l^{-1} **ii)** 13
e) i) 10^{-9} mol l^{-1} **ii)** 9

18 TCP: 10^{-12} mol l^{-1}; handwash: 10^{-6} mol l^{-1}; cleaning fluid: 10^{-3} mol l^{-1}; lemonade: 10^{-10} mol l^{-1}.

19 **a) i)** Ethanoic acid is a weak acid or is partially dissociated into ions.
ii) Sulphuric acid has 2 mol of hydrogen ions per mol of acid.
b) Z contains 0.02 mol of H^+ ions. The alkali added contains 0.01 mol of OH^- ions. After reaction there will be 0.01 mol of H^+ in 100 cm^3 of solution, so $[H^+] = 10^{-1}$ mol l^{-1}.

20 **a)** Increases
b) $[H^+] = 10^{-2}$ mol l^{-1}, $[OH^-] = 10^{-12}$ mol l^{-1}
c) $CHCl_2COOH + KOH \rightarrow CHCl_2COOK + H_2O$
d) <7

21 **a)** pH 1.0
b) A solution which is partially dissociated into ions and contains hydrogen ions.
c) i) Nitric acid reacts faster.
ii) They will be the same.

22 **a) i)** H_2SO_3 **ii)** $NaHSO_3$
b) SO_3^{2-} ions combine with H^+ ions from water. Water equilibrium produces more OH^-.
c) It will be lower.

CHAPTER 17

SECTION A

1 A 2 D 3 B 4 B
5 C 6 A 7 D

8 **a)** $Mn^{2+} + 2H_2O \rightarrow MnO_2 + 4H^+ + 2e^-$;
Mn^{2+} is a reducing agent.
b) $VO_3^- + 6H^+ + 3e^- \rightarrow V^{2+} + 3H_2O$;
VO_3^- is an oxidising agent.
c) $H_2O_2 \rightarrow O_2 + 2H^+ + 2e^-$;
H_2O_2 is a reducing agent.
d) $Ti + H_2O \rightarrow TiO^{2+} + 2H^+ + 4e^-$;
Ti is a reducing agent.
e) $2BrO_3^- + 12H^+ + 10e^- \rightarrow Br_2 + 6H_2O$;
BrO_3^- is an oxidising agent.

9 a) $Cr \rightarrow Cr^{3+} + 3e^-$
b) $SO_3^{2-} + H_2O \rightarrow SO_4^{2-} + 2H^+ + 2e^-$

10 a) $Br_2 + 2e^- \rightarrow 2Br^-$
b) $H_2O_2 + 2H^+ + 2e^- \rightarrow 2H_2O$

11 a) $2Fe^{3+} + Sn^{2+} \rightarrow 2Fe^{2+} + Sn^{4+}$
b) $6I^- + Cr_2O_7^{2-} + 14H^+ \rightarrow 3I_2 + 2Cr^{3+} + 7H_2O$

12 a) $3C_2H_5OH + Cr_2O_7^{2-} + 8H^+ \rightarrow 3CH_3CHO + 2Cr^{3+} + 7H_2O$
b) i) $CH_3CHO + H_2O \rightarrow CH_3COOH + 2H^+ + 2e^-$
ii) 0.33

13 a) $2S_2O_3^{2-} \rightarrow S_4O_6^{2-} + 2e^-$
b) Blue to colourless
c) 0.0585 mol l^{-1}

14 a) $MnO_4^- + 8H^+ + 5e^- \rightarrow Mn^{2+} + 4H_2O$
b) The permanganate solution also acts as an indicator, turning from purple to colourless.
c) i) Transfer 20.0 cm^3 of the hair bleach by pipette to a 500 cm^3 standard flask. Add pure water up to the mark on the flask. Stopper the flask and mix contents thoroughly.
ii) 1.54 mol l^{-1}

15 a) Equilibrium moves to the right, thus increasing the concentration of ozone.
b) i) 1.123×10^{-4}
ii) 2.69×10^{-5} cm^3.

SECTION B

1 B	2 B	3 B	4 C
5 D	6 A	7 C	8 A

9 a) 119 s b) 4.2 A c) 58.8 kg

10 a) X: pH rises since OH^- produced.
Y: pH falls since H^+ produced.
b) 79.6 cm^3

11 a) i) 192990 C ii) 96495 C
b) 6.03×10^{23}

12 a) i) 111.8 C ii) 0.037 g iii) 13.9 cm^3
b) i) 193000 C produces 1 mole of each gas, so 111.8 C produces equal volumes of each gas, at same temperature and pressure.
ii) Chlorine is more soluble in water.
c) i) 7 cm^3 ii) pH decreases.

13 a) 5×10^{-6}
b) i) Starch
ii) 0.008
iii) Stir or mix the solution.

CHAPTER 18

1 B	2 C	3 D	4 B	5 D
6 A	7 C	8 B	9 A	10 D
11 D	12 D	13 A	14 B	15 D

16 a) A: $^{233}_{91}Pa$ B: $^{225}_{89}Ac$ C: $^{209}_{83}Bi$
b) i) 8 ii) 5
c)

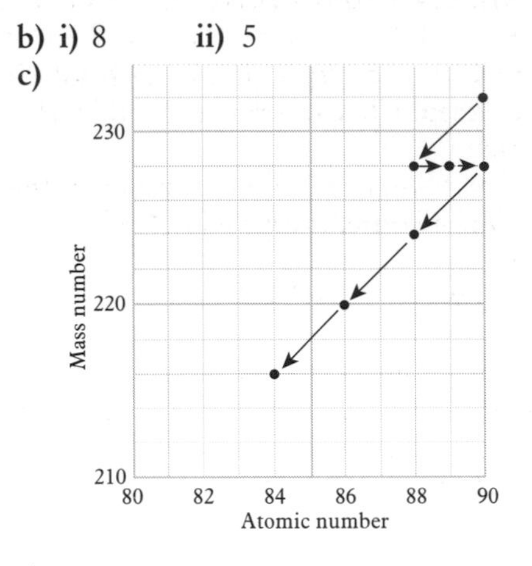

d) 4n + 2

17 a) i) ^{206}Bi ii) At^-
b) Mass number is only important in nuclear changes; a) ii) describes a chemical change.

18 a) $^{27}_{13}Al + ^{1}_{0}n \rightarrow ^{24}_{11}Na + ^{4}_{2}He$
b) $^{32}_{16}S(n,p)^{32}_{15}P$

19 a) i) Yes ii) No
b) i) Yes ii) No
c) i) No ii) No
d) i) Yes ii) Yes

20 a) Nuclear fusion
b) α emission
c) 268 and 264

21 a) Gamma rays, cosmic rays, thoron, medical uses.
b) $^{220}_{86}Rn \rightarrow ^{216}_{84}Po + ^{4}_{2}He$
c) 220 s
d) No effect

22 a) X: $^{1}_{0}n$ b) Y: $^{3}_{1}H$

23 a) $^{226}_{88}Ra \rightarrow ^{222}_{86}Rn + ^{4}_{2}He$
b) 3.82 days
c) 3.8×10^{15}
d) 6.25×10^{-9}
e) 6.08×10^{23}

24 a) Fears of possible nuclear accident.
b) Use of sea water for cooling.
c) Small harbour, cement works, road and rail links nearby.

PRESCRIBED PRACTICAL ACTIVITIES

1 a) Volume and concentration of each solution used except KI(aq); temperature of reaction mixture.

b) The concentrations of all solutions in the reaction mixture, except KI(aq), are constant.

c) 29.0 s (rate = 0.0345 s^{-1})

d) i) 0.017 s^{-1}

ii) Error in measuring the volume of a solution or the reaction was carried out at a higher temperature.

e) Rate of reaction is directly proportional to the concentration of KI(aq) or rate increases as [KI] increases.

2 a) Purple to colourless.

b) Volume and concentration of each solution used.

c) x = 0.028 s^{-1} ; y = 19.2 s

d)

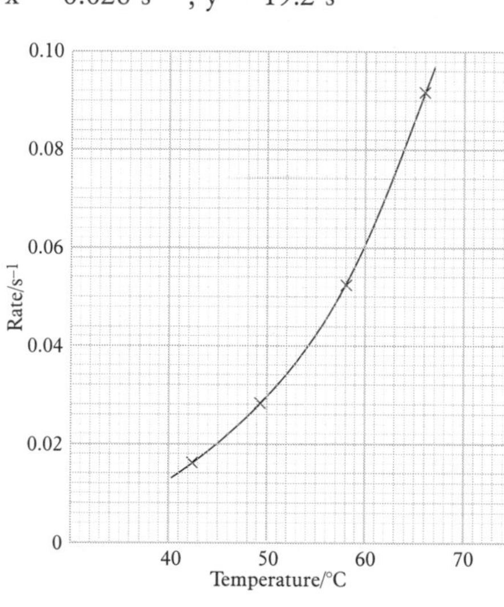

e) i) 0.042 s^{-1} **ii)** 65 °C

3 a) Diagram similar to Fig 9 page 15.

b) $CH_3OH(l) + \frac{3}{2} O_2(g) \rightarrow CO_2(g) + 2H_2O(l)$

c) −535 kJ mol^{-1}

d) Incomplete combustion

4 a) Extinguish the flame.

b) A ketone or alkanone.

c) i) $Ag^+ + e^- \rightarrow Ag$

ii) An aldehyde or alkanal.

d) Missing words: orange, blue-green, butanoic acid.

e) i) Fehling's (or Benedict's) solution.

ii) Copper(I) oxide, Cu_2O.

5 a) Concentrated sulphuric acid.

b) To condense volatile reactants and/or products.

c) i) An immiscible layer forms on the surface of the solution.

ii) Excess acid reacts releasing carbon dioxide.

d)

```
     O                 H   H   H   H
     ‖                 |   |   |   |
H—C—O—H  +  H—O—C—C—C—C—H  ⟶
                       |   |   |   |
                       H   H   H   H

               O     H   H   H   H
               ‖     |   |   |   |
          H—C—O—C—C—C—C—H  +  H2O
                     |   |   |   |
                     H   H   H   H
```

6 a) Potato 'discs'.

b) Water bath, thermometer

c) Rate at which gas bubbles through the water in the trough.

d) Rate increases with temperature to a maximum and then decreases at higher temperatures.

e) Replace the water in the test tube with solutions of different pH.

f) Use the same volume and concentration of $H_2O_2(aq)$, 5 cm^3 of each other solution and same number of potato discs. Keep temperature constant.

7 a) i) Enthalpy of solution of KOH(s).

ii) To minimise heat loss to surroundings.

iii) 5 °C

b) −53.5 kJ mol^{-1}

c) i) Known mass of KOH(s) is added to HCl(aq) of known volume and concentration. Temperature of solution is measured before and after addition of solid.

ii) $KOH(s) + HCl(aq) \rightarrow KCl(aq) + H_2O(l)$

iii) -105.75 kJ mol^{-1}

8 a) One mole of hydrogen.

b) Current in amps, time in seconds.

c) $2H^+ + 2e^- \rightarrow H_2$

d) Its molar volume under experimental conditions of temperature and pressure.

e) 193000 C

9 a) Dissolve the tablet in pure water (about 100 cm^3) and pour solution into a 250 cm^3 standard flask. Rinse beaker and add washings to flask. Make solution up to the mark with water, stopper flask and mix contents thoroughly.

b) Missing words: pipette, conical flask, starch, burette, blue, concordant titres.

c) 0.498 g

PRACTICE PRELIM: UNITS 1 and 2

SECTION A

1 C	11 C	21 B	31 B
2 B	12 D	22 A	32 D
3 B	13 A	23 A	33 D
4 C	14 D	24 C	34 B
5 B	15 A	25 A	35 B
6 A	16 C	26 A	36 A
7 A	17 B	27 C	37 D
8 D	18 C	28 D	38 C
9 D	19 D	29 C	39 C
10 B	20 B	30 D	40 B

SECTION B

1 a) Electronegativity.
b) It increases.
c) Second electron to be removed is nearer the nucleus or comes from an inner energy level.

2 a) Diagram similar to Fig 5 page 25.
b) The layers of atoms can move over each other.
c) 5.02×10^{18}.

3 a) i) Heterogeneous.
ii) Fat is more saturated or has fewer double bonds.
b) i) Glycerol or propane-1,2,3-triol.
ii) Soap.

4 a) i) It has delocalised electrons.
ii) $C_6H_3OCl_3$
iii)

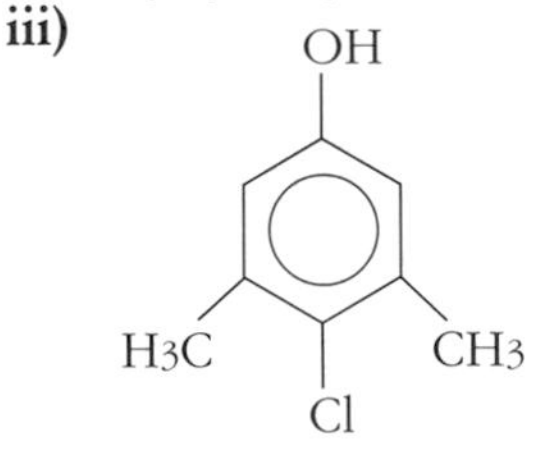

b) To make petrol.

5 a) i)

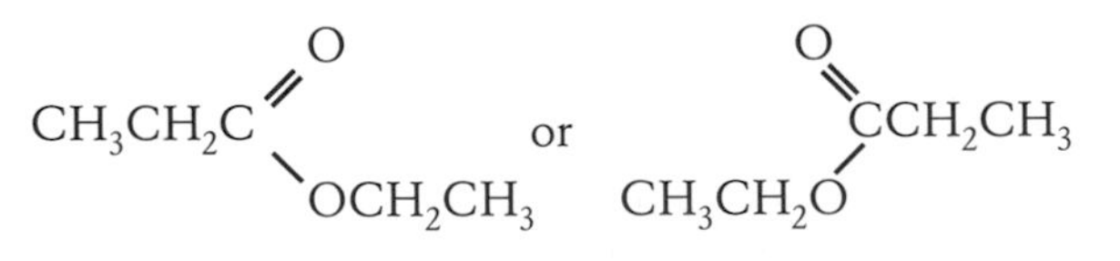

or correct expanded structural formula.

ii) Diagram similar to Fig 1 page 65.
b)

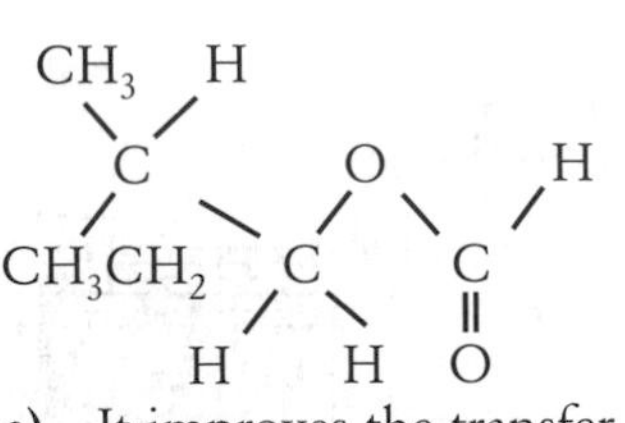

6 a) It improves the transfer of heat to the water.
b) The use of oxygen (instead of air) ensures complete combustion.
c) -1366 kJ mol^{-1} [cmΔT = 29.09 kJ, then divide by n = 0.0213].

7 a) Phosphorus and hydrogen have the same electronegativity,i.e. 2.2.
b) 35 cm^3.
c) 0.12 litres or 120 cm^3.

8 a) i) The carbonyl group is at the end of the carbon chain or has a hydrogen atom attached to it.
ii) Blue (solution) to orange-red (precipitate).
iii) It increases, showing that oxidation has occurred.
iv) Propanoic acid.
b) i) Addition.
ii)

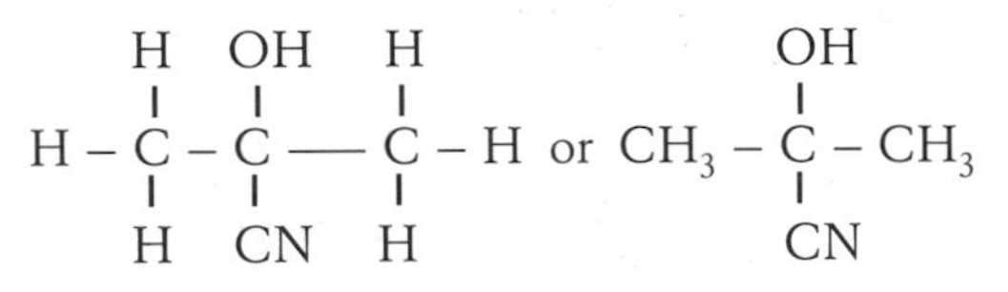

9 a) i)

$$\text{HOOC–}C_6H_4\text{–COOH}$$

(benzene ring with two C(=O)–O–H groups in 1,3-positions)

ii) 1,4-diaminobenzene.
b) Photoconductivity.

10 Number of moles of Zn = 0.5.
Number of moles of HCl = 0.8, which reacts with 0.4 mol of Zn.
Zn is in excess, so add more acid.

11 a) i) Cyclopentane.
ii) It is a catalyst.
iii) It has a branched-chain structure or is less liable to auto-ignite or ‘knock’.
b) Covalent network.

12 a) & b)

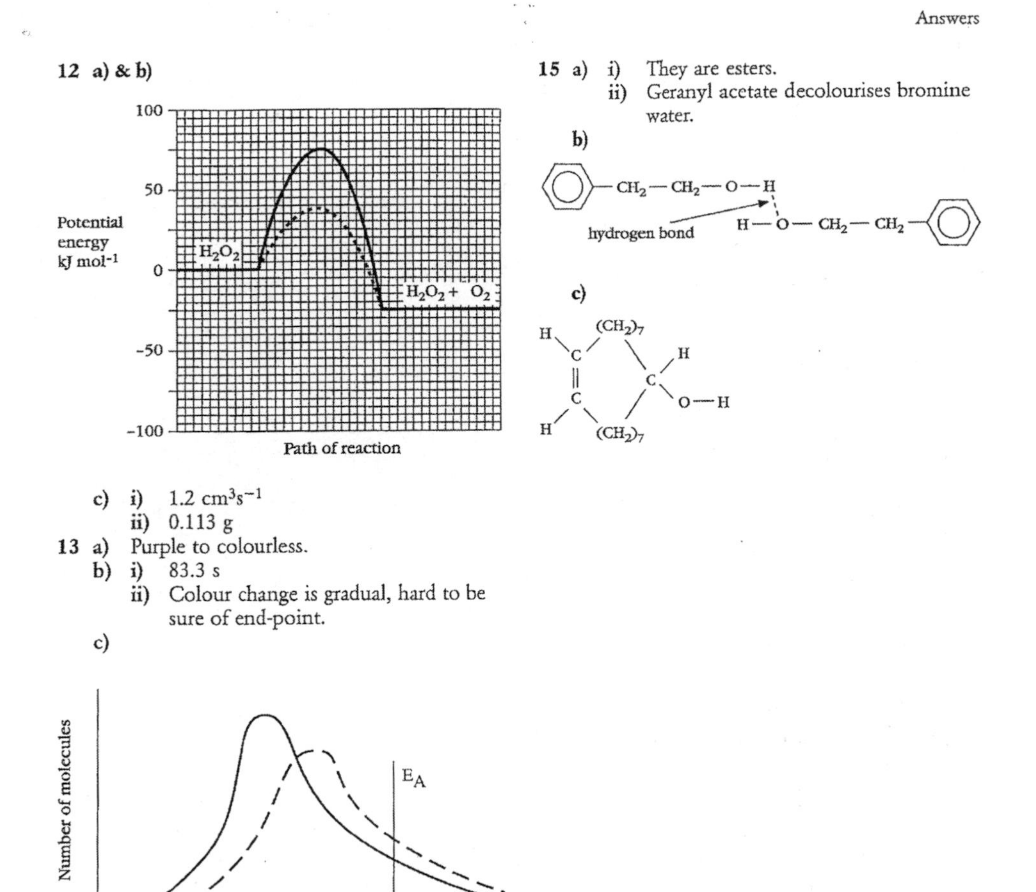

c) i) 1.2 cm^3s^{-1}
ii) 0.113 g

13 a) Purple to colourless.
b) i) 83.3 s
ii) Colour change is gradual, hard to be sure of end-point.
c)

14 a) i) Hydrolysis.
ii) Tertiary.
iii) 74.9% [Theoretical yield = 7.68 g].
b) To make thermosetting polymers or methanoic acid.

15 a) i) They are esters.
ii) Geranyl acetate decolourises bromine water.
b)

c)

12 a) & b)

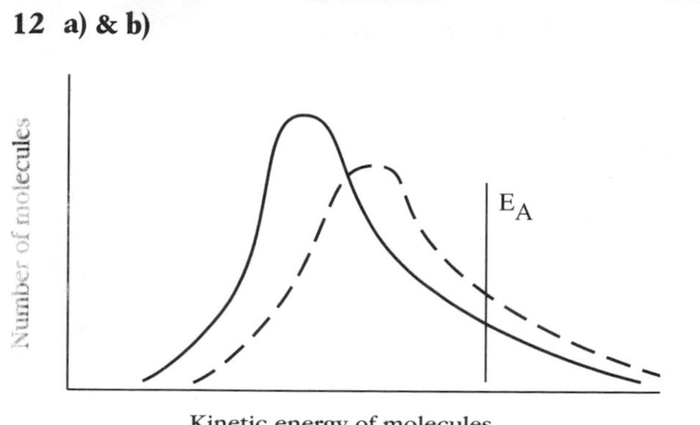

c) i) 1.2 cm^3s^{-1}

ii) 0.113 g

13 a) Purple to colourless.

b) i) 83.3 s

ii) Colour change is gradual, hard to be sure of end-point.

c)

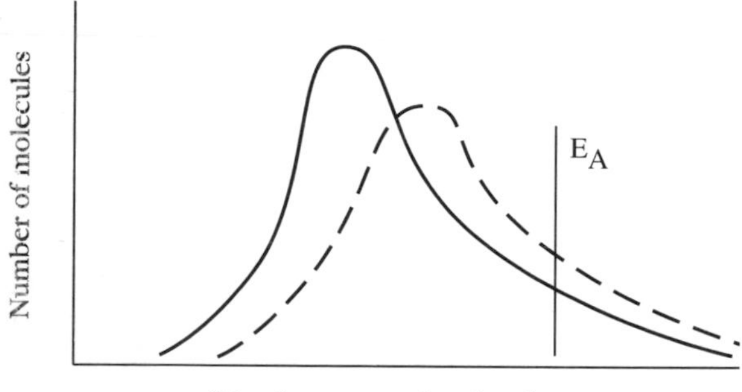

14 a) i) Hydrolysis.

ii) Tertiary.

iii) 74.9% [Theoretical yield = 7.68 g].

b) To make thermosetting polymers or methanoic acid.

15 a) i) They are esters.

ii) Geranyl acetate decolourises bromine water.

b)

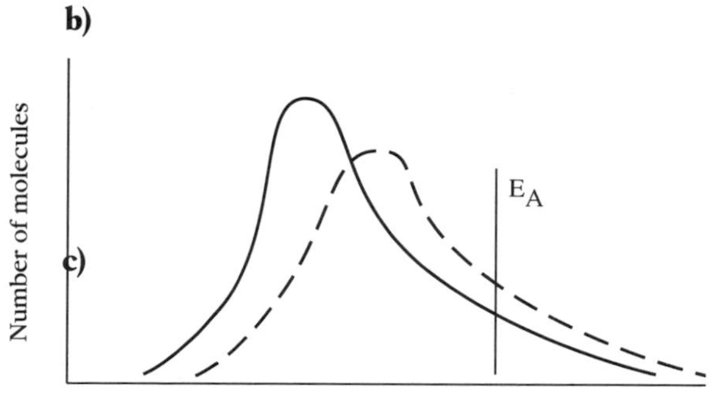

c)

UNIT 3 TEST

SECTION A

1 A	**5** B	**9** B	**13** C
2 B	**6** D	**10** A	**14** A
3 B	**7** D	**11** A	**15** D
4 C	**8** B	**12** D	**16** C

SECTION B

1 a) $C_6H_8O_6 \rightarrow C_6H_6O_6 + 2H^+ + 2e^-$.
b) i) Rinse beaker several times with water and add washings to flask.
ii) Colourless to blue.
c) 1.04 g.

2 a) ${}^{24}_{11}Na \rightarrow {}^{24}_{12}Mg + {}^{0}_{-1}e$.
b) It increases.
c) 15 hours.
d) 7 g.

3 a) -484 kJ mol^{-1}
b) Ethanoate ions combine with H^+ ions from water since ethanoic acid is a weak acid,
ie $CH_3COO^- + H^+ = CH_3COOH$
equilibrium moves to the right.
Removal of H^+ ions moves the water equilibrium to the right,
i.e. $H_2O \rightleftharpoons H^+ + OH^-$, thus giving an excess of OH^- ions, so pH > 7.

4 a) $Pd^{2+} + 2e^- \rightarrow Pd$
b) $2CO + O_2 \rightarrow 2CO_2$

5 a) Rates of forward and reverse reactions are the <u>same</u>. Concentrations of reactants and products are <u>usually different</u>.
b) i) It has a negative sign or it is exothermic.
ii) No effect.

6 a) The enthalpy change for a chemical reaction is independent of the route taken.
b) Mass or volume of water, initial temperature of water, final temperature of solution.
c) Hydrochloric acid.

7 a) Arrow upwards from NaOH formed in stage 2 to NaOH used in stage 1.
b) Acidic oxide.
c) Dehydration or decomposition.
d) 6.04×10^4 g or 60.4 kg
[$Q = 6.48 \times 10^8$ C].
e) They burn away due to reaction with oxygen produced at this electrode.

Expanding Nursing and Health Care Practice

Series Editor: Lynne Wigens

Management of Pain

A practical approach for health care professionals

Kathleen Mac Lellan

Published in 2006 by:
Nelson Thornes Ltd
Delta Place
27 Bath Road
CHELTENHAM
GL53 7TH
United Kingdom

06 07 08 09 10 / 10 9 8 7 6 5 4 3 2 1

A catalogue record for this book is available from the British Library

ISBN 0-7487-9621-5

Cover photograph by Stockbyte RF (NT)
Illustrations by Florence Production Ltd, Stoodleigh, Devon
Page make-up by Florence Production Ltd, Stoodleigh, Devon

Printed and bound by UniPrint Hungary Kft, Székesfehérvár

Contents

Important Notice

The various methods of management of pain both pharmacological and others are subjects of on-going research and debate. As such evidence changes rapidly, all health professionals should continuously update and review new evidence as it emerges. Pain management should be based on the best available evidence. Good evidence is likely to come from good systematic reviews of good clinical trials. Whilst every effort has been made in this book to ensure that best evidence is presented, health care professionals should always consult drug company literature, local and national protocols and other sources of evidence before utilising pain interventions detailed. The accuracy, currency or completeness of the information on the websites and information sources listed cannot be guaranteed and no legal liability or responsibility is assumed.

Introduction

This book aims to support health professionals during their undergraduate education and as they start out as qualified health professionals to understand and provide appropriate, effective pain management. All health professionals are in the privileged position to ensure that the patients they care for, whether in the community or in hospital, receive care that ensures that their pain is minimised and that both functional activity and quality of life are maximised. Modern health care supports the concept that patients receive better care when health professionals work as a team and provide evidence-based care. There are no benefits for patients to endure pain and indeed, as this book outlines, the prevalence of pain is the cause of many adverse effects from reduced quality of life to decreased lung volumes, myocardial ischaemia, decreased gastric and bowel motility and anxiety. Planned, strategic approaches to pain management are emphasised in the book. The evidence suggests that it is not always new approaches that are needed, that the key to effective pain management is often good and systematic use of current methods of pain management.

Chapter 1 provides the foundation for this book outlining contemporary definitions of pain and associated terms. Pain management as part of health and social policy is detailed in **Chapter 2**, which outlines socio-economic and quality-of-life issues. The management of pain has evolved over time with the introduction of specialist pain services and new therapeutic interventions including new medicines. These are described in **Chapter 3**. **Chapter 4** provides an overview of the physiology of pain including illustrations of the nervous system linking these with the processes of pain management techniques. Many factors from anxiety to culture are thought to influence the pain experience as described in **Chapter 5**. **Chapter 6** reviews the common pain assessment techniques and tools. The array of pain interventions including medications, transcutaneous electrical nerve stimulation (TENs) and behavioural therapy are detailed in **Chapter 7**. **Chapter 8** reviews complementary therapies and their role in pain management. Management of acute pain including chest pain, postoperative pain and pain in the emergency department are outlined in **Chapter 9**.

Chapter 10 provides a review of chronic pain including back pain and arthritis. The principles of palliative care are outlined in **Chapter 11**. **Chapter 12** is the concluding chapter and highlights the role of nurses and health professionals in pain management.

The most important message for you, as a health professional, emerging from this book is the **critical role of evidence-based practice**. Evidence for pain management is the subject of ongoing research and debate. As such evidence changes rapidly, all health professionals should continuously update and review developments. Good evidence is likely to come from sound systematic reviews of well-planned clinical trials. This means that you as a health professional must be able to access, understand and interpret research in order to provide evidence-based care.

1 Definitions of pain and associated terms

Learning outcomes

By the end of this chapter you should be able to:

★ Detail contemporary definitions of pain and associated terms

★ Understand the difficulties in defining pain

★ Appreciate the main ways of classifying pain

★ Outline the main components of a pain care service.

Introduction

This chapter provides an overview of the definitions of pain and associated terms. These definitions provide the foundation for the chapters that follow where more detail is provided on the particular subject areas. Numerous definitions of pain are available, all attempting to provide meaning to a complex phenomenon that has been debated and researched over time. This chapter will look at the more common definitions of pain as well as how pain is classified. These classifications can be based on a number of factors, such as intensity and duration and whether it is acute or chronic in nature. Key resources for research and information on pain are provided throughout.

Defining pain – a contemporary debate

Within the field of health care you will come across people experiencing pain in almost all clinical settings, both within acute clinical specialties and in the community. The pain experience of the older adult with arthritis in a nursing home differs from the pain of a 20-year-old who has undergone surgery for a sports injury to the knee. The sting of a wasp will cause severe pain for one adult and minimal pain in another. It is these contradictions that create the debate, difficulties and interest in defining and understanding pain.

Everyone has experienced pain at some time thus each individual has an innate sense of what pain is, based on their personal experiences. The experience of the person is influenced by past encounters with pain, physical and mental wellbeing, culture and coping mechanisms.

It is therefore reasonable to state that while much is known about the physiology of pain, the sensory and emotive aspects of pain can be difficult to interpret, separate and describe. The interactions of the mind and body in the production and perception of pain therefore are ongoing subjects for research and debate.

Defining pain is a difficult task and the definitions provided try to encompass a phenomenon with many facets. It is in the context of these considerations that this chapter provides an overview of the more common definitions of pain.

Keywords

Nociception
A neurophysiologic term which denotes activity in the nerve pathways

Pain and **nociception** can be considered two different terms. Pain is a subjective experience that accompanies nociception, but can also arise without any physiological stimuli, such as due to an emotional response. Pain can be present in the absence of tissue damage or inflammation, for instance a person can experience a headache but not necessarily have a nociceptive stimulus or tissue damage.

Background to defining pain

In order to contextualise and interpret pain definitions, a brief overview of the process of the pain experience is provided here. Chapter 4 discusses this further and provides a detailed account of the physiology of pain.

Pain is part of the functioning of the nervous system – both central and peripheral. At the peripheral nervous level, free nerve endings called nociceptors send pain messages via a series of nerve fibres or neurones (this is called nociception) to the central nervous system entering it at the spinal cord. **Synapses** occur at the dorsal horn of the grey matter of the spinal cord which comprises layers or laminae of cells.

Keywords

Synapse
Nerve impulses are conducted from one neuron to another across a synapse – a junction between two neurons

The signal then reaches the sensory cortex of the brain via the ascending pain pathways where it can be distributed to other centres in the brain for analysis and the attribution of meaning, linking it with emotions and motor activity. Pain is a key component of the body's defence system. It is part of the rapid warning system. For example, if a person accidentally touches a hot barbeque grill and burns his/her hand the nerve endings in the skin will send messages to the brain via stimulation of the nerve endings (nociceptors). These messages are relayed, via the dorsal horn of the spinal cord and the ascending pain pathways, to the brain. The response may be emotional such as crying or motor such as moving the affected area, i.e. removing the hand from the grill. The individual therefore experiences the pain in a subjective manner. The extent of the pain is directly related to the size and depth of the burn and the individual's personal emotional response to the pain.

Reflective activity

Pain is a common experience yet defining it is elusive.
Consider this statement; can you define pain in a way that will allow you to manage your patients' pain?
What are the pros and cons of this statement?

Definitions of pain

Pain refers to a category of complex experiences, not to a specific sensation that varies only along a single-intensity dimension. There are numerous aspects to pain. Pain can be described as a private and internal sensation that cannot be directly observed or measured. A person's experience of pain is influenced by past experience, family attitudes, culture, meaning of the situation, attention, anxiety, suggestion and other factors unique to the individual. Pain appears to have three main dimensions: **sensory**, **emotional** and **intensity**. So, it is understandable that people suffering a similar pain experience that pain differently.

Keywords

Pain dimensions:

Sensory
The nature of the pain, e.g. ache, sharp, burning

Emotional
Pain feelings, e.g. frustration, anger, fed up, depression, annoyance

Intensity
Strength or severity of the pain, e.g. on a scale of 1–10 with 1 being no pain and 10 severe pain

One of the more common definitions of pain that has emerged is McCaffrey's: 'Pain is whatever the experiencing patient says it is, existing whenever he says it does'. This 1972 definition is a 'catch all' that could describe pain in any setting of any severity. The difficulty with this definition is that it is almost an approach or philosophy to pain management. It does, however, provide a strong foundation for the need for individual pain assessment and management. A weakness is that McCaffrey's 1972 definition relies on the person being able to describe his/her own pain. Many people are unable to or have difficulty describing their own pain for example patients receiving assisted respiratory ventilation or confused patients.

Another well-known definition is supplied by the International Association for the Study of Pain (1994):

> Pain is an unpleasant sensory and emotional experience associated with actual or potential tissue damage, or described in terms of such damage.
>
> Notes accompanying definition:
>
> Note 1: The inability to communicate verbally does not negate the possibility that an individual is experiencing pain and is in need of appropriate pain-relieving treatment.
>
> Note 2: Pain is always subjective. Each individual learns the application of the word through experiences related to injury in early life. Biologists recognise that those stimuli which cause pain are liable to damage tissue. Accordingly, pain is the experience we associate with actual or potential tissue damage. It is unquestionably a sensation in a part of the body, but it is also always unpleasant, and therefore is also an emotional experience.
>
> Experiences which resemble pain but are not unpleasant e.g. pricking should not be called pain. Unpleasant abnormal

> experiences (dysesthesias) may also be pain but not necessarily so because, subjectively, they may not have the usual qualities of pain.
>
> Many people report pain in the absence of tissue damage or any likely pathological cause and usually this happens for psychological reasons. There is usually no way to distinguish this experience from that due to tissue damage. If a person regards their experience as pain caused by tissue damage, it should be accepted as pain. This definition avoids tying pain to a physiological stimulus. Activity induced in the nociceptor and nociceptive pathways by a noxious stimulus is not pain, which is always a psychologic state, even though we may well appreciate that pain most often has a proximate physical cause.

This definition given by the International Association for the Study of Pain, although lengthy, is quite helpful in teasing out the complex nature of pain. It attempts to draw together the many facets of the pain experience, and it encompasses both the nociception and the emotional aspects of pain.

Crombie *et al.* (1999) provide the following definition:

> Pain is an unpleasant sensory and emotional experience associated with actual or potential tissue damage or described in terms of such damage. Acute pain is associated with acute injury or disease. Chronic pain is defined as pain that has persisted for longer than three months or past the expected time of healing following injury or disease. Patients with cancer may suffer from both acute and chronic pain. Epidemiological studies have revealed widespread unrelieved pain throughout society.

This definition is quite practical in that it describes the pain sensation and also gives a sense of the meaning of the much utilised categories of pain: acute and chronic.

Dame Cicely Saunders is responsible for coining the term 'total pain' which captures not only the physical but also the psychological, social and spiritual components of suffering, as developed and illustrated by Robert Twycross (2003). Embracing the concept of total pain helps to remind health care professionals that the personal experience of pain is far deeper than just the physical pain.

The true sense of what pain is begins to emerge as we read the variety of available definitions. Each definition, on its own, adds to an understanding of the phenomenon of pain. It is clear that pain is not a simple phenomenon.

Over to you

At your next journal club, clinical supervisory session, or in-service session review your management of the last patient you dealt with who presented with pain. Present to your colleagues how you assessed that pain, how you managed it and how you communicated this to the patient and others involved in the care. Detail two of the definitions of pain outlined above and discuss whether they supported your decision making.

Classification of pain

The definitions of pain previously discussed give a sense of the broadness of the term pain which can be used to encompass many facets of the same phenomena. Pain is classified in a number of ways according to the type of pain, source of pain, speed of transmission of nerve signals or associated problems/pains. These classifications are detailed further in Chapter 4. Figure 1.1 illustrates these classifications.

Category of pain

Pain is described as having two main categories: acute and chronic.

Acute pain subsides as healing takes place, i.e. it has a predictable end and it is of brief duration, usually less than six months. Acute pain often means sudden severe pain. An example would be postoperative pain felt after surgery.

Chronic pain is prolonged. It is pain that persists beyond the usual healing phase of the disease process. An example is low back pain that persists beyond 3–6 months.

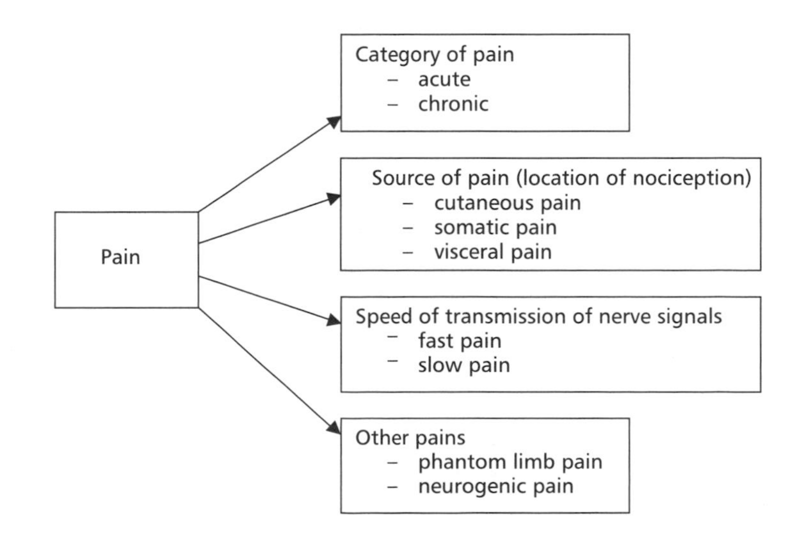

Figure 1.1 Classification of pain

Source of pain

The source or site of pain provides another classification of pain. The source of the pain influences a number of the reactions of the body to the pain.

Cutaneous pain is produced by stimulation of the pain receptors in the skin. It can be accurately localised due to the large number of receptors in the skin. An example would be the pain from a first-degree burn injury.

Somatic pain is produced by stimulation of pain receptors in the deep structures, i.e. muscles, bones, joints, tendons, ligaments. Unlike cutaneous pain, somatic pain is dull, intense and prolonged. An example could be the pain caused by an ankle fracture.

Visceral pain is that produced by stimulation of pain receptors in the viscera. Visceral nociceptors are located within body organs and internal cavities. It is poorly localised and often radiates to other sites. An example of this type of pain is myocardial ischaemia which is often felt in the left upper arm or shoulder.

Speed of transmission of nerve signals

Two types of nerve transmit pain signals from the peripheral nervous system to the central nervous system: A-delta and C fibres. See Figure 1.2.

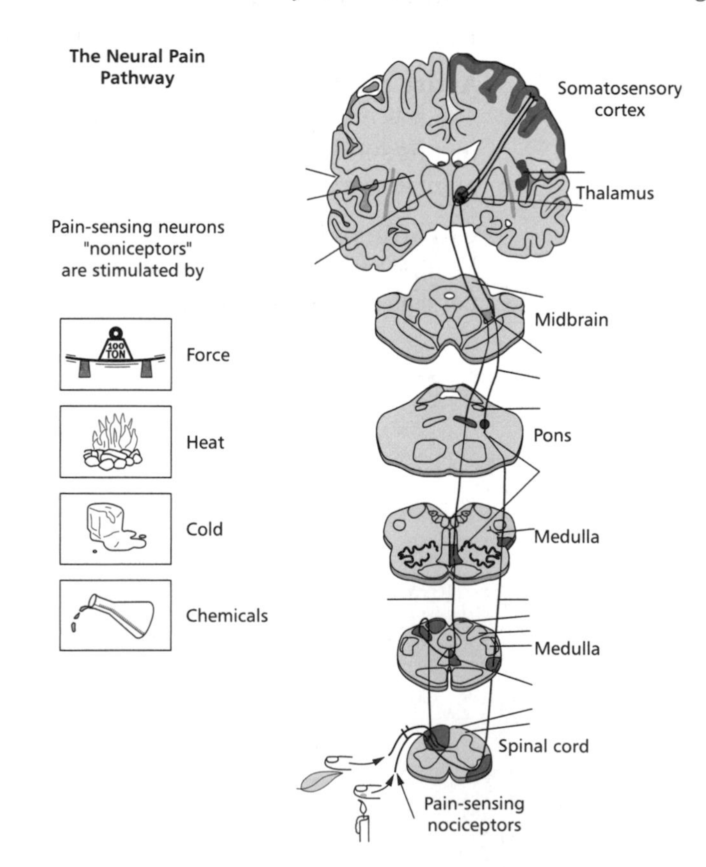

Figure 1.2 Line diagram of pain pathways

Fast pain is transmitted by A-delta fibres. It occurs very rapidly, usually within 0.1 seconds after the stimulus is applied, and is not felt in the deeper tissues of the body. This type of pain is also known as sharp, fast and pricking pain. An example is pain felt from a needle puncture or a knife cut into the skin.

Slow pain is transmitted by C fibres. C fibres take slightly longer than A-delta fibres to transmit pain and terminate over a wide area of the brain stem and thalamus. These slow pain signals are not relayed to the somatosensory cortex and are difficult to localise. An example of this type of pain is a toothache.

Other pains

There are other pains or associated problems that do not fit neatly into the above classifications yet are important to understand. Some examples are given here.

Phantom limb pain is the sensation of pain from an absent or inert limb – an experience sometimes discussed by amputees or quadriplegics.

Neuropathic pain or neuralgia can occur as a result of injury or disease to the nerve tissue itself. An example this might be trigeminal neuralgia.

Mr Jones is a 69-year-old with peripheral vascular disease. He returns from theatre with a right below-knee amputation. During the night he complains of a tingling sensation and pain in his right toe.

Consider how you would explain to Mr Jones why he has such a pain.

Pain management

Pain management is an overall term used to describe the care provided for those in pain. Pain management encompasses pain assessment, pain treatment, pain interventions, pain therapies, pain audit, staff and patient education and any other activities involved in providing care and services for those with pain. Pain management approaches are tailored to the clinical situations involved. Aspects of pain management are discussed throughout this book, with pain audit detailed in Chapter 2 and pain interventions examined in Chapter 7.

Over to you

How would you define pain management for your clinical setting?

Keywords

Intensity
The strength of the pain, e.g. how severe the pain is

Quality
The character of the pain, e.g. throbbing, aching

Location
The site of the pain, e.g. right upper arm

Pain assessment

Pain assessment can be defined as the determination of the **intensity**, **quality**, **location** and duration of pain. In relation to pain assessment a number of terms are used to describe the methods and tools. The following is a brief overview of the terms. Chapter 6 provides more detail on these.

Pain assessment tools tend to be either single or multidimensional (see Figures 1.3 and 1.4).

VRS: comprises four to five words descriptive of pain, of which the patient picks the word which best describes their pain. Each word has a score and the patient's intensity score is the number associated with the word they choose as most descriptive of their pain level. For example, words like mild, moderate or severe pain could be used with each word getting a score.

VAS: consists of a straight line, the ends of which are defined in terms of the extreme limits of the pain experience. The number corresponding to the patient's mark on the line is their pain intensity score.

NRS: this is a variation of the VAS and consists of asking the patient to rate their pain on a numerical scale 0–10. The number on the scale that the patient chooses is their pain intensity score.

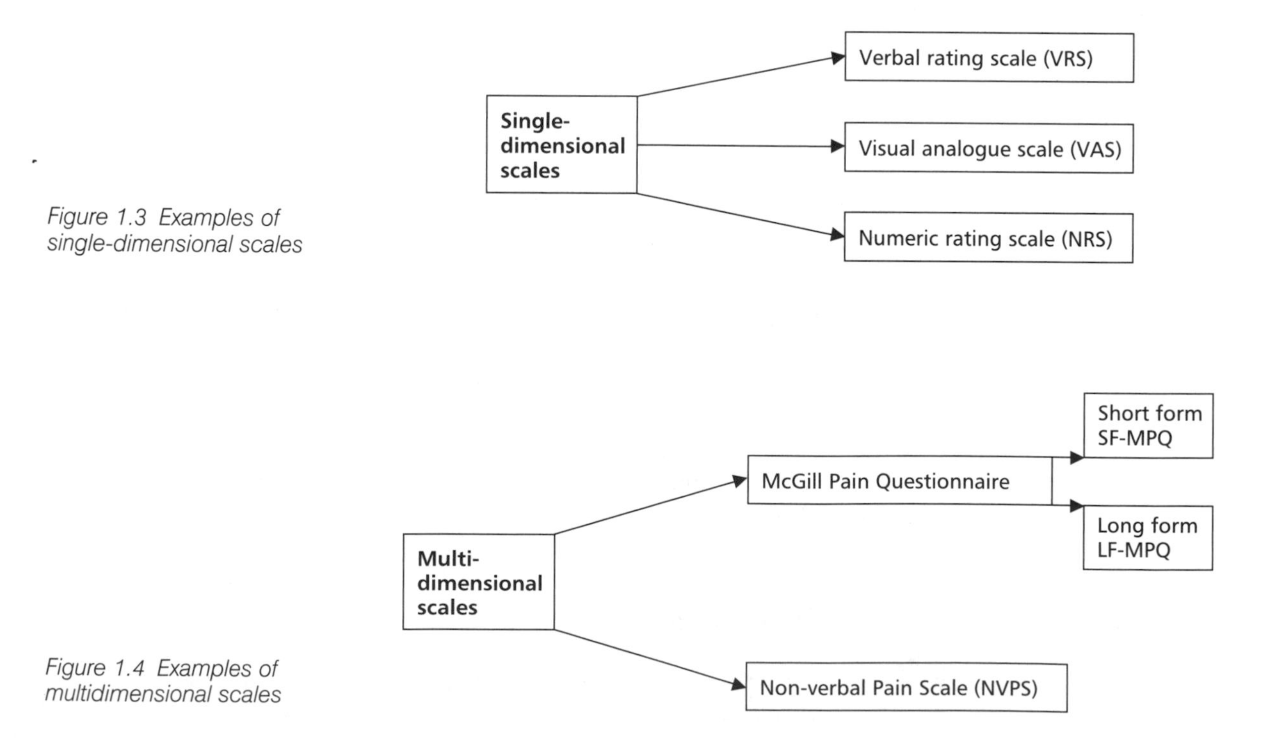

Figure 1.3 Examples of single-dimensional scales

Figure 1.4 Examples of multidimensional scales

MPQ: The McGill Pain Questionnaire (MPQ) both long (LF-MPQ) and short (SF-MPQ) forms are multidimensional scales involving subscales, which represent different aspects of pain.

NVPS: attempts to capture all aspects of pain multidimensional scales for those patients who cannot verbalise their pain experience.

Pain services

The breadth of care required to manage pain has provided numerous challenges to health care professionals. In the early 1980s, it began to emerge that the need to have dedicated staff to support pain management was critical if progress was to be made in the everyday management of pain. It was evident that pain could be targeted better by recognising that particular areas of pain practice were becoming too specialised for generalist practitioners to manage effectively. A number of terms began to emerge under the broad term 'pain service' such as pain teams, acute pain services, pain clinics, chronic pain services, pain management programmes and palliative care services. A **pain service** is an umbrella term to encompass the various services that have emerged such as pain teams, pain clinics etc.

Keywords

Pain service
A dedicated facility for specific classifications of pain that provides diagnosis, treatments and interventions as appropriate. The facility is staffed by a multidisciplinary team with committed time and resources. The pain service has a philosophy of continuous quality improvement and education and empowerment of other health care staff

A number of desirable characteristics emerge which help in defining pain services.

It should be noted that various expert organisations have standards and detailed requirements for such services. Specialised services in acute pain, chronic pain and palliative care have emerged. Even within the specific area of pain care, sub-specialties have developed, e.g. neuropathic pain, migraine. The critical issue was the need to focus specialised knowledge and skills towards this important aspect of patient care, and the speciality in question. These developments allow for dedicated time and resources from health professionals to support and develop more effective processes for pain management. Chapter 3 discusses the evidence base for acute pain services and the key components of chronic pain services.

Palliative care services

The World Health Organization describes palliative care as an approach that improves the quality of life of patients and their families facing the problems associated with life-threatening illness, through the prevention and relief of suffering by means of early identification and impeccable assessment and treatment of pain and other problems, physical, psychosocial and spiritual (Sepulveda *et al.*, 2002). Implicit in palliative care is the provision of relief from pain and other distressing symptoms.

Everyone facing a life-threatening illness will need some degree of supportive care in addition to treatment for their condition. The National Institute for Clinical Excellence (NICE, 2004: 18) describes supportive care for people with cancer as follows: 'Supportive care helps the patient and their family to cope with their condition and treatment of it – from pre-diagnosis, through the process of diagnosis and treatment, to cure, continuing illness or death and into bereavement. It helps the patient to maximise the benefits of treatment and to live as well as possible with the effects of the disease. It is given equal priority alongside diagnosis and treatment.'

Supportive care should be fully integrated with diagnosis and treatment. The elements it encompasses are shown in Figure 1.5.

Chapter 11 provides in-depth material in relation to palliative care.

A number of qualities emerge as defining characteristics of pain services as shown in Figure 1.6.

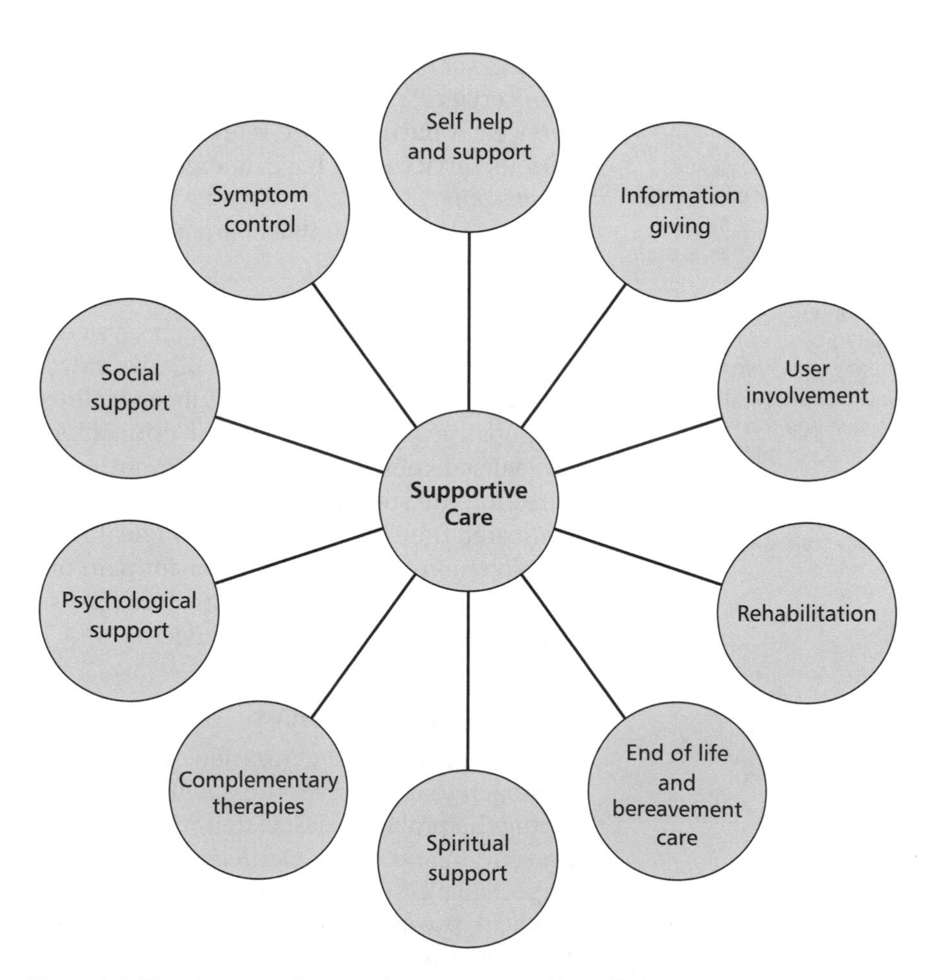

Figure 1.5 The elements of supportive care required in palliative services

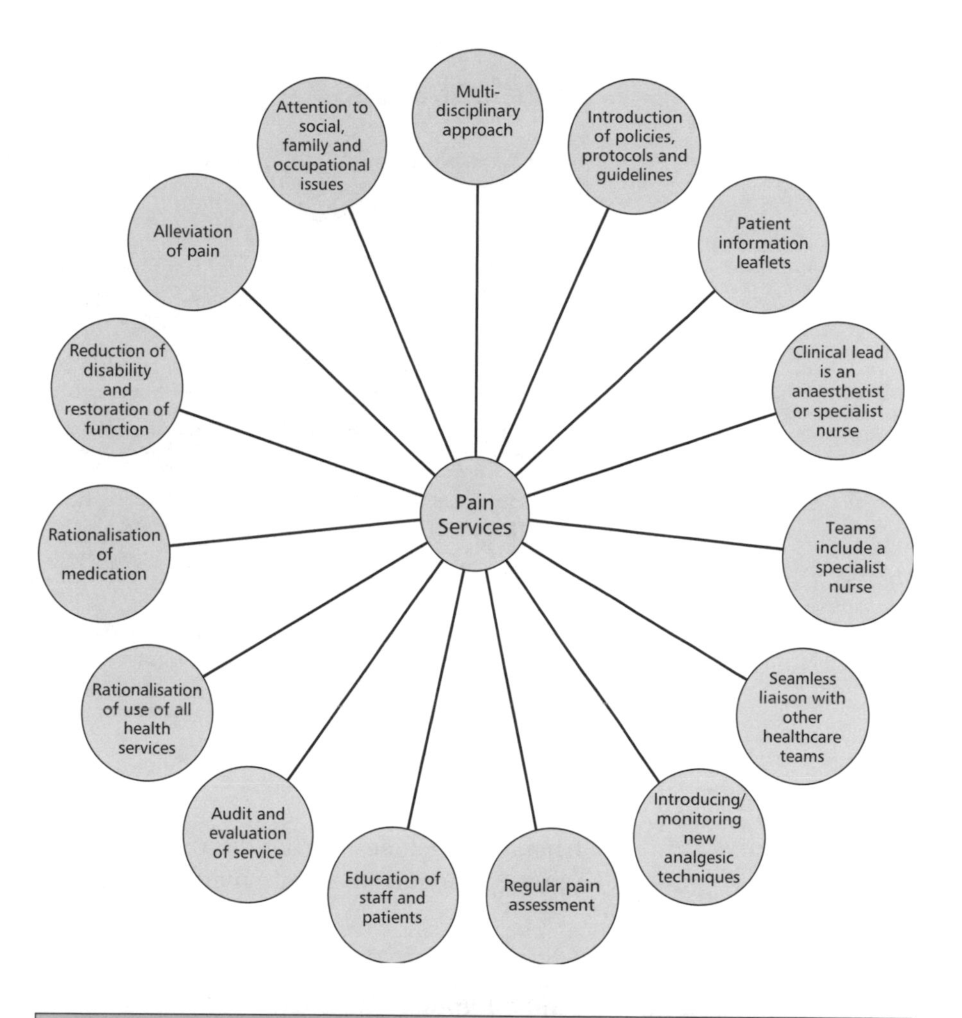

Figure 1.6 Essential requirements of pain services

Evidence base

Go to this website: www.iasp-pain.org and review International Association for the Study of Pain: desirable characteristics for pain treatment facilities (1990).

When you are reviewing the characteristics of pain treatment facilities consider how you treat pain in your clinical setting. Is there anything that could enhance how this pain is managed?

Evidence-based resources for pain and its management

The management of pain and its associated definitions are the subject of ongoing intense research and debate. As such evidence changes rapidly, all health professionals should continuously update and review developments.

Evidence base

The following are some key resource websites and journals that you should look at.

Key resources

- The Royal College of Anaesthetists, 48/49 Russell Square, London WC1B 4JY. www.rcoa.ac.uk
- The Pain Society, 21 Portland Place, London W1B 1PY. www.britishpainsociety.org
- International Association for the Study of Pain (IASP), 909 NE 43rd St., Suite 306, Seattle, WA 98105–6020, USA. www.iasp-pain.org
- Bandolier Internet site which accompanies *Bandolier's Little Book of Pain* (Moore *et al.*, 2004). www.ebandolier.com
- *Journal of Pain Medicine and Palliative Care*
- *Pain* (the journal)
- *Journal of Pain and Symptom Management*
- *Clinical Journal of Pain*

Chapter 2 includes further discussion in relation to evidence-based practice and its role in pain management.

Key points

Top tips

- Pain is a complex phenomenon and, therefore, difficult to define
- Pain is classified in a number of ways: according to type of pain, source of pain, speed of transmission of nerve signals or other pains
- Specialised services in acute pain, chronic pain and palliative care have emerged
- Pain management encompasses pain assessment, pain treatment, pain interventions, pain therapies, pain audit, staff and patient education and any other activities involved in providing care and services for those with pain.
- A pain service is a dedicated facility for a specific classification of pain which provides diagnosis, treatment and intervention as appropriate. The facility is staffed by a multidisciplinary team with committed time and resources. The pain service has a philosophy of continuous quality improvement and education and empowerment of other health care staff
- Evidence for pain changes rapidly – health professionals need access to quality evidence-based journals and websites

Pain affecting quality of life

Liz is a housewife who has two children, aged 11 and 16. The children are very sporty and the 11-year-old is an excellent hockey player. They usually play in the local grounds but also have to travel to neighbouring counties for matches. Her partner is a builder and labourer who travels with a building contractor all over the country. He is generally only home at weekends. Liz is very active in the local community and has been very involved in running a number of fundraising events for the school. She used to spend two mornings a week on voluntary duty in the local older persons' home. Liz has always been very popular in the community and, with a good sense of fun, has been a welcome addition to local events. Two years ago a car travelling at high speed ran into the back of Liz's car. She suffered a severe whiplash neck injury and had to have intensive physiotherapy for more than a year. She was on nonsteroidal anti-inflammatory medications for 12 months.

After the accident, Liz had to give up her voluntary work. She is unable to be as involved in community events as she would have liked. Her children are getting older and becoming more independent. Liz's neck has improved a lot although she continues to have chronic aches and pains in her neck. Liz's doctor has advised her not to go back to work in the older persons' home as this could exacerbate her neck pain. Liz has started to become less outgoing and has begun to avoid the local social events. If she is feeling stressed the neck pain becomes more acute and she has started to get regular headaches. Her doctor can't find any cause for the headaches. Liz's quality of life is much poorer than it was two years ago.

- What are the main features of Liz's pain?
- How would you define her pain?
- What is the source of Liz's pain?

Rapid recap

1. What are main dimensions of pain?
2. What is the main message for health care professionals coming from the more common definitions of pain?
3. Which parts of the nervous system are concerned with the transmission of pain?
4. List five desirable characteristics of a pain care service.

References

Crombie I.K., Croft P.R., Lindon S.J., Resche L. and Von Korff M. (eds.). (1999) *Epidemiology of Pain*. IASP Press, Seattle.

IASP Task Force on Taxonomy (1994) *Classification of Chronic Pain*, 2nd edn, (eds. Merskey H. and Bogduk N.) IASP Press, Seattle.

McCaffrey M. (1972) Pain in the context of nursing care. In: *Nursing Management of the Patient with Pain.* J.B. Lippincott Company, Toronto p. 8.

The National Institute for Clinical Excellence (2004) *Guidance on Cancer Services. Improving Supportive and Palliative Care for Adults with Cancer: The Manual* (available at www.nice.org.uk/pdf/csgspmanual.pdf).

Sepulveda C., Marlin A., Yoshida T. and Ulrich A. (2002) Palliative Care: The World Health Organization's Global Perspective. *Journal of Pain Symptom Management.* **24**(2): 91–96.

Twycross, R. (2003) *Introducing palliative care*, 4th edn. Radcliffe Medical Press, Oxon.

2 Health and social policy issues in pain management

Learning outcomes

By the end of this chapter you should be able to:

★ Understand why pain should form part of health and social policy
★ Detail the prevalence of pain
★ Discuss integrated approaches to pain management
★ Explain how evidence-based practice and clinical auditing are involved in effective pain management.

Introduction

This chapter will help you to consider the broader context of pain. By taking into account the implications of national and international collaboration on standards and guidelines, pain as a public health issue is addressed. Evidence-based approaches and processes are provided as the foundation for all health professionals to ensure best practice. The concepts of quality and integrated approaches to pain management are also considered.

Pain as a public health issue

Pain has become a world public health issue. Both acute and chronic pain are often poorly managed, for many reasons, including culture, attitude, education and logistics. The gap between an increasingly sophisticated knowledge of pain and its treatment and the effective application of knowledge is large and widening (Brennan and Cousins 2004). Under-treated severe pain is associated with a number of harmful effects including decreased lung volumes, myocardial ischaemia, decreased gastric and bowel motility and anxiety (Macintyre and Ready, 2002). The economic and social cost of pain is high.

The economic burden of pain may be evaluated by considering the following:

- direct costs (medical expenditure, such as the cost of prevention, detection, treatment, rehabilitation, long-term care and ongoing, medical and private expenditure)
- indirect costs (lost work output attributable to a reduced capacity for activity, lost productivity, lost earnings, lost opportunities for family members, lost earnings of family members and lost tax revenue)
- intangible costs (psychosocial burden resulting in reduced quality of life, such as job stress, economic stress, family stress and suffering).

Pain management is of proven benefit in improving the quality of life of patients. When untreated, pain can cause helplessness, depression, isolation, family breakdown and inappropriate disability (CSAG, 2000). For example, one study of chronic pain demonstrated that lower levels of quality of life were associated with loss of efficiency in the workplace and eventual absenteeism (Lamers *et al.*, 2005). Annual costs of chronic pain in Canada are in excess of $10 billion based on direct care costs. This does not include the less quantifiable costs of quality of life and productivity (Chronic Pain Association of Canada, 2004). Pain is clearly an issue that should form part of health and social policy.

Enhancing quality of life

Mary is 42 and has had rheumatoid arthritis for 10 years. She suffers constant pain and has considerable reduced mobility. She has had to reduce her working hours in the local library to 15 hours per week and attends her general practitioner monthly.

Mary is admitted to the local hospital emergency department with acute pain in her right knee.

- How has Mary's quality of life been affected?
- As a health professional, what could you do to enhance Mary's quality of life?

International collaboration is an expanding feature of public health. Increasingly, it is recognised that public health issues transcend national boundaries, and The International Association for the Study of Pain and the World Health Organization (WHO) have led the way on raising global awareness of poorly relieved acute pain as a major factor contributing to the delayed recovery of health and function after surgery and trauma. They have also highlighted the immense burden of chronic pain unrelated to cancer. The WHO's influence on governments worldwide and on national health care programmes and policies is seen to lay the ground work for better management of acute and chronic pain and improved health-related quality of life in a number of major diseases and conditions that involve pain (Bond and Breivik, 2004).

The WHO has had immense impact on health care policies and in changing the culture of pain relief for cancer pain and symptom control (Bond and Breivik, 2004). The WHO analgesic ladder was a major landmark in the management of pain (see Chapter 7 for further information on the analgesic ladder). It recommended cheap, widely available and effective analgesics. It was a strong simple message that changed the culture of cancer pain management. It can be seen from the World Cancer Report that worldwide medical use of morphine is increasing significantly (Stewart and Kleihues, 2003).

The core purposes of public health centre around:

- health promotion including the wider public health
- improving quality of clinical standards and
- protection of public health and management of risk (Hunter, 2003).

The common understanding of public health and the implications of pain and its management place pain as a broad public health issue worldwide. This understanding increases the imperative for health care providers to review current processes and management techniques for pain.

Clinical Caseload

You need to be constantly reviewing the way you undertake pain care. For instance, we have recently changed our way of managing the administration of oral morphine solution. Previously the Trust had treated all concentrations of Oramorph as controlled drugs even though there was no legal requirements to treat low concentrations like 10mg in 5ml in this way. Information obtained from the Department of Health, the Royal Pharmaceutical Society and the Nursing and Midwifery Council helped the pain care team to convince the Trust that deregulating Oramorph would be acceptable. We worked with the pharmacists, doctors and nurses to get policies in place that were acceptable to everyone. The acute pain team introduced a local education programme delivered through individual face-to-face sessions (doctor to doctor, nurse to nurse) rather than through seminars. This ensured that the process had minimal impact on clinical commitments, and that the tutorials were designed to reflect the likely concerns and anxieties of professionals. This now means that pain relief when required is given more speedily as there is no delay due to the controlled drug procedures, and what was great was that this change in pain care management has been reflected in improved patient satisfaction with their pain care treatment.

The right to pain relief and the consideration that poor management of pain constitutes negligence have led to a vibrant debate among key stakeholders. Indeed Haugh (2005) describes case studies where patients and families sue health care providers for inadequate pain management. The International Association for the Study of Pain suggests that there is no single 'right' to pain relief; instead there is a constellation of 'rights', each with a variable degree of legal enforceability. Reform in current pain management will depend on a combination of approaches including education for health care undergraduates, adoption of universal pain management standards by professional bodies, the promotion of legislative reform, liberalisation of national policies on opioid availability, reduction of cost of analgesics, domestic pain treatment forums and continued lobbying of the major stakeholders (Brennan and Cousins, 2004).

Prevalence of pain

It will be obvious, anecdotally, how prevalent pain is in most health care and community settings. This prevalence is not always measured; however, there are some studies which detail the statistics in various care

Keywords

Prevalence of pain
The proportion of individuals in a population who have pain at a specific instant (Prevalence = number of individuals with pain/total population at a given point in time)

Incidence of pain
The proportion of individuals in a population who get pain during a specified period of time (Incidence = number of new individuals with pain/total population at risk of pain)

settings. Comparisons of **prevalence** and **incidence** can be difficult due to differing definitions, methods of measurement and reporting.

Over to you

Consider the patient profile within your current clinical setting.Is pain prevalence audited? If not complete this simple quick task:

- Estimate the prevalence of pain in your clinical area.
- Take a simple numeric rating scale and do a point prevalence for 10 patients (this will take about five minutes).

Are the pain levels what you expected? How do they compare to national and international statistics?

Pain in hospitals

Various surveys have measured the extent and prevalence of pain in hospital settings. In 1994 a survey of 5150 hospital patients found that 61% suffered pain, 87% of whom had severe or moderate pain (Bruster *et al.*, 1994). In another survey 80–89% of patients surveyed were experiencing pain with 45.8–57.2% of patients reporting moderate to severe pain, yet 90% of patients were satisfied or very satisfied with the pain management provided. These were hospital patients from both medical and surgical areas (Comley and DeMeyer, 2001).

Postoperative pain studies demonstrate that patients still suffer moderate to severe pain after their surgery (Weis *et al.*, 1983; Donovan *et al.*, 1987; Melzack *et al,*. 1987; Kuhn *et al.*, 1990; Owen *et al.*, 1990; Wilder-Smith and Schuler L., 1992; Watt-Watson *et al.*, 2001; Mac Lellan, 2004). The issue of persistent acute pain following tissue damage after surgery or chronicity of acute pain is relevant. For example, Bay-Nielson (2001) highlight that, one year after inguinal hernia repair, pain is common (28.7%) and is associated with functional impairment in more than half of those with pain.

Pain is cited as the number one reason for attending emergency departments. Cordell *et al.* (2002) on reviewing patients' charts over one week (n = 1665) found a prevalence of 52.2% of pain as the chief complaint for visits to emergency departments (United States). Chest pain accounted for 2–4% of all new attendances at emergency departments per year in the UK (Herren and Mackway-Jones, 2001).

Palliative care – prevalence of pain

Pain in palliative care includes both cancer pain and non-malignant pain.

Cancer pain

More than 11 million people are diagnosed with cancer every year. It is estimated that there will be 16 million new cases every year by 2020. Cancer causes 7 million deaths every year – or 12.5% of deaths worldwide (World Health Organization, 2002).

Ross (2004) claims that, in the United States, cancer pain occurs in one-quarter of all patients at time of diagnosis, and two-thirds of patients receiving anti-cancer therapy report pain. As the stage of the disease advances the number of patients experiencing pain increases. Three quarters of patients hospitalised with advanced cancer report unrelieved pain. However, it is still important to note that cancer and pain are not synonymous (Twycross, 2003).

Non-malignant pain

The prevalence of pain in non-malignant conditions is often underestimated. Approximately 60% of patients with advanced diseases suffer troublesome pain and this figure is similar for cardiac disease, neurological disorders and AIDS. Unfortunately, the chances of patients being treated for pain due to non-malignant conditions is reduced (Regnard and Kindlen, 2004).

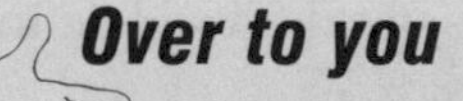

Over to you

Brian is 70 and he has been retired for 10 years. He has suffered from angina over the past five years and if he walks for more than 25 minutes he gets tightness in his chest and a dull pain. He finds this very frightening and has started to stay in the house rather than do his usual walk to the shops in the morning. His general practitioner in consultation with the local hospital cardiologist agrees that his heart disease does not warrant major intervention and is ideally controlled with medication.

Do you think that there are many others in the community like Brian with pain and reduced qualify of life? Go to the internet/Medline or CINAHL and search for 'prevalence of angina in the community'.

Chronic pain prevalence

Chronic pain tends to be considered as pain without apparent biological value that has persisted beyond the normal tissue healing time (usually taken to be three months).

Common conditions that cause chronic pain include: low back pain, headache, arthritis and peripheral neuropathy. Chronic pain has a significant effect on the lives of sufferers and their families, as it will effect patients' mood, social relationships and quality of life (Veillette *et al.*, 2005).

Ferrell (1995) reports that 62% of nursing home residents have pain. A study by the American Geriatrics Society (AGS) Panel found that 25–50% of those living in the community suffer significant pain problems (AGS Panel on Chronic Pain in Older Persons, 1998). Elliot *et al.* (1999) in a UK survey of 29 general practices (5036 patients, 72% response rate) report that 50.4% of patients self report chronic pain. Back pain and arthritis were the most common complaints. Unsurprisingly, with ongoing pain, increased pain intensity is associated with decreased quality of life (Laursen *et al.*, 2005).

Palmer *et al.*, (2000) report on two prevalence surveys of back pain at an interval of 10 years. Measurements included low back pain and low back pain making it impossible to put on hosiery. Over a 10-year interval the one-year prevalence of back pain, standarised for the age and sex distribution, rose from 36.4% to 49.1%. There was no increase in the prevalence of symptoms sufficient to prevent people putting on hosiery. A recent Canadian study (Veillette *et al.*, 2005) identified that between 18% and 29% of Canadians experience chronic pain.

The challenges of reviewing studies on measuring prevalence of chronic pain are numerous due to variations in population sampled, methods used to collect data and criteria used to define chronic pain. Standard definitions for chronic pain are not available. Harstall and Ospina (2003) call for a research agenda to conduct concurrent, prospective, epidemiological studies to estimate the chronic pain prevalence using clear, standardised definitions and well-validated and reliable collection tools.

Table 2.1 presents data from nine studies which are part of a systematic review by Harstall and Ospina (2003) highlighting the prevalence of pain and also some of the issues in measuring chronic pain.

Table 2.1 The prevalance of pain

Definition used	Number of studies	Population	% pooled prevalence
International Association for Study of Pain definition (chronic pain)	5 studies (1993–2001)	Adults	11% among adults
American College of Rheumatology definition (chronic widespread pain)	3 studies (1993–2000)	Adults	11.8% among adults (7.2% male and 14.7% female)
International Association for Study of Pain definition (chronic pain)	1 study (1997)	Elderly general population aged 65 years and older	50.2%

Recommendations and guidelines

A number of key agencies have emerged as leaders in providing resources, guidelines and standards in relation to pain management. The common theme emerging from these statements is that pain is a common and often under-treated phenomenon. Management of pain should become a key objective of health providers with an evidence-based, high-quality service provided. Multidisciplinary and multi-modal methods of pain management should be utilised. Key to this is robust undergraduate education on pain and continuing professional development for all health professionals.

Evidence base

The following timeline details some of the key position statements that have emerged over time.

Timeline

- **2003** The Royal College of Anaesthetists and The Pain Society, the British Chapter of the International Association for the Study of Pain, *Pain Management Services – Good Practice.*
- **2000** Services for Patients with Pain, *Clinical Standards Advisory Group (CSAG).* London, Department of Health.
- **2000** American Pain Society, *Pain assessment and treatment in the managed care environment. A position statement from the American Pain Society.*
- **1999** National Health and Medical Research Council, Australia. *Acute Pain Management: The Scientific Evidence.*
- **1995** American Pain Society. Quality Improvement Guidelines for the Treatment of Acute Pain and Cancer Pain (American Pain Society Quality of Care Committee). *JAMA.* **274**(23): 1874–80.
- **1992** Acute Pain Management Guideline Panel. *Acute Pain Management: Operative or Medical Procedures and Trauma. Clinical Practice Guideline.* AHCPR Pub. No. 92–0032. Rockville, MD: Agency for Health Care Policy and Research, Public Health Service, U.S. Department of Health and Human Services. Feb. 1992.
- **1991** International Associations for the Study of Pain, Task Force on Acute Pain. (Ready and Edwards, 1992): *Clinical Practice Guideline for Acute Pain Management.* IASP.
- **1990** Royal College of Surgeons and Anaesthetists, Commission on the Provision of Surgical Services, *Report on the Working Party on Pain after Surgery.*

In 2000, the clinical standards advisory group (CSAG) published a report on services for patients with pain. This group, which is now disbanded, was set up under statute as an independent source of expert advice to the UK health ministers and to the NHS on standards of clinical care for, and access to and availability of services to, NHS patients with acute and chronic pain. Key recommendations in relation to various types of adult pain are summarised in Table 2.2 below.

The common thread in the recommendations is effective communications, rapid access to appropriate services, dedicated staff and services and education for staff.

The Royal College of Anaesthetists and The Pain Society, the British Chapter of the International Association for the Study of Pain (2003) suggest that integration with primary care teams is essential for maximum patient benefit. They further state that pain management provides outstanding potential for fruitful working across the boundaries between primary and secondary care. They recommended that new and innovative models for supporting care should be supported and developed. For instance, Hewlett *et al.* (2000), in a randomised controlled trial with two years of follow up, describe a shared care system of hospital follow-up for patients with rheumatoid arthritis. Patient and general practitioner initiated shared care was compared to traditional rheumatologist initiation hospital care in patients with established rheumatoid arthritis. The study concluded that a shared system of hospital follow-up reduced both pain and use of health care resources.

Table 2.2 Key recommendations of the CSAG on the management of pain in adults (Source: CSAG, 2000)

Chronic pain	Postoperative and other acute pain	Accident and Emergency (A&E) departments
Objective of pain management services should include prevention of the development of chronicity and the disability that follows. GPs should have early access to appropriate investigations and specialist secondary care services when needed.	All patients undergoing potentially painful procedures should have access to the services of an acute pain team if necessary. The components of an acute pain team are described.	There should be greater emphasis on effective pain management in patients attending A&E departments. Actions to achieve this are described.

Community physiotherapist

Integrated care between the hospital and community means that the patient can get rapid access to care or treatments needed without delays and bureaucracy. For example, I had a young lad last week who was attending for a sports injury to the knee. After two treatments I decided he would benefit from a visit to the pain clinic in the local hospital. This was easily arranged though the shared care processes set up between my services, the entire community services and the local hospital. It involved a brief phone call to the nurse co-ordinating the clinic and an electronic appointment was sent for the next week. Such rapid access meant that the patient was able to continue my treatments and also benefit from a specialised service provided at the hospital.

The American Pain Society (1995) advised key processes for improving acute pain management which include:

- recognition and timely, prompt treatment
- making information about analgesics readily available
- developing policies for advanced analgesic techniques
- the goal of continuous improvement examining the process and outcomes of pain management.

Audit and quality of pain management practices

In order to improve the quality of care, ongoing monitoring of practice utilising clinical audit tools is required. Audit is concerned with the monitoring of current practice against standards. Clinical audit is at the heart of clinical governance. It is there to improve the quality of patient care and clinical practice.

Clinical audit of pain will establish the baseline performance for a clinical area/service. Once this is completed, the data can be reviewed and the process of identification of opportunities for improvement of practice can begin. Practice can be monitored over time. The pain service can be benchmarked with national and international standards. All of this supports proactive planned service-planning processes. This is a continuous cycle supporting continuous quality improvement (see Figure 2.1).

What should be included in the clinical audit?

Clinical audit is used to improve all aspects of patient care. Table 2.3 identifies examples of these areas for pain management.

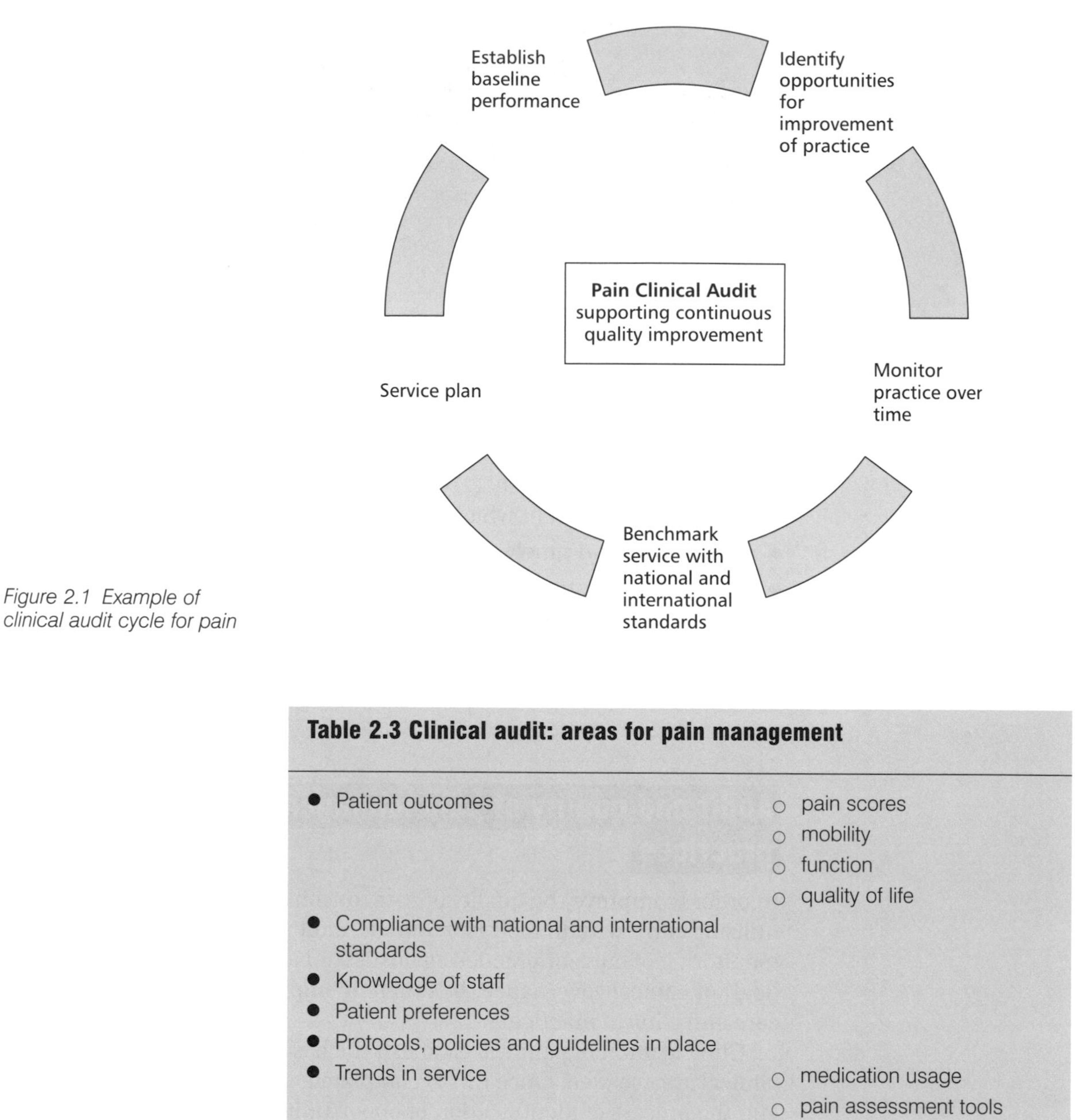

Figure 2.1 Example of clinical audit cycle for pain

Table 2.3 Clinical audit: areas for pain management

• Patient outcomes	○ pain scores ○ mobility ○ function ○ quality of life
• Compliance with national and international standards	
• Knowledge of staff	
• Patient preferences	
• Protocols, policies and guidelines in place	
• Trends in service	○ medication usage ○ pain assessment tools ○ quality of documentation

Keywords

Systematic review
A widely used method to synthesise the literature; it includes a comprehensive search, critical appraisal of the quality of selected studies and pooled analysis either qualitative or quantitative, of the results

Evidence-based approaches to pain management

Pain management, like any aspect of patient care, should be based on the best available evidence. Moore *et al.* (2004) are clear that good evidence is likely to come from robust **systematic reviews** of sound clinical trials. This means that health professionals must be able to access, understand and interpret research in order to provide evidence-based practice. As technology, skills and knowledge grow, nurses and health care professionals need to develop systems to ensure they are accessing, interpreting and utilising the best available evidence. Health care organisations need to consider multiple strategies to facilitate and promote evidence-based practice. Managerial support, facilitation and a culture that is receptive to change are essential (Gerrish, 2004).

Evidence base

Table 2.4 details current sources of research evidence.

Table 2.4 Sources of research evidence

Source	Research category	Quality checked – Pre-appraised	Sources
Evidence-based journals	Systematic reviews Randomised controlled trials Case control studies Cohort studies Qualitative research	Yes	Evidence-based nursing (http://ebn.bmjjournals.com/) Evidence-based medicine (http://ebm.bmjjournals.com/)
Cochrane database	Systematic reviews	Yes	www.cochrane.org/index0.htm
Medline CINAHL	– Research – Opinions – Discussions – Literature reviews	No	Check your local library

View the following websites for evidence on pain management

www.jr2.ox.ac.uk/bandolier/booth/painpag/index2.html
www.jr2.ox.ac.uk/bandolier/painres/MApain.html – systematic reviews with pain as an outcome
www.cochrane.org
www.jr2.ox.ac.uk/cochrane/

Go to one of the above websites and search for a systematic review on pain prevalence in the community.

Organisational reponsibility

Organisations have responsibility in a number of areas to support improvement of pain management practices. National and international standards and guidance form the platform for organisational responsibility (see Figure 2.2).

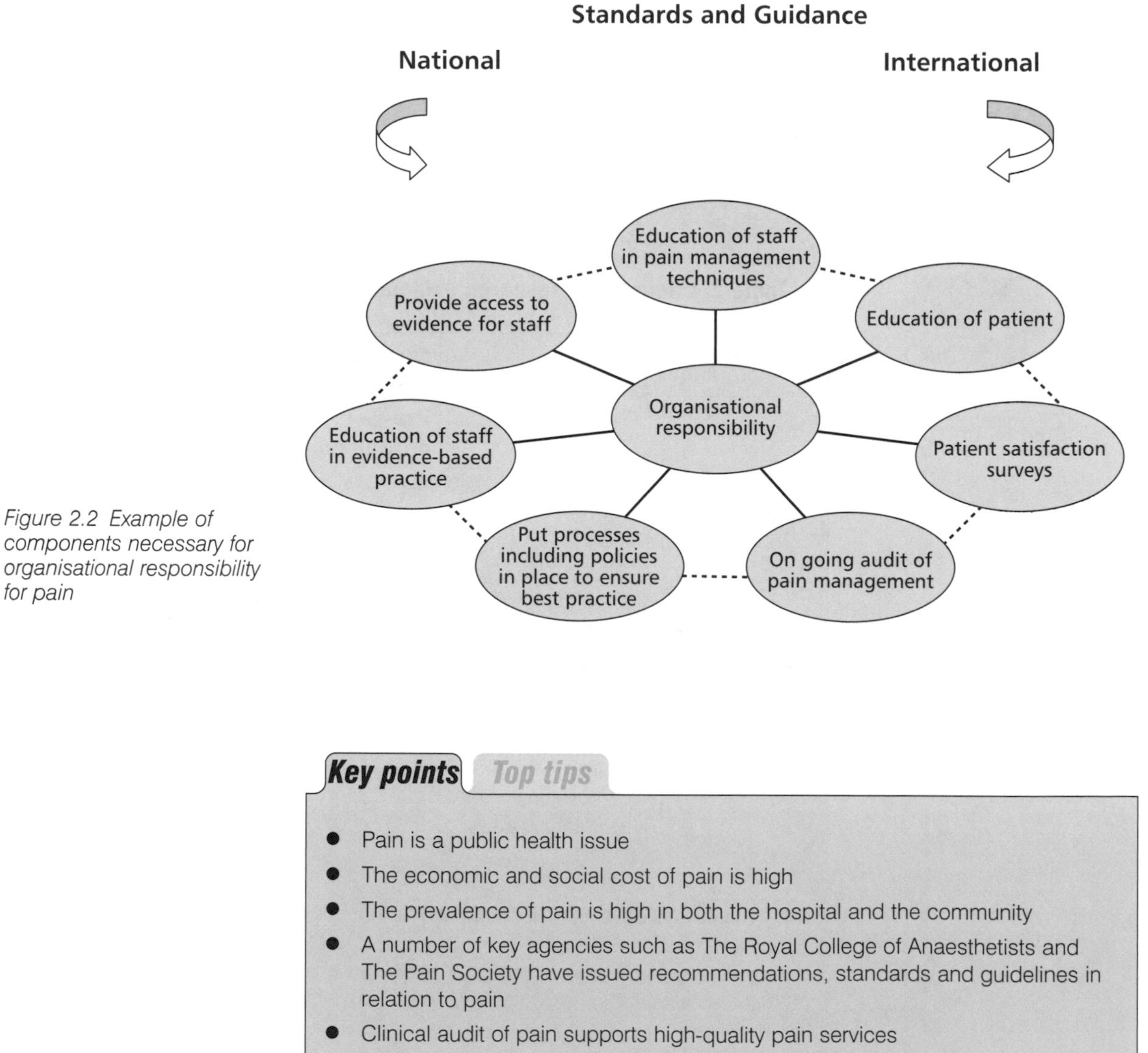

Figure 2.2 Example of components necessary for organisational responsibility for pain

Key points Top tips

- Pain is a public health issue
- The economic and social cost of pain is high
- The prevalence of pain is high in both the hospital and the community
- A number of key agencies such as The Royal College of Anaesthetists and The Pain Society have issued recommendations, standards and guidelines in relation to pain
- Clinical audit of pain supports high-quality pain services
- Evidence-based approaches should be used to manage pain

Pain as an indicator of quality of care

St Margaret's is a small general hospital with 200 beds. It has two surgical and four medical wards. There is a critical care area that manages acutely ill patients from both medicine and surgery. There is a minor injuries unit, but the hospital does not operate an accident and emergency department. The hospital performs routine surgery such as appendectomies and cholecystectomies. It has a day unit for endoscopies.

St Margaret's has just appointed a new clinical director and a new director of nursing. Both are keen to prove their worth and make St Margaret's a centre of excellence. They realise that, while the staff all work hard and provide excellent care, there has been little investment in new technologies, equipment or professional development of staff. They decide to do an audit of some of the main indicators of quality of care including pain in order to benchmark care.

A three-day audit of pain in the hospital is shown in Table 2.5.

The clinical director and director of nursing sit down to review their entire audit. Their eyes fall on the pain prevalence. It strikes them both that pain prevalence in the surgical ward is rather high.

- Why is pain a good indicator of the quality of patient care?
- What do you interpret from the table of pain prevalence?
- Why do you think there is more pain at night in the minor injuries unit than during the day?

Table 2.5 St Margaret's three-day audit of pain

Location	Mean pain prevalence (using visual analogue scale)
Medical wards	Daytime: 4.5 Night-time: 3.8
Surgical wards	Daytime: 7.6 Night-time: 7.8
Day unit	Daytime: 2.1
Minor injuries unit	Daytime: 4.5 Night-time: 5.3

Rapid recap

1. How do you define prevalence of pain?
2. How would you describe the economic burden of pain?
3. What are the sources of research evidence?

References

AGS Panel on Chronic Pain in Older Persons (1998) The management of chronic pain in older persons: AGS Panel on Chronic Pain in Older Persons. American Geriatrics Society. *Journal of American Geriatric Society*, **46**: 635–512.

American Pain Society Quality of Care Committee (1995) Quality improvement guidelines for the treatment of acute and cancer pain. *JAMA*, **274**(23): 1871–1880.

Bay-Nielson M., Perkins F.M. and Kehlet H. (2001) Pain and functional impairment 1 year after inguinal herniorrhaphy: A Nationwide Questionnaire Study. *Annals of Surgery*, **233**(1): 1–7.

Bond M. and Breivik H. (2004) Why pain control matters in a world of killer diseases. *Pain Clinical Updates*, XII(4).

Brennan F. and Cousins M.J. (2004) Pain relief as a human right. *Pain Clinical Updates*, XII(5).

Bruster S., Jarman B., Bosanquet N., Weston D., Erens R. and Delbanco T.L. (1994) National survey of hospital patients. *BMJ*. **309**(6968): 1542–1546.

Chronic Pain Association of Canada (2004) Pain Facts. www.chronicpaincanada.com

Clinical Standards Advisory Group(CSAG) (2000) Services for Patients with Pain. London, Department of Health.

Comley A.L. and DeMeyer E. (2001) Assessing patient satisfaction with pain management through a continuous quality improvement effort. *Journal of Pain and Symptom Management*, **21**(1): 27–40.

Cordell W.H., Keene K.K., Giles B.K., Jones J.B., Jones J.H. and Brizendine E.J. (2002) The high prevalence of pain in emergency medical care. *American Journal of Emergency Medicine*, **20**(3): 165–169.

Donovan M, Dillon P. and McGuire L. (1987) Incidence and characteristics of pain in a sample of medical-surgical inpatients. *Pain*, **30**: 69–78.

Elliot A.M., Smith B.H., Penny K.I., Cairns Smith W. and Chambers W.A. (1999) Epidemiology of chronic pain in the community. *Lancet*, **354**: 1248–1252.

Ferrell B.A. (1995) Pain evaluation and management in a nursing home. *Annals Internal Medicine*, **123**: 681–687.

Gerrish K. (2004) Promoting evidence-based practice: an organizational approach. *Journal of Nursing Management*, **12**: 114–123.

Harstall C. and Ospina M. (2003) How prevalent is chronic pin? *Pain Clinical Updates*, XI(2).

Haugh R. (2005) Hospitals and clinicians confront a new imperative: Pain management. *Hospitals and Health Networks*, **79**(4): 51–56.

Herren K.R., and Mackway-Jones K. (2001) Emergency management of cardiac chest pain: a Review. *Emergency Medical Journal*, **18**: 6–10.

Hewlett S, Mitchell K., Haynes J., Paine T., Korendowych E. and Kirwan J.R. (2000) Patient-initiated hospital follow-up for rheumatoid arthritis. *Rheumatology*, **39**: 990–997.

Hunter D.J. (2003) Public Health Policy. In *Public Health for the 21st Century. New Perspectives on Policy, Participation and Practice* (eds. Orme J., Powell J., Taylor P.) Open University Press, McGraw-Hill Education, Berkshire.

Kuhn S., Cooke K., Collins M,. Jones M. and Mucklow J.C. (1990) Perceptions of pain relief after surgery. *BMJ*, **300**: 1687–1690.

Lamers L.M., Meerding W.J., Severens J.L. and Brouwer W.B.F. (2005) The relationship between productivity and health-related quality of life: An empirical exploration in persons with low back pain. *Quality of Life Research*, **14**: 805–813.

Laursen B.S., Bajaj P., Olesen A.S., Delmar C. and Arendt-Nielsen L. (2005) Health related quality of life and quantitative pain measurement in females with chronic non-malignant pain. *European Journal of Pain*, **9**: 267–275.

Mac Lellan K. (2004) Postoperative pain: Strategy for improving patient experiences. *Journal of Advanced Nursing*, **46**: 179–185.

Macintyre P.E. and Ready L.B. (2002) *Acute Pain Management, A Practical Guide*, 2nd edn. W.B. Saunders, Philadelphia.

Melzack R, Abbott F.V., Zackson W., Mulder D.S. and Davis M.W.L. (1987) Pain on a surgical ward: a survey of the duration and intensity of pain and the effectiveness of medication. *Pain*, **29**: 67–72.

Moore A., Edwards J., Barden J. and McQuay H. (2004) *Bandolier's Little Book of Pain*. Oxford University Press, Oxford.

Owen H., McMillan V. and Rogowski D. (1990) Postoperative pain therapy: a survey of patients' expectations and their experiences. *Pain*, **41**: 303–307.

Palmer K.T., Walsh K., Bendall H., Cooper C. and Coggon D. (2000) Back pain in Britain: Comparison of two prevalence surveys at an interval of 10 years. *BMJ*, **320**: 1577–1578.

Regnard C. and Kindlen, M. (2004) What is pain? In *Helping the Patient with Advanced Disease* (ed. Regnard C.). Radcliffe Medical Press, Oxford.

Ross E. (2004) *Pain Management*. Hanley & Belfus, Philadelphia.

Royal College of Anaesthetists and The Pain Society, the British Chapter of the International Association for the Study of Pain (2003) *Pain Management Services – Good Practice.* The Royal College of Anaesthetists and The Pain Society, London.

Spranger M.A.G., de Regt E.B., Andries F., van Agt H.M., Bijl R.V., de Boer J.B. *et al.* (2000) Which chronic conditions are associated with a better or poorer quality of life? *Journal of Clinical Epidemiology*,. **53**: 895–907.

Stewart B.W. and Kleihues P. (eds) (2003) World Cancer Report. IARC Press, Lyon.

Twycross R. (2003) *Introducing Palliative Care,* 4th edn. Radcliffe Medical Press, Oxford.

Veillette Y., Dion D., Altier N. and Choiniere M. (2005) The treatment of chronic pain in Quebec: a study of hospital-based services offered within anesthesia departments. *Canadian Journal Anesthesia*, **52**(6): 600–606.

Watt-Watson J.H., Stevens B., Garfinkel P., Streiner D. and Gallop R. (2001) Relationships between nurses' pain knowledge and pain management outcomes for their postoperative cardiac patients. *Journal of Advanced Nursing*, **36**: 535–545.

Weis O.F., Sriwatanakul K., Alloza J.L., Weintraub M. and Lasagna L. (1983) Attitudes of patients, housestaff and nurses toward postoperative analgesic care. *Anesthesia and Analgesia*, **62**: 70–74.

Wilder-Smith C.H. and Schuler L. (1992) Postoperative analgesia: pain by choice? The influence of patient attitudes and patient education. *Pain*, **50**: 257–262.

World Health Organization (2002) *National Cancer Control Programmes: Policies and Managerial Guidelines,* 2nd edn. WHO, Geneva.

3 Modern developments in pain management

Learning outcomes

By the end of this chapter you should be able to:

- ★ Describe developments in the approaches to pain management
- ★ Outline modern, contemporary approaches to pain management
- ★ Understand how service improvements in pain management link to evidence-based care.

Introduction

This chapter begins by describing the developments in pain management over the past few decades. The major landmarks in the history of pain management are highlighted including the development of new advanced medications. Technology has had a major impact on pain interventions, and technologies such as patient-controlled analgesia are detailed. Future developments that are currently being researched are referred to. The importance of pain services and the evidence base for their development are outlined.

History

The nature of pain has puzzled society for centuries. Pain has been and remains the subject of medicine, philosophy, religion and politics. As an age-old problem it is interesting and challenging that society and, indeed, health professionals continue to struggle with its understanding, physiological basis and subsequent management. The past 20 years have seen new drugs being developed, many of which are refinements of the traditional approaches to pain. The development of more refined nonsteroidal anti-inflammatories such as the new COX-2 inhibitors or research currently being conducted on the use of intravenous paracetamol are examples. New methods of delivery of medications have emerged, such as patient-controlled analgesia and epidural analgesia. Pain physiology is better understood, but many questions still remain.

Pain relief has been used since the late 1600s, with new developments emerging over time. Here are some of the key developments in pain management over time (source: Meldrum, 2003).

- In 1680 laudanum, a mix of opium and sherry, was introduced by Thomas Sydenham
- In 1868 W.T.G. Morton demonstrated anaesthesia
- Throughout the 19th century opiates were standard treatment for acute pain. There was much concern about addiction
- During World War I, Rene Leriche introduced the concept of nerve blocks
- In 1953 John Bonica published *The Management of Pain* bringing together available information on pain treatments
- In the 1950s and 1960s nerve block clinics were set up
- 1950s Dame Cecily Saunder's concept of palliative care and 'total pain' was founded
- 1965 Melzack and Wall published the *Gate Control Theory of Pain*
- In the 1970s the concept of endogenous neurochemical reactions to pain was debated and researched
- 1972 Formation of International Association of Pain
- 1982 Development of the World Health Organization *analgesic ladder*
- 1980–2006 Clinical drug research
- 1980–2006 Dedicated acute and chronic pain services emerged
- 1990–2006 Multidisciplinary multi-modal approaches to pain. Pain management standards emerging from interest groups
- 2001 Patients and families sue health care provider for inadequate pain management (source: Haugh, 2005).

Impact of technology on pain management

Technology has revolutionised modes of analgesic administration allowing for smaller, more regular amounts of analgesia to be administered in a safe and efficient manner. Patients have been enabled to take more control over pain relief. It is suggested that refinements in systemic opioid administration and epidural analgesia have by far the greatest impact on the quality of acute pain care (Carr *et al.* 2005). Patient-controlled analgesia (PCA) using traditional intravenous opioids, the newer concept of the fentanyl hyrdrochloride patient-controlled transdermal system (PCTS) and epidural analgesia are reviewed below.

Patient-controlled analgesia (PCA)

Active involvement of the patient in the management of postoperative pain by the use of PCA is now a well-established part of clinical practice. The PCA system consists of a syringe pump and a timing device. The patient activates the system by pressing a button, which causes a small dose of analgesia to be delivered into the venous circulation. Simultaneously, a lockout device is activated, ensuring another dose cannot be delivered until a pre-set time. PCA is now over 30 years old and is in widespread use.

PCA systems have the following features: bolus demand dose, lockout interval, background infusion and maximum dose. The bolus demand dose is a predetermined dose which is delivered by the machine when the patient presses the button. Drugs commonly used in PCA systems include morphine, pethidine and fentanyl.

The literature details numerous studies on the effects, advantages and disadvantages of PCA systems. Advantages such as ease of use, time saving and patient control are all cited. Ballantyne *et al.* (1993) published a **meta-analysis** of initial randomised controlled trials of postoperative patient-controlled analgesia compared with conventional intramuscular analgesia. They included 15 randomised controlled trials. Their conclusions were that:

- Patient preference strongly favours PCA over conventional analgesia
- Patients using PCA obtain better pain relief than those with conventional analgesia, without an increase in side effects
- Favourable effects of PCA upon analgesic usage and length of hospital stay did not attain statistical significance (the authors suggest that this may be due to the low numbers involved in the randomised controlled trials).

Keywords

Meta-analysis

A meta-analysis uses *statistical methods* to combine the results from a number of previous experiments or studies examining the same question, in an attempt to summarise the totality of evidence relating to a particular issue. Meta-analysis includes a qualitative component (applies predetermined criteria of study quality) and a quantitative component (integration of numerical information) (Cochrane Collaboration, 2001).

The debate with regard to the efficacy of PCA continues in the literature with conflicting evidence emerging (Passchier *et al.*, 1993; Thomas *et al.*, 1995; Williams, 1996; Snell *et al.*, 1997). The overall incidence of potentially life-threatening complications with PCA is reported as low (Fleming and Coombes, 1992; Sidebotham and Schug, 1997). However, Choiniere *et al.* (1998) reported PCA costs as higher and found no clinical advantages with its use.

The main advantage of a patient-controlled analgesia is that it is a system that is designed to accommodate the wide range of analgesic requirements that can be anticipated when managing acute pain. Control over pain relief is cited in a number of studies as an advantage (Chumbley *et al.*, 1998).

Macintyre (2001), in a review of the literature, suggests that PCA can be a very effective and safe method of pain relief and may allow easier individualisation of therapy compared to conventional methods of opioid analgesia. The success lies in how well it is used. He suggests that if similar attention was given to other methods of opioid administration, conventional methods of analgesia could be as effective as PCA in many patients. Larijani *et al.* (2005) suggest that successful postoperative pain management using PCA is difficult to achieve on a consistent basis unless treatment is individualised.

Clinical Caseload . . .

If successful pain management utilising PCA is difficult to achieve on a consistent basis unless treatment is individualised, think about what you can do to support individualised treatment on a busy surgical ward.

Patient-controlled transdermal analgesia (PCTS)

A fentanyl hyrdrochloride patient-controlled transdermal system (PCTS) is under development as an alternative to PCA. It delivers small doses of fentanlyl by iontoporesis with electro transport delivery platform technology. The system uses a low intensity direct current to move fentanyl from a hydrogel reservoir into the skin, where it then diffuses into the local circulation and is transported to the central nervous system. The self-adhesive unit, which is about the size of a credit card, is worn on the patient's upper arm or chest (unlike PCA it does not have intravenous tubing or pump etc.) (Viscusi *et al.*, 2004). A patient-controlled transdermal analgesic delivery system provides the patient with a locus of control akin to the PCA. It is suggested that PCTS does not require the personnel and resource requirements of a conventional intravenous PCA and is as effective (Carr *et al.*, 2005).

In a randomised controlled trial Chelly *et al.*, (2004) demonstrated that fentanyl hydrochloride 40 μg was superior to placebo for the management of acute postoperative pain after surgery. Viscusi *et al.* (2004) assessed whether the transdermal PCTS system is equivalent to a standard morphine IV PCA regimen in postoperative pain management. The authors conclude that the pain control is equivalent and that the incidence of opioid-related side-effects similar.

Studies comparing PCTS with conventional postoperative treatments are needed.

Further research and development is required to evaluate the efficacy and feasibility of PCTS for potential widespread use.

Over to you

Research on patient-controlled transdermal systems is in its infancy. Go to Medline and CINAHL and see if you can access any new research on such systems.

Epidural analgesia

The spinal epidural space extends from the sacral hiatus to the base of the skull. Drugs injected into the epidural space can block or modulate afferent impulses and cord processing of those impulses. Both epidural local anaesthetics and epidural opioids can produce analgesia (McQuay, 1994).

Epidural analgesia is indicated for the provision of postoperative analgesia (Mallet and Bailey, 1996). Block *et al.* (2003) conducted a meta-analysis of efficacy of postoperative epidural analgesia including 100 studies (1966–2002). They concluded that epidural analgesia, regardless of analgesic agent, location of catheter, and type and time of pain assessment, provided better postoperative analgesia compared with parenteral opioids.

Wheatley *et al.* (2001) consider the efficacy and safety of epidural analgesia in patients recovering from major surgery, based on a computerised search of the literature from 1976 to 2000. The authors conclude that continuous thoracic epidural analgesia with a low-dose local anaesthetic-opioid combination has the potential to provide effective dynamic pain relief for many patients, even those undergoing major upper abdominal or thoracic procedures. However, there are major factors to consider in the safe and effective ward-based use of epidural analgesia including the need for staff training and regular monitoring of vital signs. Furthermore, the authors say that it is not uncommon for epidurals to be ineffective in terms of dynamic pain relief on a surgical ward for these reasons.

Keywords

Presurgical nociceptive blockage
This is the blockage of pain nerve fibre transmission prior to surgery

Pre-emptive analgesia

The concept of pre-emptive analgesia prior to surgery to reduce analgesic consumption and post surgical pain has been under discussion, review and research for a number of decades. The principle behind pre-emptive analgesia is that giving an intervention prior to surgery prevents or decreases surgical pain by preventing central sensitisation by administering a **presurgical nociceptive blockage**. This means that before surgery an analgesia is given in order to try to prevent nerve sensitisation once surgery commences.

To date, however, research reports have shown modest or disappointing results in showing reduced surgical pain (Ochroch *et al.*, 2003; Wnek *et al.*, 2004). However, it is suggested that nonsteroidal anti-inflammatories will have an increasing role in pre-emptive analgesia. Aida (2005) states that pre-emptive analgesia is challenging and that to be successful it must be deep enough to block all nociception, wide enough to cover the entire surgical area and prolonged enough to last throughout surgery and even into the postoperative period.

Moiniche *et al.* (2002) report on a systematic review of 80 double-blind **randomised controlled trials** (RCTs) designed to compare the role of timing of analgesia, i.e. preoperative versus intraoperative or

Keywords

Randomised controlled trials
RCTs are quantitative, comparative, controlled experiments in which investigators allocate at random two or more interventions to a series of individuals and measure specified outcomes. In a double-blind trial neither the researchers nor the patients know whether they are receiving the control or the intervention

postoperative initiation of analgesia. They conclude that the current evidence available reveals a lack of evidence for any important effect (rather than evidence for lack of effect) with pre-emptive analgesia.

Further research and development in this area will demonstrate whether there is a clinical justification for widespread use of pre-emptive analgesia.

Role of intravenous paracetamol and nitroxparacetamol

The use of paracetamol in the management of postoperative pain has been a source of interest for some time. Power (2005) discusses its use and proposes the introduction of an intravenous preparation. He reports that the analgesic and anti-inflammatory properties and safety advantages of a nitric oxide (NO)-releasing form may represent significant advances in the use of oral paracetamol. It is suggested that the intravenous formulation may have a safety advantage over the oral form by producing more predictable plasma paracetamol concentrations in the immediate postoperative period. Further research will demonstrate whether this will have any clinical advantage.

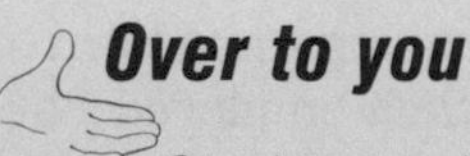

Over to you

Speak to your local pain care service team, and find out their views on any recent or future development in pain care and its associated technological advances.

Pain services

The need to have dedicated staff to support pain management has emerged over the past three decades. It was becoming evident that pain management was complex, required specialised staff and dedicated education and training. Each particular pain area was creating its own evidence base and process for pain management. Pain management could be better targeted by recognising that there is a need to view particular pain areas as specialisms. Acute pain, chronic pain and palliative care are now viewed as specialist areas of pain care and within these areas sub-specialties have emerged, e.g. neuropathic pain, migraine. The need for pain treatment facilities has emerged and there are a number of terms utilised to describe the types of pain services developed. This section details the development of acute pain services, chronic pain services and palliative care services.

Acute pain services

Worldwide acute pain services are developing in response to increasing recognition that the solution to the problem of inadequate postoperative pain relief lies not so much in development of new techniques but in the development of a formal organisation for better use of existing techniques (Rawal and Berggren, 1994). In 1988, Ready *et al.* drew attention to the potential role of an acute pain service by developing an anaesthesiology-based postoperative pain management service. They said, 'just as chronic pain management has become a special area of medical practice, treatment of acute pain deserves a similar commitment by practitioners with special expertise.' Following systematic review of acute pain teams McDonnell *et al.* 2003a describe them as multidisciplinary teams who assume day-to-day responsibility for the management of postoperative pain. Team members generally include specialist nurses, anaesthetists and pharmacists. Acute pain services and acute pain teams are in the main providing similar services. It appears that significant investment has been put into these approaches by some services and that it takes time and organisational commitment to set up a good service.

Windsor (1996) reported that 42.7% of UK hospitals had a multidisciplinary acute pain service in place by 1994. In 1995, 44% of UK hospitals reported having some form of acute pain service (Harmer *et al.*, 1995). Of these, 28% employed a dedicated pain nurse and 16% had specific 'fixed' consultant sessions allocated to acute pain. By 2003, this number had almost doubled, with 84% of acute English hospitals performing inpatient surgery for adult patients, having an acute pain team (McDonnell *et al.*, 2003a). Warfield surveyed 300 US hospitals in 1995. She found that 42% of hospitals had acute pain management programmes and that an additional 13% had plans to establish an acute pain management programme. A survey of acute pain services in Canada showed that 53% of hospitals were operating such a service in 1993. (Zimmermann and Stewart, 1993).

Key points Top tips

Acute pain teams/acute pain services usually encompass the following elements:

- introduction of policies, protocols and guidelines
- education of staff and patients
- audit and evaluation of service
- development of patient information leaflets
- introduction of regular pain assessment

- introduction and monitoring of new analgesic techniques
- clinical lead is an anaesthetist or specialist nurse
- services/teams include a specialist nurse
- seamless liaison with other health care teams

Evidence base

Benefits of acute pain service/acute pain team

Many benefits are ascribed to acute pain services and acute pain teams. There are numerous published studies, using both research and audit techniques, which report the benefits of such services. The following presents a sample of published work.

Timeline

- **1991** Wheatley *et al.* describe the benefits, risk and resource implications of providing an acute pain service indicating that their daily audit of the service and feedback to the anaesthetic and surgical colleagues had led to the delivery of a more consistent standard of postoperative care.
- **1992** Gould *et al.* outline the effect of sequential changes in a policy for controlling pain after surgery. Their objective was to observe the effects of introducing an acute pain service to the general surgical wards of a large teaching hospital. Main results showed a reduction in median visual analogue pain scores 24 hours after surgery for pain during relaxation, pain on movement and pain on deep inspiration.
- **1993** Schug and Torrie, in relation to the safety of acute pain services, reported no serious complications in morbidity and mortality in a study of 3016 patients receiving care from an acute pain service.
- **1997** Mackintosh and Bowles evaluated the impact of a nurse-led acute pain service. The evaluation focused on three areas of concern: preoperative information, patients' self-reported levels of pain and analgesics prescribed. The results of evaluation indicate significant reductions in reported levels of pain and patterns of analgesic prescribing.
- **1997** Pesut and Johnson evaluated the effectiveness of an acute pain service in Canada. No significant differences in mean pain scores were noted following the introduction of the pain service.
- **1997** Tsui *et al.* audited 2509 patients under the care of an acute pain service. They documented all side-effects and complications. They concluded that a standard monitoring and management protocol, an experienced nursing team and a reliable acute pain service coverage is mandatory for the safe use of modern analgesic techniques.
- **1998** Tighe *et al.*, following the introduction of an acute pain service whose emphasis was on multimodal pain therapy, reported that the service significantly improved inpatient perception of pain relief upon return of consciousness after anaesthesia and for two days postoperatively. They report

no changes in the incidence of emetic sequelae. They found a significant improvement in patient satisfaction and sleep pattern in hospital.

- **1998** Harmer reported on the effect of education, assessment and a standardised prescription on postoperative pain management following a clinical audit of 15 hospitals in the United Kingdom. He reported that following the introduction of the above there was an overall reduction in the percentage of patients who experienced moderate to severe pain at rest from 32% to 12%. The incidence of severe pain on movement decreased from 37% to 13% and moderate to severe pain on deep inspiration from 41% to 22%. He also noted decreases in the incidence of nausea and vomiting.
- **2003** McDonnell *et al.* report a systematic review and meta-analysis of 15 studies. The research team reported that the difference in quality between the studies made meta-analysis difficult. They conclude that there is insufficient robust research to assess the impact of acute pain teams (APTs) on postoperative outcomes of adult patients or on the processes of postoperative pain relief.

Reflective activity

Consider the timeline of evidence presented above. Read McDonnell *et al.* (2003b) and contemplate how future evaluations of acute pain services/acute pain teams should be undertaken.

Chronic-pain services

Many patients with chronic pain continue to have symptoms and may develop psychological distress, overall disability and increased dependency on family members and social services even though there are many advances in pain-relieving techniques. Pain management programmes have been developed to enhance the patient's physical performance and help them cope more effectively with their pain (Nurmikko *et al.*, 1998). Flor *et al.* (1992), following meta-analysis evaluating the efficacy of multidisciplinary treatments for chronic pain, suggest that patients treated in multidisciplinary pain clinics show improvements in pain and psychological functioning compared to conventional treatments. Guzman *et al.* (2001) conducted a systematic review of 10 trials in relation to multidisciplinary rehabilitation for chronic low back pain. The authors suggest that there was strong evidence that intensive multidisciplinary biopsychosocial rehabilitation with functional restoration improves function when compared with inpatient or outpatient non-multidisciplinary treatments and moderate evidence that such rehabilitation reduces pain. Ospina and Harstall (2003) support this through analysis of the literature, concluding that the evidence for the effectiveness of multidisciplinary pain programmes is strong for chronic low back pain. However, they state the evidence is inconclusive for widespread pain, neck and shoulder pain. Good practice in chronic-pain management suggests the provision of core services for chronic-pain in all district general hospitals and most specialist hospitals (The Royal College of Anaesthetists and The Pain Society 2003).

Veillette *et al.* (2005) report on a Canadian survey of the availability of hospital-based anaesthesia departments with chronic non-cancer pain services. The survey indicates that 73% of anaesthesia departments offer chronic non-cancer pain services, with 26% providing some form of multidisciplinary assessment and treatment. However, 4500 patients were waiting for their first appointment to see a pain consultant, with 67% of these waiting for nine months or more.

There is limited published empirical work on the prevalence or effectiveness of chronic-pain services. It emerges, however, that chronic pain services should include the following as detailed below.

Key points Top tips

Chronic-pain services usually encompass the following elements:

- introduction of policies, protocols and guidelines
- education of staff and patients
- audit and evaluation of service
- alleviation of pain
- reduction of disability and restoration of function
- rationalisation of medication
- rationalisation of use of all health services
- attention to social, family and occupational issues
- multidisciplinary approach

Palliative care

The history of palliative care spans a number of centuries. From the beginning of the nineteenth century, charitable endeavours led to the creation of 'hospices', meaning a place of rest and recovery for pilgrims. In 1842, Jeanne Garnier formed 'L'Association des Dames du Calvaire' in Lyon, France and her influence led to six other 'Calvaire' hospices in Paris and New York by the end of the century.

In 1879, Our Lady's Hospice for the Dying in Harold's Cross, Dublin, Ireland, opened thanks to the ambition of Mother Mary Aikenhead, the Superior of a religious Order – the Irish Sisters of Charity. Here, religious Sisters offered care, shelter, comfort and warmth to the destitute, chronically frail and to those whose condition was beyond cure. Many consider this to be the first institutional provision of its kind beyond the mainland of Europe (O'Brien and Clark, 2005). Healy (2004) reports that consequently the Irish Sisters of Charity founded London's first hospice, St Joseph's Hospice in Hackney. It was in this very London hospice in the 1950s that modern palliative care was born.

Dame Cicely Saunders was responsible for establishing the discipline and the culture of palliative care. During her lifetime, she worked as a nurse, medical social worker, volunteer and, eventually, as a physician. In 1967, Dame Cicely Saunders founded the first modern hospice, St Christopher's Hospice in London. Here, she promoted an integrated approach that recognised all of the difficulties that the dying have to face and helped them in meeting these challenges, but without pain. Since then, palliative care has developed and evolved worldwide. In the UK, Palliative care is provided by two distinct categories of health and social-care professionals:

- Those providing the day-to-day care to patients and carers in their homes and in hospitals
- Those who specialise in palliative care (consultants in palliative medicine and clinical nurse specialists in palliative care, for example) supported by a multidisciplinary palliative care team.

The National Council for Palliative Care's website indicates the recent provision of specialist palliative care services in England, Wales and Northern Ireland. As at January 2004, there were:

- 196 specialist inpatient units providing 2730 beds, of which 19.9% were NHS beds.
- 341 home care services – this figure includes both primarily advisory services delivered by hospice and NHS-based community palliative care teams as well as other more sustained care provided in the patient's home. Some 17% of referrals were at time of diagnosis.
- 324 hospital-based services.
- 237 day-care services.
- 273 bereavement support services.

(Source: The National Council for Palliative Care, 2004)

Key points Top tips

Palliative care aims to:

- affirm life and regard dying as a normal process
- provide relief from pain and other distressing symptoms
- integrate the psychological and spiritual aspects of patient care
- offer a support system to help patients live as actively as possible until death
- offer a support system to help the family cope during the patient's illness and in their bereavement

Integrated approaches to pain management

This chapter title hints at a panacea of modern developments that will revolutionise the way all types of pain are managed. Yet many of the modern developments are messages highlighting the importance of listening to the patient, assessing pain regularly, and using multidisciplinary, multimodal approaches. Above all, the modern message is to take pain seriously, seriously enough to have dedicated services, protocols and guidelines, and have staff and patient education in place.

Organisational approaches to pain management should be co-ordinated, planned, audited and firmly rooted in an evidence-based process utilising policies and guidelines. It would appear that improvement in one part of the pain process will have greater impact if an integrated, multidisciplinary approach is taken. This means that there should be a dependent interrelationship between pain assessment, pain history, pain management, knowledge of patients and staff and clinical audit. It is only by adopting this modern approach that the quality of pain will improve for patients.

The knowledge and attitudes of patients and health care staff has become a subject of increasing interest in pain management. The context of pain, culture, past experiences and family all influence patient perception of pain. Figure 3.1 illustrates this.

Figure 3.1 Influences on the quality of pain management

Increasingly, nurses have taken on more specialist and advanced practice roles in pain management. They are seen as the lead in some pain teams (Mackintosh, 1997). Titles such as pain resource nurses (Holley *et al.*, 2005), nurse specialists, pain nurse consultants have all emerged. Nurses are seen are key members of the multidisciplinary team in managing pain (CSAG 2000). The Pain Society (The British Chapter of the IASP) (2002) has produced recommendations for nursing practice in pain management. These guidelines support a career pathway for nurses in pain management and outline competencies for novice, intermediate and higher levels of practice. They were developed in response to the development of nursing at a range of levels in pain management across acute and chronic settings. They can be accessed at www.britishpainsociety.org.

The US agency for health care policy and research (AHCPR, 1992) recommends an integrated approach to pain management and states that pain control options should include:

- cognitive-behavioural interventions, relaxation, distraction and imagery: these can be taught preoperatively and can reduce pain, anxiety and the amount of drugs needed for pain control
- systematic administration of nonsteroidal anti-inflammatory drugs (NSAIDs) or opioids using the traditional 'as needed' schedule or round-the-clock administration
- patient-controlled analgesia (PCA) usually means self-medication with intravenous doses of an opioid; this can include other classes of drugs administered orally or by other routes
- spinal analgesia, usually by means of an epidural opioid and/ or local anaesthetic injected intermittently or infused continuously
- intermittent or continuous local neural blockade (examples of the former include intercostal nerve blockade with local anaesthetic or cryoprobe; the latter includes infusion of local anaesthetic through an interpleural catheter)
- other forms of analgesia – such as transcutaneous electrical nerve stimulation.

Key points Top tips

- Pain management has developed significantly over time
- Pain management has developed into various specialisms: acute, chronic and palliative care
- Multidisciplinary teams and multimodal approaches support quality patient care
- New technologies continue to develop which will support the development of analgesic management in clinical areas

An audit of pain prevalence to inform practice development

Derek is the clinical nurse manager in one of St Margaret's two surgical wards (see case study in Chapter 2). The ward manages patients post routine surgery such as appendectomies and cholecystectomies. St Margaret's new clinical director and director of nursing have just completed a pain audit as part of an overall quality of care audit in the hospital. They are keen to benchmark care and are very aware that St Margaret's has not invested in new technologies, equipment or professional development of staff.

Derek reviews the data for his ward: daytime mean pain prevalence is 7.6 and night-time mean pain prevalence is 7.8 using a visual analogue scale. Derek decides to call a ward meeting to discuss it. He quickly realises from feedback from his staff that the mean pain prevalence presented to him is so crude that it is not indicative of the true picture. He decides to investigate further and take a sample of 10 patients and measures their pain score every six hours after surgery for three days. Table 3.1 details the results.

When Derek and his staff review the results, they see they have a pain prevalence peak at 12 hours post surgery. One of the staff nurses suggests that, at this stage, the patients have recovered from their anaesthetic, and analgesia given in theatre has worn off. After this, patients are given intramuscular analgesia on an as-needed basis or if there is a patient-controlled analgesia pump available they will be given that. There are only five PCA pumps for the surgical wards so not all patients will get a PCA pump.

- Do you think that the pain scores in Table 3 compare favourably to the national average post routine surgery?
- What do you think is the issue with the analgesic regime utilised on the ward?
- If you were in Derek's shoes what would you do next?

Table 3.1 Record of pain scores taken six-hourly for three days

Day	Time post surgery	Mean pain score
Day 1	6 hrs	3.4
	12 hrs	8.2
	18 hrs	6.5
	24 hrs	7.6
Day 2	30 hrs	6.6
	36 hrs	6.8
	42 hrs	5.4
	48 hrs	5.6
Day 3	54 hrs	4.7
	60 hrs	3.2
	66 hrs	4.6
	72 hrs	4.1

Rapid recap

1 Name two new technologies that have emerged in relation to analgesic administration in the last 30 years.
2 What are the main elements of a pain service?
3 What new roles in pain management are emerging for nurses?

References

Agency for Health Care Policy and Research, Public Health Service (AHCPR) Acute Pain Management Guideline Panel (1992) *Acute Pain Management: Operative or Medical Procedures and Trauma. Clinical Practice Guideline*. AHCPR Pub. No. 92–0032. Department of Health and Human Services, Rockville, Maryland.

Aida S. (2005) The challenge of pre-emptive analgesia. *Pain Clinical Updates IASP*, XIII(2).

Ballantyne J.C., Carr D.B., Chalmers T.C., Keith B.G., Angelillo I.F. and Mosteller F. (1993) Postoperative patient-controlled analgesia: Meta-analyses of initial randomised control trials. *Journal of Clinical Anethesia*, **5**: 182–193.

Bennett R.L., Batenhorst R.L., Bivins B.A., Bell R.M., Bauman T., Graves D.A. *et al*. (1982) Drug use patterns in patient controlled analgesia. *Anaesthesiology*, **57**(3): A210.

Block B.M., Liu S.S., Rowlingson A.J., Cowan A.R. and Wu C.L. (2003) Efficacy of postoperative epidural analgesia. *JAMA*, **290**(18): 2455–2463.

Carr D.B., Reines D., Schaffer J., Polomano R.C. and Lande S. (2005) The impact of technology on the analgesic gap and quality of acute pain management. *Regional Anesthesia and Pain Medicine*, **30**(3): 286–291.

Chelly J.E., Grass J., Houseman T.W., Minkowitz H. and Pue A. (2004) The safety and efficacy of a fentanyl patient-controlled transdermal system for acute postoperative analgesia: A multicenter, placebo-controlled trial. *Anesthesia and Analgesia*, **98**: 427–433.

Choiniere M., Rittenhouse B.E., Perreault S., Chartrand D., Rousseau P., Smith B. and Pepler C. (1998) Efficacy and costs of patient-controlled analgesia versus regularly administered intramuscular opioid therapy. *Anesthesiology*, **89**: 1377–1388.

Chumbley G.M., Hall G.M. and Salmon P. (1998) Patient controlled analgesia: an assessment of 200 patients. *Anaesthesia*, **53**: 216–221.

Clinical Standards Advisory Group (CSAG) (2000) *Services for Patients with Pain*, Department of Health, London.

Cochrane Collaboration (2001) Definitions www.vichealth.vic.gov.au/cochrane/overview/definitions.htm.

Fleming B.M. and Coombes D.W. (1992) A survey of complications documented in a quality control analysis of patient controlled analgesia in the postoperative patient. *Journal of Pain and Symptom Management*, **7**(8): 463–469.

Flor H., Fydrich T. and Turk D.C. (1992) Efficacy of multidisciplinary pain treatment centres: a meta-analytic review. *Pain*, **49**: 221–230.

Gould T.H., Crosby D.L., Harmer M., Lloyd S.M., Lunn J.N., Rees G.A.D. *et al*. (1992) Policy for controlling pain after surgery: effects of sequential changes in management. *BMJ*, **305**: 1187–1193.

Guzman J., Esmail R., Karjalainen K., Malmivaara A., Irvin E. and Bombardier C. (2001) Multidisciplinary rehabilitation for chronic low back pain: systematic review. *BMJ*, **322**:1511–1516.

Harmer M. and Davies K.A. (1998) The effect of education, assessment and a standardised prescription on postoperative pain management. *Anaesthesia,* **53**: 424–430.

Harmer M., Davies K.A. and Lunn J.N. (1995) A survey of acute pain services in the United Kingdom. *BMJ*, **311**: 360–361.

Haugh R. (2005) Hospitals and clinicians confront a new imperative: Pain management. *Hospitals and Health Networks*, **79**(4): 51–56.

Healy, T. (2004) *Our Lady's Hospice: 125 Years of Caring in Ireland.* A. & A. Farma, Dublin.

Holley S., McMillan S.C., Hagan S.J., Palacios P. and Rosenberg D. (2005) Training pain resource nurses: Changes in their knowledge and attitudes. *Oncology Nursing Forum*, **32**(4): 843–848.

Larijani G.E., Sharaf I., Warshal D.P., Marr A., Gratz I. and Goldberg M.E. (2005) Pain evaluation in patients receiving intravenous patient controlled analgesia after surgery. *Pharmacotherapy*, **25**: 1168–1173.

Macintyre P.E. (2001) Safety and efficacy of patient-controlled analgesia. *British Journal of Anaesthesia*, **87**(1): 36–46.

Mackintosh B.A. and Bowles S. (1997) Evaluation of a nurse-led acute pain service. Can clinical nurse specialists make a difference? *Journal of Advanced Nursing*, **25**: 30–37.

Mallett J. and Bailey C. (1996) Epidural Analgesia in *The Royal Marsden NHS Trust Manual of Clinical Nursing Procedures*, 4th edn. Blackwell Science, Oxford, p.259.

McDonnell A., Nicholl J. and Reid S.M. (2003a) Acute pain teams in England: current provision and their role in postoperative pain management. *Journal of Clinical Nursing,* **12**: 387–393.

McDonnell A., Nicholl J. and Reid S.M. (2003b) Acute pain teams in England: a systematic review and meta-analysis. *Journal of Advanced Nursing*, **41**(3): 261–273.

McQuay H.J. (1994) Epidural Analgesics. In Melzack R. and Wall P. *The Textbook of Pain.* Churchill Livingston, Edinburgh, pp.1025–1033.

Meldrum M.L. (2003) A capsule history of pain management. *JAMA*, **290**(18): 2479–2475.

Moiniche S., Kehlet H. and Berg Dahl J. (2002) A qualitative and quantitative systematic review of preemptive analgesia for postoperative pain relief: the role of timing of analgesia. *Anesthesiology*, **96**(3): 725–741.

National Council for Palliative Care. (2004) Palliative care explained. Availaible at: www.ncpc.org.uk/palliative_care.html.

Nurmikko T.J., Nash T.P. and Wiles J.R. (1998) Recent advances: control of chronic pain. *BMJ*, **317**: 1438–1441.

O'Brien, T. and Clark, D. (2005) A national plan for palliative care – the Irish experience. In: *Palliative Care in Ireland* (eds. Ling J. and O'Siorain, L.). Open University Press, Berkshire.

Ochroch E.A., Mardini I.A. and Gottschalk A. (2003) What is the role of NSAIDs in pre-emptive analgesia? *Drugs*, **63**(24): 2709–2723.

Ospina M. and Harstall C. (2003) Multidisciplinary pain programmes for chronic pain: Evidence from systematic reviews. Alberta Heritage Foundation for Medical Research, *Health Technology Assessment*, HTA 30 (Series A): 1–48.

Pain Society, The British Chapter of the International Association for the Study of Pain (2002) *Recommendations for Nursing Practice in Pain Management*. The Pain Society, London.

Passchier J., Rupreht J., Koenders M.E.F., Olree M., Luitweiler R.L. and Bonke B. (1993) Patient controlled analgesia leads to more postoperative pain relief, but also to more fatigue and less vigour. *Acta Anaesthesiology Scandinavia*, **37**: 659–663.

Pesut B. and Johnson J. (1997) Evaluation of an acute pain service. *Canadian Journal of Nursing Administration*, Nov–Dec: 86–107.

Power I. (2005) Recent advances in postoperative pain therapy. *British Journal of Anaesthesia*, **95**(1): 43–51.

Rawal N. and Berggren L. (1994) Organisation of acute pain services: a low cost model. *Pain*, **57**:117–123.

Ready B.L., Oden R., Chadwick H.S., Benedetti C., Rooke G.A., Caplan R *et al*. (1988) Development of an anaestiology-based postoperative pain management service. *Anaesthesiology*, **68**(1): 100–106.

Royal College of Anaesthetists and The Pain Society, the British Chapter of the International Association for the Study of Pain (2003) *Pain Management Services – Good Practice.* Royal College of Anaesthetists and The Pain Society, London.

Schug S.A. and Torrie J.J. (1993) Safety assessment of postoperative pain management by an acute pain service. *Pain*, **55**: 387–391.

Sidebotham D. and Schug S.A. (1997) The safety and utilisation of patient controlled analgesia. *Journal of Pain and Symptom Management*, **14**(4): 202–209.

Snell C.C., Fothergill-Bourvonnais F. and Durocher-Henriks S. (1997) Patient controlled analgesia and intramuscular injections: a comparison of patient pain experiences and postoperative outcomes. *Journal of Advanced Nursing*, **25**: 681–690.

Thomas V., Heath M., Rose D. and Flory P. (1995) Psychological characteristics and the effectiveness of patient controlled analgesia. *British Journal of Anaesthesia*, **74**: 271–276.

Tighe S.Q.M., Bie J.A., Nelson R.A. and Skues M.A. (1998) The acute pain service: effective or expensive care? *Anaesthesia*, **53**: 382–403.

Tsui S.L., Irwin M.G., Wong C.M.L., Fung S.K.Y., Hui T.W.C., Ng K.F. *et al*. (1997) An audit of the safety of an acute pain service. *Anaesthesia*, **52**: 1042–1047.

Veillette Y., Dion D., Altier N. and Choiniere M. (2005) The treatment of chronic pain in Quebec: a study of hospital-based services offered within anesthesia departments. *Canadian Journal of Anesthesia*, **52**(6): 600–606.

Viscusi E.R., Reynolds L., Chung F., Atkinson L.E. and Khanna S. (2004) Patient-controlled transdermal fentanyl hydrochloride vs intravenous morphine pump for postoperative pain. *JAMA*, **291**(11): 1333–1341.

Warfield C.A. and Kahn C.H. (1995) Acute pain management programs in U.S. hospitals and experiences and attitudes among U.S. adults. *Anaesthesiology*, **83**: 1090–1094.

Wheatley R.G., Madjel T.H., Jackson I.J.B. and Hunter D. (1991) The first year's experience of an acute pain service. *British Journal of Anaesthesia*, **67**: 353–359.

Wheatley R.G., Schug S.A. and Watson D. (2001) Safety and efficacy of postoperative epidural analgesia. *British Journal of Anaesthesia*, **87**(1): 47–61.

Williams C. (1996) Patient controlled analgesia: a review of the literature. *Journal of Clinical Nursing*, **5**: 139–147.

Windsor A.M., Glynn C.J. and Mason D.G. (1996) National provision of acute pain services. *Anaesthesia*, **51**: 228–231.

Wnek W., Zajaczkowska R., Wordliczek J., Dobrogowski J.and Korbut R. (2004) Influence of pre-operative ketoprofen administration (preemptive analgesia) on analgesic requirement and the level of prostaglandins in the early postoperative period. *Polish Journal of Pharmacology*, **56**: 547–552.

Zimmermann D.L. and Stewart J. (1993) Postoperative pain management and acute pain service activity in Canada. *Canadian Journal of Anaesthesia*, **40**(6): 568–575.

4
Pain physiology

> **Learning outcomes**
> By the end of this chapter you should be able to:
> ★ Describe the functions of the nervous system relative to pain
> ★ Detail the pain pathways
> ★ Explain the importance of the pain pathway for modulating pain
> ★ Appreciate the link between pain classification and the physiology of pain.

Introduction

In order to manage pain effectively health professionals should have a firm understanding of the physiology of pain including the concept that each individual perceives their pain uniquely. This provides the basis for management and modulation of pain. This chapter provides an overview of the central and peripheral nervous system as they relate to pain. The important structures in understanding the physiological mechanisms of pain are provided. As each of these is described, possibilities for modulation of pain at this stage of the pain process will be presented.

Pain pathways

Traditionally, the pain pathway is viewed as consisting of a three-neuron chain that transmits pain signals from the periphery to the cerebral cortex (Cross, 1994). This is the peripheral nociceptor, the spinal cord and the supra-spinal (brain) levels.

At the level of the peripheral nervous system, free nerve endings called nociceptors send pain messages via a series of nerve fibres (neurones) to the central nervous system entering it at the spinal cord. Neurones are highly complex cells sensitive to changes in the environment. They consist of nerve endings, axons (the long fibre of a nerve cell that carries messages) and dendrites (short arm-like protuberances that are branced like a tree from the axon). The speed at which a nerve impulse travels is determined by the diameter of the axon and the presence of myelin. Myelin is a sheath around the axons and dendrites of peripheral neurons. The greater the diameter and the presence of myelin the faster the nerve impulse travels. Synapses (specialised junctions between nerve cells that are bridged by neurotransmitters) occur at the dorsal horn of the grey matter of the spinal cord. These comprise layers or laminae of cells.

The nerve signal then reaches the sensory cortex of the brain via the reticular formation. The reticular formation is a massive neural area of white and grey matter that has a role in the control of body posture, musculoskeletal reflex activity and general behavioural states. It is here that the brain analyses the pain, attributes meaning and links this with emotions and motor activity. The ascending pain pathways comprise:

- neospinothalamic tract (fast ascending fibres)
- paleospinothalamic tract (slow ascending fibres) (also called spino-reticulo-diencephalic tract)
- other ascending pain pathways.

The descending pathway consists of descending fibres passing from brainstem to spinal cord, inhibiting incoming sensations of pain. See Figure 4.1.

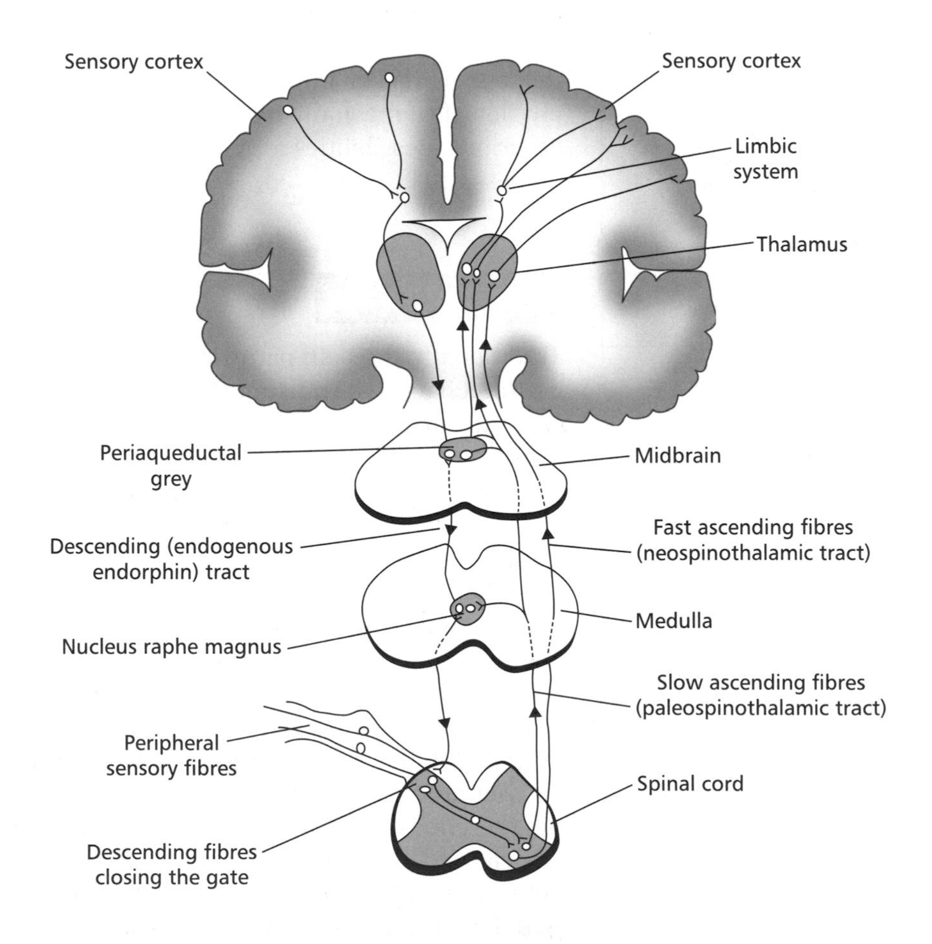

Figure 4.1 Pain pathways. Reproduced with permission from Davis B.D. (2002) Caring for people in pain *p. 37. London, Routledge, part of the Taylor & Francis Group.*

The nervous system

The physiology of pain is a function of the nervous system. It is therefore necessary to understand its anatomy and physiology in order to comprehend the physiological mechanisms of pain. The nervous system consists of two major components:

- the central nervous system (CNS) and
- the peripheral nervous system (PNS).
  (See Figure 4.2)

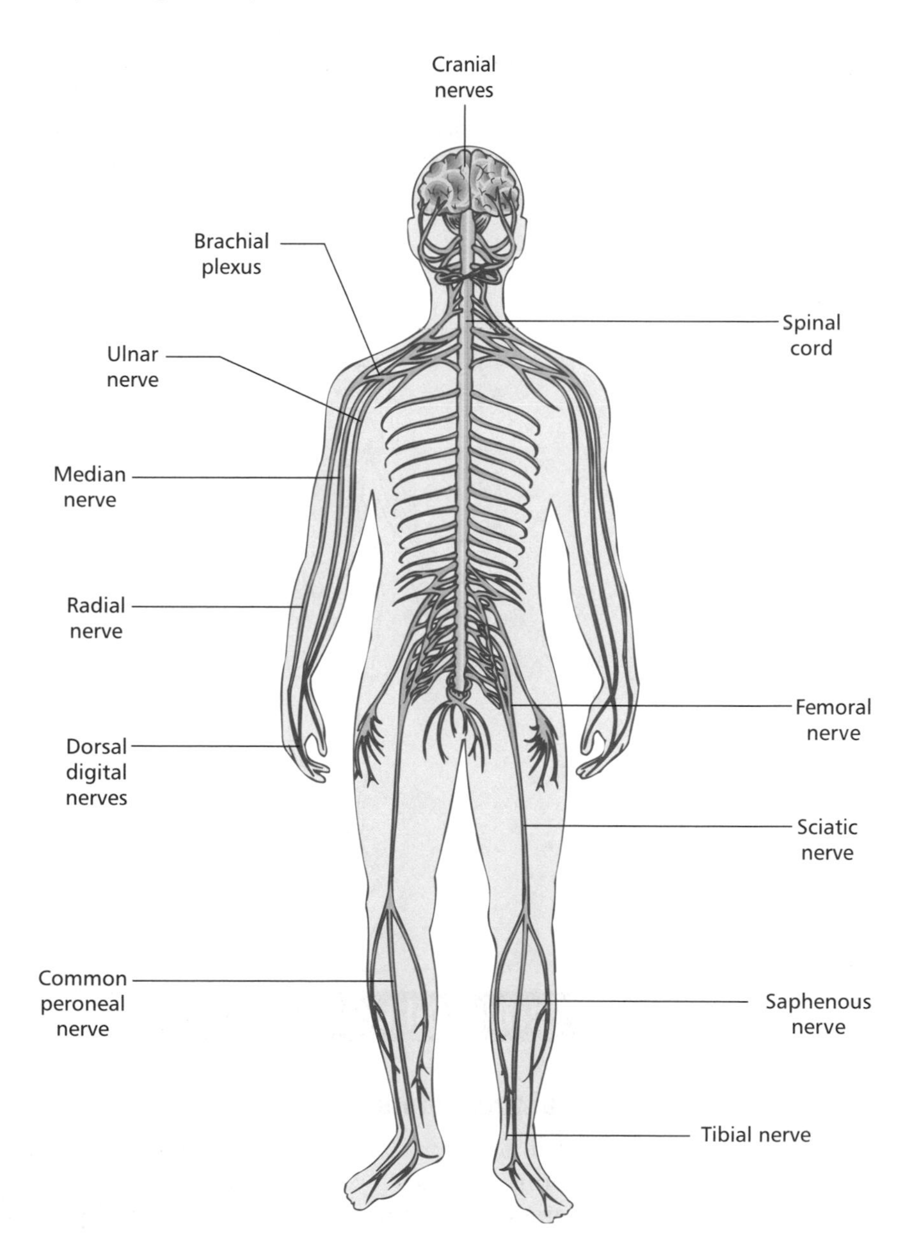

Figure 4.2 The central and peripheral nervous sytem

The CNS, which consists of the brain and spinal cord, receives and processes information.

The PNS consists of 12 pairs of cranial nerves arising from the brain and 31 pairs of nerves arising from the spinal cord. Two types of nerves, afferent and efferent, service the PNS. Afferent nerves form the sensory division of the PNS and carry sensory data including pain data from receptors in the skin, muscles, joints and internal organs to the CNS. Efferent nerves carry nerve impulses from the CNS to the rest of the body making up the motor division of the PNS. Voluntary movement of the body occurs in response to efferent nerves which serve the musculoskeletal system. Involuntary responses occur in response to the autonomic nervous system which supplies the glands, blood vessels, gastrointestinal, respiratory system and other internal organs.

The skin is innervated (stimulation of a muscle or an organ by nerves) by particular nerves and these skin areas are called dermatomes – see Figure 4.3.

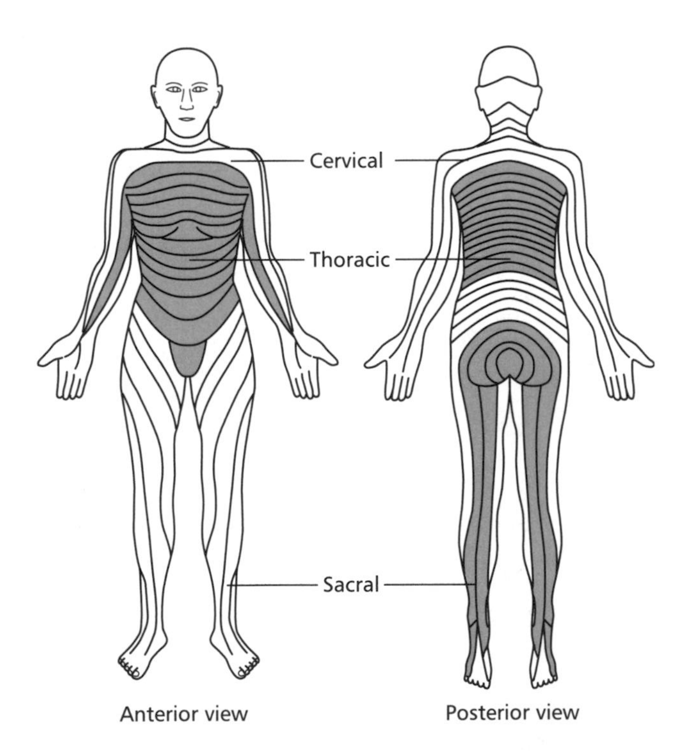

Figure 4.3 Dermatomes

Peripheral nervous system

Pain receptors

Pain receptors are called nociceptors. These are free nerve endings found in almost every tissue of the body including the skin, synovium of joints, artery walls and internal organs. Synovium is a thin layer of tissue which lines the joint space. Nociceptors respond to any type of stimulus if it is strong enough to cause tissue damage. Excessive stimulation of a sense organ causes pain. When stimuli for sensations such as touch, heat and cold reach a certain threshold they stimulate the sensation of pain. Additional stimuli for pain receptors include excessive distension or dilation of a structure, prolonged muscular contractions, muscle spasms, inadequate blood flow to an organ, or the presence of certain chemical substances.

Nociceptors are specific for painful stimuli, responding to damaging or potentially damaging mechanical, chemical and thermal stimuli. Although the actual stimulus for nociceptors is not known, it is assumed that chemicals released from cells damaged by the pain stimulus, such as histamine or bradykinin, activate the nociceptors. Pain receptors perform a protective function by identifying changes that may endanger the body.

Nociceptors are located at the distal end of sensory neurones (see Figure 4.4). They transmit messages to the spinal cord's dorsal horn via nerve fibres.

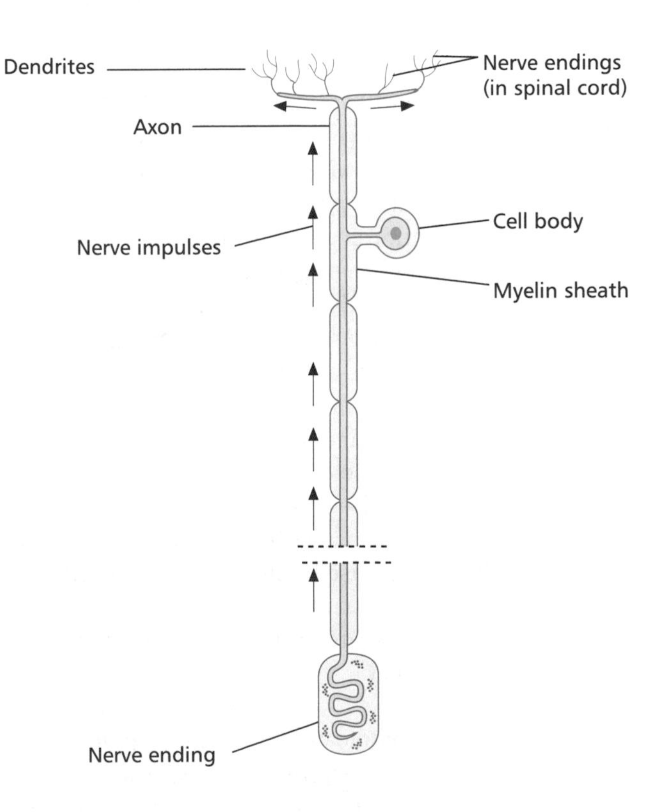

Figure 4.4 A sensory neurone

Nerve fibres are classified by size and according to whether they originate in skin or muscle.

There are two main types of nerve fibre involved in the transmission of pain (nociceptors): A-delta fibres and C-polymodal fibres. Large-diameter fibres (A-beta) are found in the skin and muscle and respond to touch. A-beta fibres are not directly involved in the transmission but are important when discussing the Gate Control Theory (see page 55).

A-delta nociceptors (nerve fibres)

A-delta nociceptors are small-diameter fibres which are myelinated and conduct the transmission of pain rapidly. The myelin sheath speeds up the process of nerve transmission. A-delta nociceptors are found mainly in and just under the skin with some present in muscle. They are activated by noxious stimuli such as pressure, surgery and ischaemia. They are known as high-threshold mechanoreceptors. Some respond to heat and are known as mechanothermal nociceptors. Fast pain occurs very rapidly, usually within 0.1 second after the stimulus is applied, and is not felt in the deeper tissues of the body. This type of pain is also known as sharp, fast and pricking pain. Fast pain is a well-localised, pinprick sensation that results from activating the nociceptors on the A-delta fibres.

C-polymodal nociceptors (nerve fibres)

C-polymodal fibres are small-diameter fibres which are not myelinated and conduct the transmission of pain slower. They are called polymodal because of their ability to respond to mechanical, chemical or thermal stimulus. Slow pain begins after a second or more and then gradually increases in intensity after a period of several seconds or more. It is referred to as burning, aching, throbbing and slow pain. An example is the pain associated with toothache. Slow or delayed pain is a poorly localised, dull, burning sensation that results from activating the nociceptors on the C fibres. This activation is due to the effects of substances released and triggered by damaged cells (see below).

The two types of pain sensation elicit different reflexes. Fast pain evokes a withdrawal reflex and a sympathetic response including an increase in blood pressure and a mobilisation of body energy supplies. Slow pain produces nausea, profuse sweating, a lowering of blood pressure and generalised reduction in skeletal muscle tone (Wang, 1991).

Chemicals at the site of pain

Prostaglandins, histamine, bradykinin and serotonin are chemicals found at the site of pain and thought to excite nociceptors thus increasing the pain sensation (see Table 4.1).

Table 4.1 Chemicals found at the site of pain

Chemical	Action
Prostaglandins	Very specialised fatty acids. They are found at the site of pain/injury. They can cause vasodilation or vasoconstriction, muscle contraction or relaxation, and increase intensity of pain.
Histamine	Released at the site of injury and causes vasodilation and oedema.
Bradykinin	Released upon tissue injury and is present in inflammatory exudates. It evokes a response in nociceptors.
Serotonin	Found at site of pain-activating nociceptors.

Hyperalgesia

Nociceptors can be sensitised so that they continue to send pain messages long after the stimulus is removed. Hyperalgesia is the phenomenon of increased sensitivity of damaged areas to painful stimuli and is categorised as follows:

- primary hyperalgesia occurs within the damaged areas
- secondary hyperalgesia occurs in undamaged tissues surrounding the area.

Modulation of pain at the nociceptor

The aim of modulating pain at the nociceptor is to prevent transmission of the pain signal at the site of pain. Pain at the site is exacerbated by the presence of certain chemicals, such as prostoglandins or bradykinin which is released naturally by the body as part of the inflammatory response. Pain perception is diminished by the reduction or eradication of these chemicals. The drug class of nonsteroidal anti-inflammatory drugs (NSAIDs) act at the site of pain reducing the presence of prostaglandins. NSAIDs have the ability to inhibit prostaglandins, one of the sensitisers of nociceptors. This is why it is clinically indicated in certain circumstances to use NSAIDs early to prevent sensitisation and the consequent augmentation of pain. NSAIDs have numerous side-effects however. COX-2 inhibitors (which are specific NSAIDs selectively inhibiting the cyclo-oxygenase enzyme COX-2) have fewer side-effects than traditional non-selective NSAIDs. NSAIDs (such as Nimesulide (Aulin/Mesulid), which are COX-2 inhibitors, are particularly useful for inflammatory pains such as rheumatoid arthritis.

Peripheral effects of opioids are now being explored. The use of opioid drugs, whose function is to bind with specialised opioid receptors, reduces excitability of nociceptors, and inflammatory peptide release from nerve endings (Power, 2005).

Modulation of pain at the nerve fibre

The aim of modulating pain at the nerve fibre is to prevent the transmission of pain signals along the nerve cell. Nerve conduction depends on sodium ion channels in the axon. Sodium channels help to maintain a neuron's resting state (when sodium channels are closed) and its activation (when sodium channels open). Local anaesthetic agents are absorbed across the membrane of the neurone and block the sodium channels. This means that the electrical impulse cannot be transmitted. For example, a local anaesthetic agent such as lignocaine injected prior to inserting sutures for a traumatic wound in the emergency department is very useful for preventing pain.

Central nervous system

The CNS consists of the brain and spinal cord. A number of neurotransmitters are found in both the brain and spinal cord that are thought to stimulate or modulate the perception of pain. Neurotransmitters facilitate, excite or inhibit post-synaptic neurons. They establish the lines of contact between brain cells (see Table 4.2).

In addition, neuropeptides are present within the brain and spinal cord. They are chemical messengers in the brain. Most act primarily to modulate the response of or the response to a neurotransmitter. (see Table 4.3).

Table 4.2 Neurotransmitters involved in the sensation of pain

Neurotransmitters	Action
Substance P	Found in sensory nerves, spinal cord pathways and parts of the brain associated with pain: stimulates perception of pain. Endorphins may exert their pain-inhibiting properties by suppressing release of substance P.
5-hydroxytryptamine (serotonin)	Modulates pain when acting in the CNS
Glutamine	Likely to be major transmitter in afferent A and C fibres – this occurs in the spinal cord after nerve injury. Glutamine is likely to be a key transmitter in central sensitisation. NMDA (N-methyl-D-aspartine) is a receptor for glutamine. It is found in the dorsal horn. Ketamine is an NMDA antagonist.

Table 4.3 Neuropeptides

Neuropeptide	Action
Enkephalins	Concentrated in the thalamus, hypothalamus, parts of the limbic system and spinal cord pathways that relay pain impulses. They inhibit pain impulses by suppressing substance P. It is suggested that enkephalins are the body's natural painkillers. They act by inhibiting impulses in the pain pathway and by binding to the same receptors in the brain as morphine.
Endorphins	Concentrated in the pituitary gland. They also function by inhibiting substance P. Like enkephalins they have morphine-like properties that suppress pain.
Dymorphin	Found in the posterior pituitary gland, hypothalamus and small intestine. May be related to controlling pain.

Spinal cord and brain

Once pain receptors are stimulated the impulse they discharge travels to the spinal cord and on to the brain via the pain pathways. In the spinal cord the grey matter is divided into 10 laminae. The dorsal part is divided into five laminae (i–v), components of which deal with most incoming pain fibres. A-delta and C fibres enter laminae i and ii. A-beta fibres enter laminae iii and iv. A mixture of A-beta and A-delta enter lamina v. Links between laminae are achieved and maintained through chemical neurotransmitters (Melzack and Wall, 1982). It is in laminae i and ii that the Gate Control Theory mechanism is observed.

In the spinal cord, information on pain is received by cells in the dorsal horn and is passed on to higher centres in the brain along tracts in the spinal cord.

Pain fibres ascending into the cerebrum from the spinal cord may end in a number of sites, particularly the reticular formation, the thalamus and the cerebral cortex. In the reticular formation, the pain stimuli may evoke arousal, changes in heart rate, blood pressure, respiration and other activities. The appreciation or conscious awareness of pain is found in the thalamus and cerebral cortex.

Gate Control Theory

This theory was proposed in 1965 by Melzack and Wall to describe the transmission of pain through the dorsal horn in the spinal cord. Pain impulses must pass through the substantia gelantosia cells, which are present in the dorsal horn of grey matter, in order to travel to the brain and here there is some integration of sensory stimuli. The theory proposes that a neural mechanism in the dorsal horns of the spinal cord acts like a gate which can increase or decrease the flow of nerve impulses from the peripheral fibres to the central nervous system – see Figure 4.5.

This theory is of critical importance because it describes the mechanisms of transmission and modulation of nociceptive signals and it recognises pain as a psycho-physiological phenomenon resulting from the interaction between physiological and psychological events. The Gate Control Theory provided a framework for examining the interactions between local and distant excitatory and inhibitory systems in the dorsal horn and has generated much debate and discussion leading to many research studies. Melzack and Wall expanded the conceptualisation of pain from a purely sensory phenomenon to a multidimensional model that integrates the motivational-affective and cognitive-evaluative components with the sensory-physiological one (Turk and Melzack 2001). Dickenson (2002) suggests that the Gate Control Theory was a leap of faith but that it was right. The concepts of convergence and modulation reduced the emphasis on destruction of pathways and led to the idea that pain could be controlled by modulation. The capacity of pain signalling and modulating systems to alter in different circumstances has changed thinking about pain control.

The Gate Control Theory is a very important milestone in understanding the physiology of pain. The fact that it is relatively recent (1965) emphasises the newness of our understanding of pain physiology.

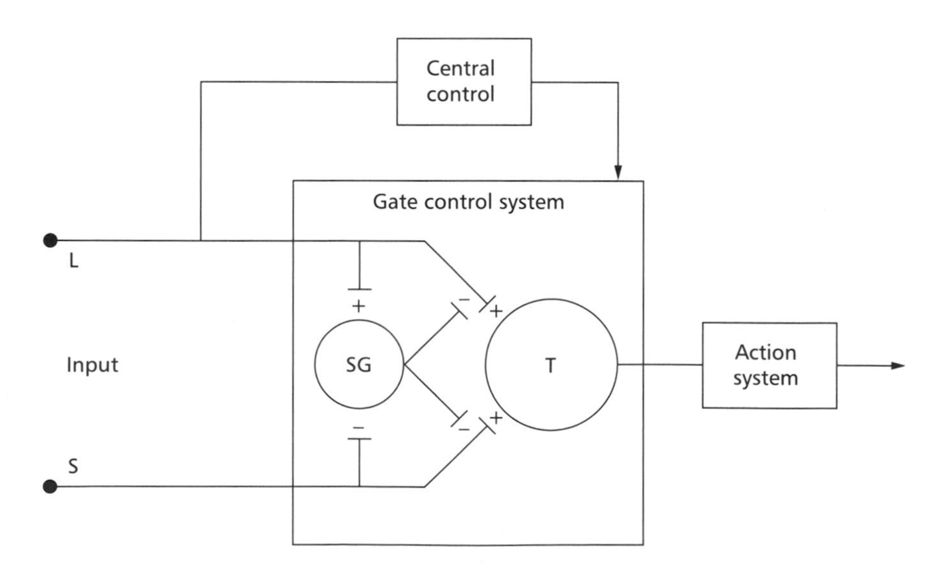

Figure 4.5 Gate Control Theory (adapted from Melzack and Wall, 1965).

Modulation of pain using the Gate Control Theory

The degree to which the gate increases or decreases sensory transmission is determined by the relative activity in large-diameter (A-beta) and small-diameter (A-delta and C) fibres and by descending influences from the brain. Entry into the CNS can be visualised as a gate, which is opened by pain-generated impulses and closed by low-intensity stimuli such as rubbing or electrical stimulation (TENS). Furthermore, it can be closed by endogenous opioid mechanisms which can be activated from the brain or peripherally by acupuncture or by gentle rubbing, massage, electrical stimulation and hot or cold therapies. The theory suggests that descending influences consist of attention, anxiety, anticipation and past experience – all of which may exert control over sensory input (Melzack and Wall, 1982).

Modulation of pain within the spinal cord and brain

The aim of modulating pain within the spinal cord is to prevent the transmission of pain signals to the brain. The Gate Control Theory, as described above, suggests processes for such modulation.

These include the inhibition of the release of chemical transmitters at the various synapses within the dorsal horn of the spinal cord and brain as a result of the uptake of naturally occurring morphine-like substances by specialised receptors on relevant neurones. The use of opioid drugs whose function is to bind with these specialised receptors will reduce pain perception. There are three main types of opioid receptor (mu μ, delta δ and kappa κ). For example, the use of morphine in a patient-controlled analgesia pump (PCA) post surgery is extremely effective for pain relief. The morphine acts as an **agonist** for the opioid receptor. Naloxone is an **antagonist**.

Agonists
Drugs that bind to and stimulate opioid receptors

Antagonists
Drugs that bind to but do not stimulate opioid receptors and may reverse the effect of opioid agonist

Classification of pain

Pain can be classified in a number of ways according to type of pain, source of pain, speed of transmission of nerve signals or other pains (see Figure 4.6).

Categories of pain

Pain is usually classified into two major categories, i.e. acute and chronic, based on speed of onset, quality and duration of the sensation.

Acute pain

A distinguishing characteristic of acute pain is that it subsides as healing takes place, i.e. it has a predictable end and it is of brief duration, usually less than six months. Acute pain usually means sudden severe pain. Postoperative pain is an example. Acute pain is characterised by a

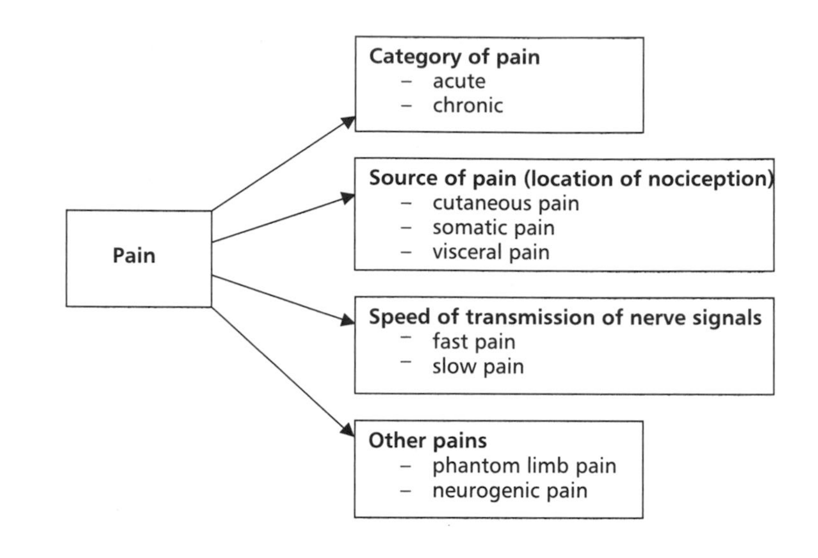

Figure 4.6 Classification of pain

well-defined time of onset and is associated with both subjective and objective signs indicating activation of the autonomic nervous system.

Chronic pain

Chronic pain is prolonged. It is pain that persists beyond the usual healing phase of the disease process. Chronic pain is characterised by patient distress rather than pain (Hardy, 1997).

Pain types

Pain can be classified into cutaneous, deep somatic or visceral pain according to the site of stimulation.

Cutaneous pain is produced by stimulation of the pain receptors in the skin. It can be accurately localised due to the large number of receptors in the skin. Touch and vision greatly increase localisation of cutaneous pain.

Somatic pain is produced by stimulation of pain receptors in the deep structures, i.e. muscles, bones, joints, tendons, ligaments. Unlike cutaneous pain, deep somatic pain is dull, intense and prolonged. It is usually associated with autonomic stimulation, e.g. sweating, vomiting and changes in heart rate and blood pressure. Pain from deeper structures can also initiate reflex contraction of nearby muscles, e.g. muscle spasm associated with bone fractures. The adequate stimuli for deep somatic pain include:

- mechanical forces, e.g. severe pressure on a bone, traction of muscle or ligament
- chemicals such as venom, acids or alkalis
- ischaemia, e.g. muscle ischaemia such as angina.

Visceral pain is produced by stimulation of pain receptors in the viscera. Visceral nociceptors are located within body organs and internal cavities. Visceral pain is caused by disorders of internal organs such as the stomach, kidney, gallbladder, urinary bladder, intestines and others. These disorders include distension from impaction or tumours, ischaemia and inflammation (Al-Chaer and Traub, 2002).

In contrast to the skin, which is richly innervated by sensory neurons, the visceral receptors of the abdominal and thoracic viscera are sparsely innervated. This means that they have fewer sensory neurons than other areas such as the skin. Therefore, severe visceral pain indicates diffuse stimulation of pain receptors from a wide area of the viscus. Visceral pain:

- is poorly localised
- is often referred or radiates to other sites, e.g. pain of a myocardial infarction is classically felt centrally just behind the sternum, radiating down the left arm and up the root of the neck into the jaw – see Figure 4.7
- is often associated with autonomic disturbances, e.g. vomiting, sweating, tachycardia
- can be associated with rigidity and tenderness of nearby skeletal muscles.

Neurogenic/neuropathic pain is described as burning, electric, tingling or shooting in nature. It may be continuous or paroxysmal in

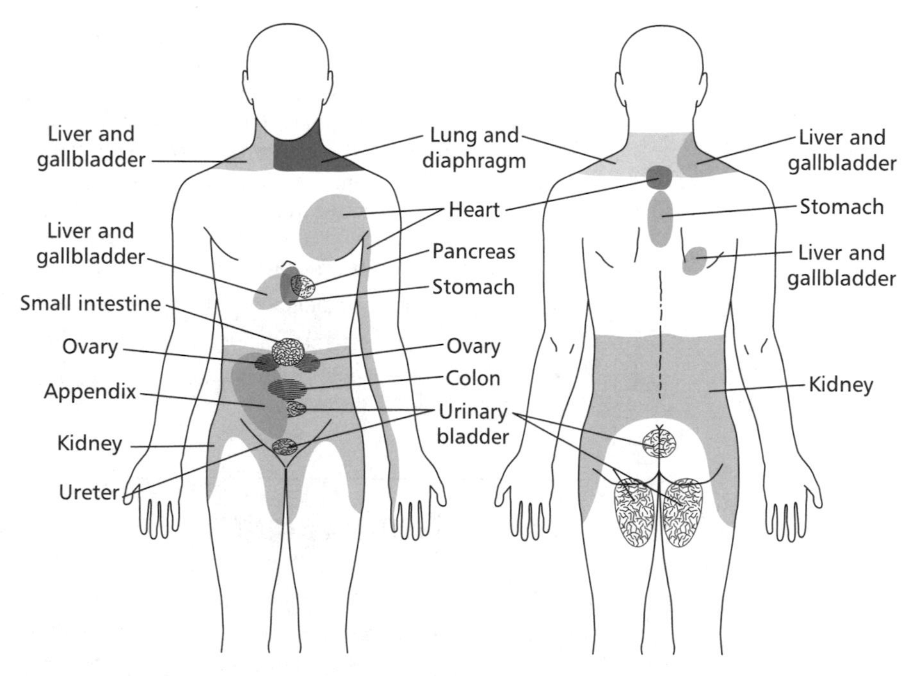

Figure 4.7 Referred pain

presentation. Neuropathic pain is produced by damage to or pathological changes to the nerve itself. Neuropathic pain, thus, does not always respond to traditional analgesic therapies of NSAIDS or opioids. It responds to co-analgesics such as tricyclics.

Key points Top tips

- Pain is a function of the nervous system, both central and peripheral
- Pain is classified in a number of ways: according to type of pain, source of pain, speed of transmission of nerve signals or other pains
- Gate Control Theory can help in understanding how some pain care treatments work
- Pain can be modulated at the site of pain, in the spinal cord and brain

Referred pain

Sue presents to her general practitioner (GP) with pain in her right side. Sue is a 42-year-old accountant. She is fit – running 45–60 minutes on the treadmill five days a week. She is slim and eats sensibly. She is studying part time for a law degree and seems to be very busy with a full life. She has had the pain for some time and is now getting anxious. The general practitioner suspects gallstones and refers Sue to the local hospital. Three weeks later, Sue is reviewed by the general surgeon who elicits a full history and performs a physical examination. Agreeing with the GP, he suspects gall stones and orders an ultrasound to confirm diagnosis. Sue is subsequently diagnosed with gall stones despite being an atypical profile. The ultrasound identified three rather large stones in the gallbladder and one in the biliary tract. She is scheduled for a laparoscopic cholecystectomy the next month.

Sue continues to have pain at home while she waits for her surgery. In addition to the pain in her side, she is also suffering from a darting pain in her right shoulder. She describes her pain as a sharp biting kind of pain, which is very acute.

Sue goes back to the general practitioner. He reassures her by telling her that her pain including that in her shoulder is classic gallstones. He prescribes her a nonsteroidal anti-inflammatory for her pain while she is waiting to go into hospital.

- Can you explain what is meant by referred pain?
- How will the nonsteroidal anti-inflammatory medication help alleviate Sue's pain?
- How would the Gate Control Theory help you manage Sue's pain?

Rapid recap

1. What are the main ascending pain pathways?
2. How do the various chemicals at the site of pain work?
3. What is the Gate Control Theory of Pain?

References

Al-Chaer E.D. and Traub R.J. (2002) Biological basis of visceral pain: Recent developments. *Pain*, **96**: 221–225.

Cross S.A. (1994) Symposium on pain management – Part 1 pathophysiology of pain. *Mayo Clinical Proceedings*, **69**: 375–383.

Dickenson A.H. (2002) Gate Control Theory of Pain stands the test of time. Editorial *British Journal of Anaesthesia*, **88**(6): 755–757.

Hardy P.A.J. (1997) The Essentials. In *Chronic Pain Management: The Essentials.* Oxford University Press, Oxford, p. 8.

Melzack R. and Wall P. (1982) The Gate Control Theory of Pain. In: *The Challenge of Pain.* 2nd edn. Penguin books, London, pp. 222–231.

Power I. (2005) Recent advances in postoperative pain therapy. *British Journal of Anaesthesia*, **95**(1): 43–51.

Turk D.C. and Melzack R. (2001) *Handbook of Pain Assessment*, 2nd edn. The Guilford Press, New York.

Wang M.B. (1991) Neurophysiology. In: *Physiology,* 2nd edn. (eds. Bullock J., Boyle J. and Wang M.B). Williams and Wilkins, Baltimore, pp. 43–44.

5 Influencing factors for the pain experience

Learning outcomes

By the end of this chapter you should be able to:

★ Understand the factors that influence the pain experience for the patient

★ Be able to discuss the emotional aspect of pain

★ Understand the importance of the role of the patient in their pain experience

Introduction

This chapter will help in understanding the factors that influence the pain experience. Pain is a personal and unique experience for the individual. While much is known about the physiology of pain the emotive and contextual aspects of pain continue to challenge our understanding. The meaning of pain to an individual and the importance of culture and knowledge of patients will be discussed. Influences on pain experiences will be described at a macro level (pain as a public health issue), meso level (organisational responses to pain) and a micro level (how the individual experiences their pain).

Meaning of pain

The cause of an individual's pain will strongly relate to how that pain is experienced by any individual, i.e. the extent of pain will relate to the level of nociception involved (see Chapter 4 for more detail on nociception). However, the context of that pain will shape the experience and subsequent management of that pain. The patients' knowledge of pain and analgesia, their expectations of pain, past experiences of pain, fear of addiction, anxiety, culture, age, lack of information and the influence of public and organisational policy and health professional responsible for care all form part of the pain experience. For those attending health services the overall service philosophy and approach to pain will be instrumental in the way pain is perceived and managed. Figure 5.1 highlights these issues.

Epidemiological studies examine pain as it occurs in groups of people or even whole populations attempting to discern the populations who suffer pain. This chapter, while attempting to review the evidence of the influences on pain, highlights the lack of empirical evidence available on the influences on pain and the need for further research in this area. Crombie (1999) supports this and details the difficulty with accessing the many studies on the epidemiology of pain due to their being published in a diverse array of journals. He stresses

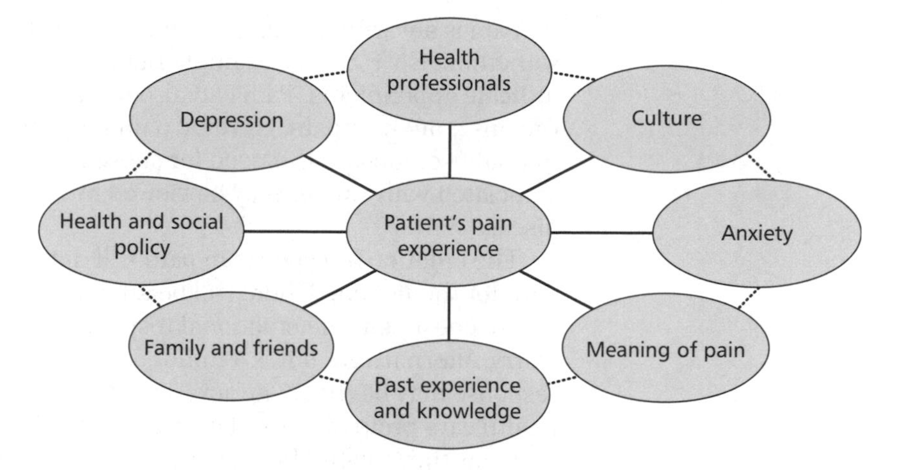

Figure 5.1 Factors that influence the pain experience

the importance of the need for more systematic reviews to grade the quality of these studies. There is great potential for the epidemiology of pain to help decide what kind and what level of health service provision should be provided and to influence health promotion. The International Association for the Study of Pain (IASP) (1999) bring together a number of high-quality research studies on the epidemiology of pain which are summarised in Table 5.1.

Table 5.1 Research studies on the epidemiology of pain. (Source: ISAP, 1999)

Factors	Key messages
Role of Psychological Factors	• Psychology contributes to the understanding of individuals' pain perception and behaviour • A psychological view enhances a holistic conception of pain • A psychological perspective may explain a patient's reactions in specific contexts
Gender Considerations in the Epidemiology of Chronic Pain	• No simple relationship in gender consideration of the epidemiology of chronic pain • Patterns differ from condition to condition • Gender-specific prevalence for most conditions varies across the life cycle
Cross-cultural Investigations of Pain	• Coping style and disability/dysfunction reflect the most variance in pain experiences across cultures
Pain in Older People	• The well-documented increase in pathological load, particularly degenerative joint and spine disease and leg and foot disorders may help explain the increased frequency of pain report in surveys of older persons • Age-related increases in overall pain prevalence does not continue beyond the seventh decade of life

Pain is a warning of danger and is often the first indication that something is wrong, for example right-sided lower abdominal pain may indicate appendicitis. Pain can also be seen as life threatening, such as the onset of chest pain. Chronic pain may indicate a potential reduction in quality of life and the need for subsequent lifestyle changes. Pain associated with cancer may be viewed by the patient as deterioration in disease status.

The emotional response to pain will depend on the meaning of the pain for the patient. There will be an emotional response. If there is severe chest pain the emotional response may be intense anxiety and worry. Alternatively, if it is a minor, superficial burn the emotional response may be that of annoyance. It is therefore important for the health care professional to be aware of the likely responses of patients and how they might affect decisions in relation to developing pain management regimes.

Various models of pain perception have evolved over the years. Linton and Skevington (1999) provide a model which demonstrates a cross-sectional view of the psychological processes involved in pain perception including attributions, coping and behaviour (Figure 5.2). The first step in the model is the noxious stimulus. The model stresses the role of appraisal and beliefs, which compliment those of coping and learning. The authors state that these may not be conscious processes. The consequences of the behaviour affect the emotional, cognitive and behavioural aspects.

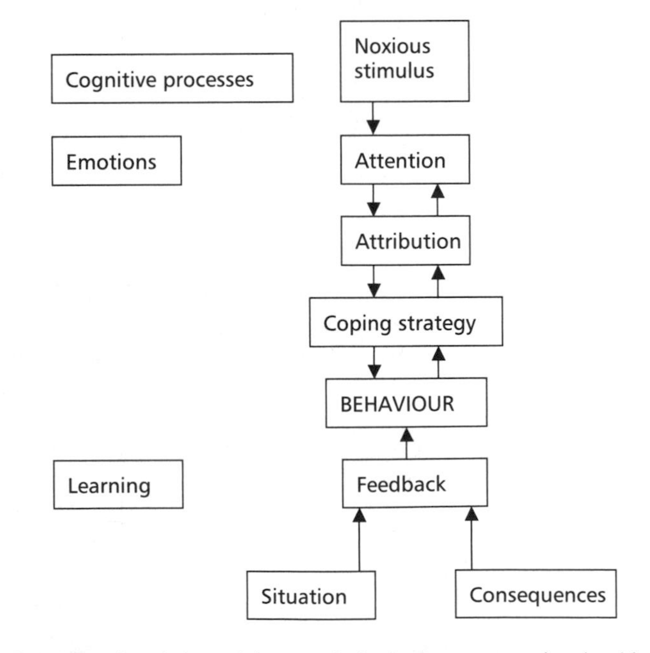

Figure 5.2 Cross-sectional view of the psychological processes involved in pain perception. Reproduced with permission from IASP Press. Originally published in Epidemiology of Pain. *Steven J Linton and Suzanne M Skevington. Editors Crombie, I.K., Croft, P.R., Linton S.J., Le Resche, L., and von Korff, M. IASP Press, Seattle.*

Linton and Skevington (1999) suggest that a psychological approach may help explain the relationship between physiological, behavioural and cognitive and emotional aspects of pain. They emphasise the reciprocal nature of the relationship, showing that pain may influence mood but mood may also affect pain perception.

Pain as a public health issue

At a macro level pain is a public health issue with international collaboration as an expanding feature of public health (further detail in Chapter 2). The influence and need for such influence from organisations such as the World Health Organization (WHO) and the IASP is critical in ensuring that pain is viewed as a public health issue and as such a global problem. The WHO's influence on governments worldwide and on national health care programmes and policies is seen to lay the ground work for better management of acute, chronic and cancer pain.

The seriousness with which governments and health service providers consider pain a health issue has considerable influence on how individuals will experience their pain. Where national protocols and guidelines exist to set templates for service providers, the more likely it is that the recipients of care will receive an adequate pain management plan.

Pain initiatives have staff and resource implications, but economic justification can be found in the reduced morbidity, faster convalescence, increased quality of life and work days, and improved satisfaction in patients who receive adequate relief from pain (Clinical Standards Advisory Group (CSAG), 2000; Macintyre and Ready, 2002). Service providers should use quality assurance procedures periodically to ensure that best practice in pain management is being carried out. The use of clinical audit described in Chapter 2 could be viewed in conjunction with evidence-based guidelines as significant influencers of change and as factors that impact on how pain is experienced.

Culture and ethnicity

Worldwide mobility has raised the profile of culture, race and ethnicity in health care in general and in pain management specifically. Culture can be defined as the beliefs, customs and behaviours of a group of individuals due to ethnicity, religion, origin or current residence (Green, 2004). Dimsdale (2000) identifies that ethnicity affects the type of illness suffered. Cultural factors related to the pain experience include pain expression, pain language, lay remedies for pain, social roles and expectations, and perceptions of the medical care system

(Lasch, 2002). Globally migration patterns challenge traditional homogeneous views of the universal problem of pain management with little research reported about pain management related to culture (Sherwood *et al.*, 2005).

Green *et al.* (2003) undertook a literature review and conclude that there are racial and ethnic disparities in pain perception, assessment and treatment across all types of pain (acute, cancer, chronic non-malignant and experimental). Sources of pain disparities are complex involving patient (e.g. patient/health care provider communication, attitudes), health care provider (e.g. decision making) and health care system (access to pain medication) factors. Blanchard *et al.* (2003) outline a number of factors that contribute to racial disparities in emergency and acute medicine care:

- poverty and access to care
- differences due to health insurance levels
- differences in the delivery of care
- physician characteristics and factors within the doctor–patient relationship.

The literature, however, reports mixed views highlighting the important need for further research in this area. Choi (2000) conducted a chart review of long-bone fractures in a UK hospital accident and emergency department with attendees from a number of ethnic backgrounds. The authors found no difference between those who received analgesia. Tamayo-Sarver *et al.* (2003) in a US study utilising vignettes (hypothetical clinical cases) identified no effect of patient race or ethnicity on physician prescription of opioids at discharge for African American, Hispanic and white patients. Tan *et al.* (2005) suggest that attitudes, beliefs, coping mechanisms and adjustments to chronic pain are comparable among non-Hispanic black and white patients comparing self reports of 128 non-Hispanic black Americans and 354 non-Hispanic white Americans. However, Bonham (2001), following a literature review on disparities in treatment of pain based on race and ethnicity, suggests that patients coming from racial and ethnic minority groups are at more risk of ineffective treatment of pain.

Over to you

Read the article by Bonham (2001): Bonham V.L. (2001) Race, Ethnicity, and Pain Treatment: Striving to Understand the Causes and Solution to the Disparities in Pain Treatment. *Journal of Law, Medicine and Ethics*, 29: 52–68.

Identify three key points from the article and bring them for discussion to your next journal club meeting or mentor/supervisor meeting.

Jones (2005), on reviewing 119 qualitative studies on end-of-life and ethnicity/race/diversity issues, stressed that a common theme emerging is a need for sensitivity to the varying expectations and the mix of involvement of patients, practitioners and families and the need for information sharing in end-of-life care. Jones also cautioned against 'cookbook' solutions to ethnicity, highlighting that there is diversity within ethnic groups themselves. Choi (2000) supports this suggesting that factors such as ethnicity, pain threshold, communication of pain to health care staff and relationships between patients and staff all influence determination of prescription of analgesia.

While it emerges that access to care, expression of pain and usage of interventions can vary among various cultures, for the nurse or health care professional managing a pain situation some common guidelines apply:

- sensitivity to individual needs of patient and their family
- utilising a comprehensive pain history and assessment appropriate to classification of pain
- constant attention to communications
- underlying philosophy of pain management 'pain is what the patient says it is'.

Reflective activity

Are there a number of different cultures represented in your patient caseload? How many of the following four principles do you use when considering pain management for your patients?

1. Sensitivity to individual needs of patient and their family
2. Utilising a comprehensive pain history and assessment appropriate to classification of pain
3. Constant attention to communications
4. Underlying philosophy of pain management 'pain is what the patient says it is'

The future may partly lie in genetic research. Kim and Dionne (2005) describe how genetic research in pain is just beginning, but that in the future it may be possible to determine treatments based on each individual's genetic profile. It may be possible to identify genes involved in the mediation of pain and genes that contribute to experimental painful stimuli.

Key points Top tips

- Understand that the pain experience is unique to each individual patient regardless of culture or ethnicity or age
- Be sensitive to individual needs
- Obtain a comprehensive pain history
- Utilise appropriate pain assessment techniques
- Ensure communication of pain assessment and management is understood by patient, family and health professionals

Team approaches

At a meso/organisational level pain is a multidimensional phenomenon which requires team approaches to its management. It requires professionals working together involving the input of doctors, anaesthetists, nurses, physiotherapists, occupational therapists and psychologists. The combined expertise of individual practitioners provides a more effective service to patients and their families. Pain management is an interactive and collaborative process involving the patient and family, nurse, doctor and other providers, as appropriate to the care setting.

The knowledge and attitudes of patients and health care staff have become an increasing subject of interest in pain management. Erroneous knowledge with regard to pain, analgesics and pain management contribute to patients not receiving best possible pain interventions. In the early 1980s it began to emerge that the need to have dedicated staff with specialised knowledge to support pain management was critical in order to improve the day-to-day management of pain. Professional standards for treatment facilities for pain set out definitions and standards for pain clinics and pain centres. There is a strong commitment to a multidisciplinary approach to diagnosis and treatment of pain. A number of statements and standards have emerged in relation to pain facilities. Chapter 3 discusses types of pain services in more detail.

Evidence-based approaches are critical to ensure that patients receive tested pain management approaches which are up to date and in line with international standards.

Modern approaches are about taking pain seriously, seriously enough to have dedicated services, protocols, guidelines, staff and patient education in place. The importance of listening to the patient, assessing pain regularly, and using multidisciplinary, multimodal approaches is advocated.

Organisational approaches to pain management should be evidence-based, planned, co-ordinated and audited utilising policies and guidelines as appropriate. Central to integration of care is a focus on the client with the most important objective being to improve continuity of care within the organisation and across providers and settings. Models of service delivery which place the client at the centre of the care continuum can help providers to reduce gaps, avoid unnecessary duplication, and ensure that clients are well supported in navigating 'the health system'. Such integration requires organisations to develop the simplest and most accessible pathways for clients. This means that there is a dependent interrelationship between pain assessment, pain history, pain management, knowledge of patients and staff, and clinical audit. By utilising this modern approach, the pain experience for patients will improve. Figure 5.3 highlights these interactions.

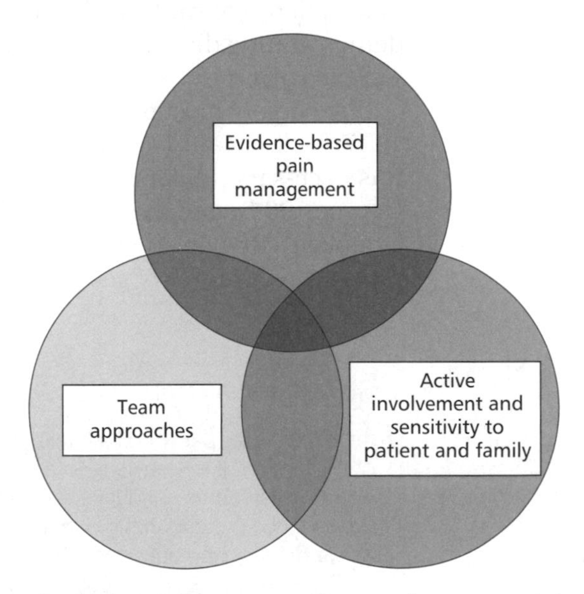

Figure 5.3 Example of interactions necessary for a modern approach to pain management

Health professional speaks

Clinical Nurse Specialist Pain Management

It is not enough any more to concentrate on looking after your patient in isolation, nurses and all the members of the health care team . . . the doctors, physiotherapists, anaesthetists must actively work together, sharing information and discussing and reviewing best case management through care pathways and audit. In addition bridges must be put in place between hospital and community to ensure that when your patient is discharged from hospital there is continuity and a smoothness in the transition of care in the community.

Patient knowledge and attitudes

At a micro level the influence that the patients themselves have on their pain management has become an increasing subject for research. It is reasonable to assume that while the patient is a recipient of care, they then have an influence on how their pain is managed. The consideration that patient expectations will influence how they communicate their pain and when they will request pain interventions has emerged. Patients expecting to have pain may make fewer demands and less reports of pain, making it a more difficult task to control. Patients believing that it is possible to become addicted to strong pain-killers while in hospital may compound these expectations (Winefield, 1990; Lavies 1992).

Many patients wait until they have severe pain before asking for medication (Owen, 1990; Winefield 1990). Juhl (1993) supports this demonstrating that only 64% of patients would always complain if they were in pain.

Pre surgery most patients anticipate a high amount of pain (Carr, 1997; Pellino 1997). Many patients believe that pain following surgery was necessary thinking they would 'just have to put up with it' (Hume,1994), and that 'you should put up with a bit of pain rather than complain' (Brydon, 1996; Scott & Hodson, 1997). Patients do not always rate the importance of good pain relief to their recovery (Laing, 1993).

Health professional speaks

Staff Nurse surgical ward

I find that older patients don't talk about pain. . . they talk about some soreness or discomfort. It is easy to ignore this and not consider it as pain and I have to explain to students and junior nurses that this is the terminology these older patients use . . . and that it is as important to manage that 'soreness' as actively as we manage the patient who complains of severe pain or that older patient will not be able to mobilise or cough which will slow his/her recovery and may mean a longer length of stay and more complications such as chest infections.

Interestingly in 2004 Fosnocht *et al.* reported that the expectations of patients attending the accident and emergency department may differ from that of other clinical settings. Their study showed that patients with pain reported a mean expectation for pain relief of 72% with 18% expecting complete relief of their pain. Expectation did not vary based on age or gender.

Patients, however, continue to have a high degree of confidence in the ability of doctors and nurses to treat pain (Scott and Hodson, 1997).

Interestingly many studies agree that patients continue to report high satisfaction with pain management in spite of high levels of pain (Miaskowski *et al.*, 1994; Jamison *et al.*, 1997; McNeill *et al.*, 1998) perhaps because they do not have sufficient knowledge of pain relief and may judge the kindness of staff rather than their way of treating pain (Bostrom *et al.*, 1997).

Increasing patients' knowledge and engaging them more actively in their own care can be the beginning of a process to improve pain management. Patients expecting to have pain make fewer demands and less reports of pain, making it a more difficult task to control. Stoic attitudes with regard to pain and fear of addiction all contribute to how patients view themselves within the health care system. It is only in the last 20 years that the need to provide appropriate information to patients has become routine. It has become evident that pain will not be managed well if the patient is not an active participant in the process. Pain standards recommend that it is part of all health professionals' history taking and assessment of pain to review patient knowledge with regard to any pain and pain relief and ensure that patients are given accurate, appropriate information in a timely manner.

Reflective activity

Consider your patient profile. How much do your patients know about pain and pain relief? Ask three patients: (a) do they think suffering pain is bad for them? (b) what painkillers are they on? and (c) are there any side effects from the painkillers?

Anxiety and depression

The role of anxiety and depression in the experience of both acute and chronic pain has been the subject of debate for some time. Anxiety, depression and pain are often linked together in the literature. However, Symreng and Fishman (2004) suggest that rigorous studies on the interactions between anxiety and pain are still in their infancy. There is increasing evidence that psychological disorders such as depression (Tsai, 2005) often coexist with or are exacerbated by chronic pain. For example, Kenefick (2004) conducted a study to examine the relationships among cognitive impairment, communication and pain in elderly nursing residents (n = 111). Results identified that cognitive impairment is significantly associated with depression, and pain increases the strength of the relationship between cognitive impairment and depression.

Carr *et al.* (2005) explored the impact of anxiety and depression on women's pain experience over time following surgery. Preoperative anxiety was found to be predictive of postoperative anxiety, with patients who experienced high levels of anxiety before surgery continuing to feel anxious afterwards. Anxiety and depression scores increased as pain increased.

Evidence base

The role of patient information and knowledge on pain perception and anxiety has been the subject of considerable research over time. It is intuitive that if patients were more knowledgeable they would be less anxious, more informed and take more control of their pain management. The research studies of Hayward (1971)and Boore (1978) are considered seminal in terms of examining the role of patient information and its effect on anxiety post surgery. One of the first studies in relation to patient information came from Hayward in 1971. He hypothesised that those patients who were given information appertaining to their illness and recovery would, when compared with an appropriate control group, report less anxiety and pain during the postoperative period. He studied two hospitals and had an informed group and a control group in each hospital. In the first hospital, although the informed group consistently had lower pain scores the only significant differences were for day five post surgery when informed patients were recording significantly less pain. In the second hospital the differences in pain between the informed group and the control group reached significance on both day four and day five post surgery. Hayward concluded that the results gave strong support to the idea that informed patients became relatively pain free more rapidly than usual. Patients in both hospitals in the informed group consistently received less analgesic medications. However, this did not reach statistical significance on all postoperative days.

Following on from this Boore (1978) tested the hypothesis that giving information about prospective treatment and care preoperatively, and teaching exercises to be performed postoperatively, would minimise the rise in biochemical indicators of stress. Pain scores were measured. There was very little difference to be seen between experimental and control groups and no significant results were obtained. Much other work has been completed since then measuring narcotic use, pain and anxiety following information given to patients. However, experimental studies on information given pre surgery do not demonstrate significant differences in level of pain experienced postoperatively (Hawkins and Price,1993; Schwartz-Barcott *et al.*, 1994; Hawkins, 1997).

More recently Lin and Wang (2005) suggest that preoperative nursing intervention for pain has positive effects for patients undergoing surgery. This experimental study provided patients with preoperative education. The study reports decreased anxiety and pain levels with a preoperative nursing intervention.

The role of anxiety, depression and patient information in the reporting and management of pain will remain topical for some time. Due to their inevitable interaction, when pain and anxiety are both present, it may be difficult to identify which one is the predisposing factor. Rigourous scientific research with larger sample sizes and more diverse patient groups is required.

Key points Top tips

- Patients' knowledge and attitudes influence how their pain is managed
- Culture, race and ethnicity are important considerations
- Education of staff is critical
- Multidisciplinary teams and multimodal approaches support quality patient care
- Governments and service managers have significant roles in ensuring quality pain management approaches

Case study

The importance of a comprehensive pain assessment

Michael is a 75-year-old farmer living in a very rural area. He runs a 150 acre farm with his wife Kate, looking after livestock, growing wheat and selling a number of market garden vegetables. They have two farm labourers who help three days a week. Michael is very 'hands on' and likes to spend about six to eight hours on the farm a day. He is very fit, although he has some arthritis but it is not so severe as to hamper his work. Michael is also very sociable and leads on the local whist games in the community hall and enjoys a few pints of beer at the weekends.

Kate has been getting a bit worried that in the last few weeks Michael doesn't seem to have much energy and falls asleep most evenings around seven o'clock which is very unlike him. He has also stopped going out to his whist games. She persuades him to visit the general practitioner. The general practitioner discusses Michael's tiredness with him. They talk about how busy the farm is. She is not unduly concerned as she perceives he is not getting any younger and maybe needs to slow down a bit. Michael leaves the general practitioner satisfied that he knew all along nothing was wrong and that Kate was worrying unnecessarily.

Michael and Kate's daughter Fiona comes to visit the next weekend. She is alarmed at how her Dad seems to be deteriorating and goes out with him for a walk. Fiona notices that every now and again her Dad puts his right hand behind him and rubs his back. She asks him if there is a problem. He says no but Fiona insists on looking at his back. She is astonished to see large fading bruises and asks him what happened. It turns out a gate had fallen on his back a few weeks beforehand. He admits to it being a bit sore saying he is not sleeping well.

Fiona immediately brings him back to the general practitioner for a full examination. Following an X-ray at the local hospital it turns out that Michael has no serious damage but has soft tissue injury. He is started on nonsteroidal anti-inflammatories and over the course of a week starts to sleep again and in three weeks has his usual energy levels back.

Kate asks him why he didn't say anything. He said that it was only a bit of pain which you should be able to put with. He knew it was going to get better anyway so why worry anyone. He said he thought he was just wasting the general practitioner's time.

1 By taking the approach that Michael's age was slowing him down meant that the general practitioner missed Michael's real diagnosis. What other questions could she have asked him?
2 How did Michael's stoic attitude affect his quality of life?

Rapid recap

1 Name three factors which form the context of patients' pain.
2 Define what is meant by culture.
3 Name two organisations who view pain as a public health issue.

References

Blanchard J.C. (2003) Racial and ethnic disparities in health: An emergency medicine perspective. *Academic Emergency Medicine*, **10**(11): 1289–1293.

Bonham V.L. (2001) Race, ethnicity, and pain treatment: striving to understand the causes and solution to the disparities in pain treatment. *Journal of Law, Medicine and Ethics*, **29**: 52–68.

Boore J. (1978) *Prescription for Recovery*. London, RCN Publication, pp. 50–56.

Bostrom B.M., Ramberg T., Davis B.D. and Fridlund B. (1997) Survey of post-operative patients' pain management. *Journal of Nursing Management*, **5**: 341–349.

Brydon C.W. and Asbury A.J. (1996) Attitudes to pain and pain relief in adult surgical patients. *Anaesthesia*, **51**: 279–281.

Carr E.C.J. (1990) Postoperative pain: patient's expectations and experiences. *Journal of Advanced Nursing*, **15**: 89–100.

Carr E.C.J. and Thomas V.J. (1997) Anticipating and experiencing post-operative pain: the patient's perspective. *Journal of Clinical Nursing*, **6**: 191–201.

Carr E.C.J., Thomas V.N.T. and Wilson-Barnet J. (2005) Patient experiences of anxiety, depression and acute pain after surgery: a longitudinal perspective. *International Journal of Nursing Studies*, **42**: 521–530.

Choi M., Yate P., Coats T., Kalinda P. and Paul E.A. (2000) Ethnicity and prescription of analgesia in an accident and emergency department: Cross-sectional study. *BMJ*, **320**: 980–981.

Clinical Standards Advisory Group (CSAG) (2000) Services for Patients with Pain. Department of Health, London.

Crombie I.K. (1999) In *Epidemiology of Pain* (eds. Crombie I.K., Croft P.R., Linton S.J., LeResche L. and Von Korff M.). IASP Press, Seattle, p. xi.

Dimsdale J.E. (2000) Stalked by the past: The influence of ethnicity on health. Presidential address. *Psychosomatic Medicine*, **62**: 161–170.

Fosnocht D.E., Heaps N.D. and Swanson E.R. (2004) Patient expectations of pain relief in the *American Journal of Emergency Medicine*, **22**(4): 286–288.

Green C.R., Anderson K.O., Baker T.A., Campbell L.C., Decker S., Fillingim R.B. *et al.* (2003) The unequal burden of pain: Confronting racial and ethnic disparities in pain. *Pain Medicine*, **4**(3): 227–294.

Green R.G. (2004) Racial Disparities in Access to Pain Treatment. *Pain Clinical Updates,* XII(6).

Hawkins R. and Price K. (1993) The effects of an education video on patients' requests for postoperative pain relief. *Australian Journal of Advanced Nursing*, **10**(4): 32–40.

Hawkins R.M.F. (1997) The role of the patient in the management of post surgical pain. *Psychology and Health*, **12**: 565–577.

Hayward J. (1971) *Information – A Prescription Against Pain*. RCN Publication, London. Series 2 Number **5**: 67–112.

Hume A. (1994) Patient knowledge of anaesthesia and peri-operative care. *Anaesthesia*, **49**: 715–718.

International Association for the Study of Pain (1999) *Epidemiology of Pain* (eds. Crombie I.K., Croft P.R., Linton S.J., LeResche L. and Von Korff M.). IASP Press, Seattle.

Jamison R.N., Ross M.J., Hoopman P., Griffen F., Levy J., Daly M. and Schaffer J.L. (1997) Assessment of postoperative pain management: patient satisfaction and perceived helpfulness. *The Clinical Journal of Pain*, **13**(3): 229–236.

Jones K. (2005) Diversities in approach to end-of-life: A view from qualitative literature. *Journal of Research in Nursing*, **10**(4): 431–454.

Juhl I.U. (1993) Postoperative pain relief, from the patients' and the nurses' point of view. *Acta Anaesthesiololgy Scandinavia*, **37**: 404–409.

Kenefick A.L. (2004) Pain treatment and quality of life. *Journal of Gerontological Nursing*, **30**(5): 22–29.

Kim H. and Dionne R.A. (2005) Genetics, pain, and analgesia. *Pain Clinical Updates*, XIII(3).

Kuhn S., Cooke K., Collins M., Jones M. and Mucklow J.C. (1990) Perceptions of pain relief after surgery. *British Medical Journal*, **300**: 1687–1690.

Laing R., Lam M., Owen H. and Plummer J.L. (1993) Perceived risks of postoperative analgesia. *Australian and New Zealand Journal of Surgery*, **63**: 760–765.

Lasch K.E. (2002) Culture and pain. *Pain Clinical Updates*. X(5).

Lavies N., Hart L., Rounsefell B. and Runciman W. (1992) Identification of patient, medical and nursing staff attitudes to postoperative analgesia: stage 1 of a longitudinal study of postoperative analgesia. *Pain*, **48**: 313–319.

Lin L.Y., Wang, R.H., (2005) Abdominal surgery, pain and anxiety: preoperative nursing intervention. *Journal of Advanced Nursing*, **51**(3): 252–260.

Lindon S.J. and Skevington S.M.(1999) In *Epidemiology of Pain* (eds. Crombie I.K., Croft P.R., Linton S.J., LeResche L. and Von Korff M.). IASP Press, Seattle, p. 27.

Macintyre P.E. and Ready L.B. (2002) *Acute Pain Management, A Practical Guide,* 2nd edn. W.B. Saunders, Elsevier Science, Philidelphia.

McNeill J.A., Sherwood G.D., Starck P.L. and Thompson C.J. (1998) Assessing Clinical Outcomes: Patient Satisfaction with Pain Management. *Journal of Pain and Symptom Management*, **16**(1): 29–40.

Miaskowski C., Nichols R., Brody R. and Synold T. (1994) Assessment of patient satisfaction utilising the American Pain Society's quality assurance standards on acute and cancer related pain. *Journal of Pain and Symptom Management*, **9**(1): 5–11.

Owen H., McMillan V. and Rogowski D. (1990) Postoperative pain therapy: a survey of patients' expectations and their experiences. *Pain*, **41**: 303–307.

Pellino T.A. (1997) Relationships between patient attitudes, subjective norms, perceived control and analgesic use following elective orthopaedic surgery. *Research in Nursing and Health*, **20**: 97–105.

Schwartz-Barcott D., Fortin J.D. and Kim H.S. (1994) Client–Nurse interaction: testing for its impact in preoperative instruction. *International Journal of Nursing Studies*, **31**(1): 23–35.

Scott N.B. and Hodson M. (1997) Public perceptions of postoperative pain and its relief. *Anaesthesia*, **52**:438–422.

Sherwood G., McNeill J.A., Hernandez L., Penarrieta I. and Peterson J.M. (2005) A multinational study of pain management among Hispanics. *Journal of Research in Nursing*, **10**(4): 403–423.

Symreng I. and Fishman S.M. (2004) Anxiety and Pain. *Pain Clinical Updates*, XII(7).

Tamayo-Sarver J.H., Hinze S.W., Cydulka R.K., Baker D.W. (2003) Racial and ethnic disparities in emergency department analgesic prescription. *American Journal of Public Health*, **93**(12): 2067–2073.

Tan G., Henson M.P., Thornby J. and Anderson K.O. (2005) Ethnicity, control appraisal, coping and adjustment to chronic pain among black and white Americans. *Pain Medicine*, **6**(1): 18–28.

Tsai P.F. (2005) Predictors of distress and depression in elders with arthritic pain. *Journal of Advanced Nursing*, **51**(2): 158–165.

Winefield H.R., Katsikitis M., Hart L.M. and Rounsefell B.F. (1990) Postoperative pain experiences: relevant patient and staff attitudes. *Journal of Psychosomatic Research*, **34**(5): 543–552.

Zinn C. (2003) Doctors told to use positive language in managing pain. *BMJ*, **326**: 301.

6 Assessment of pain and the effect of pain interventions

Learning outcomes

By the end of this chapter you should be able to:

- ★ Discuss the importance of pain assessment
- ★ Describe various approaches to assessment of pain
- ★ Detail pain assessment tools.

Introduction

This chapter looks at how the effective assessment of pain contributes towards the development of an appropriate pain management plan. Pain assessment is not a choice; it is a necessary component of the management of patients' pain in all health care settings. It is the key to decision making in relation to pain management interventions and provides evidence as to whether such interventions are effective.

Pain assessment

The key to accurate pain assessment lies in a consistent scientific approach. Appropriate assessment and management is essential in light of the many disabling effects of pain which are referred to in this and other chapters. A freshness of thinking and modern approaches are needed. Health professionals should be confident in their pain assessments and patients empowered to contribute to this approach. It should be remembered that much of the physical, psychological and social assessment of patient symptoms is highly dependent on careful, accurate health-professional skills and on patient reporting. Pain assessment should be approached with considered, reliable and valid processes. The accurate recording of patients' reported pain will provide sensitive and consistent results for the majority of the patient population (Moore *et al.* 2004).

Traditional debates continue to highlight the difficulty of assessing pain in the absence of quantitative definitive pain assessment tools in the form of physiological measures. The literature is replete with evidence of underassessment and poor documentation of pain. Little wonder then that health professionals, patients and families approach pain assessment with hesitation and a sense of suspicion. Societal anecdotes only compound the thinking by promoting the consideration that a little pain never harmed anyone. Who wants to be labelled a complainer?

A team approach

Evidence supports the concept that patient interactions linked to pain assessment should not be left to one person and that the patient benefits more from a multidisciplinary team approach utilising an integrated pathway process. An organisational system's approach to pain assessment will ensure a culture of regular appropriate pain assessment and ensure that pain assessment is part of an overall pain policy linked to organisational and national targets and guidelines.

Organisations should take pain assessment seriously and ensure that pain policies/protocols and guidelines are comprehensive enough to encompass modern approaches to pain assessment.

Nursing Practice Development Co-ordinator

We as an organisation decided to take an overall approach to pain management. We developed a hospital policy which adopted a definition of pain aligned to 'pain is what the patient says it is'; all patients now receive regular pain assessment . . . we use a Visual Analogue Scale (VAS) for surgical patients and a longer scale for chronic and cancer pain. Additionally we introduced short pain courses for all staff . . . these are multidisciplinary . . . they are provided by doctors, nurses, pharmacists, anaesthetists and physiotherapists and attended by the same groups. These are now part of induction programmes for new staff.

The individuals providing care at particular points in the patient episode have a responsibility to assess and document pain appropriately. However, approaches to deciding which methods are appropriate for assessment of pain should be based on relevant timely information and audit. This will allow for planning of necessary training and education for staff and ensure that pain assessment approaches are relevant to patient groups and case mix.

Pain assessment is an interactive and collaborative process involving the patient and family, nurse, doctor, anaesthetist, physiotherapist and other health carers as appropriate to the care setting (see Figure 6.1 below, which highlights the interactions). Such assessment will provide the basis for selecting and monitoring interventions.

A considered approach in the form of service guideline is key to successful and valid pain assessment.

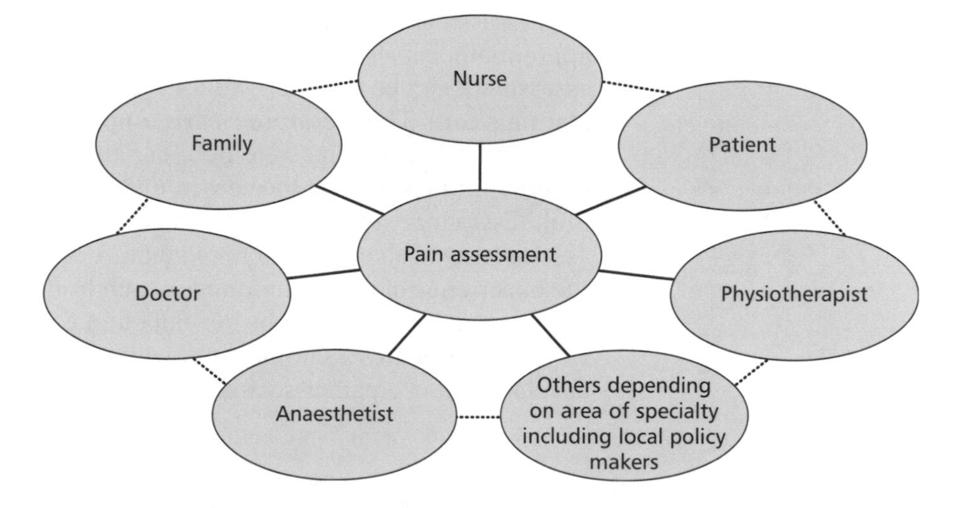

Figure 6.1 Multidisciplinary team involved in pain assessment

Reflective activity

You have just moved to a new general surgical ward in a medium-sized district hospital – there are no processes in place to record pain assessments on this ward (although you are aware that some of the other wards do undertake some pain assessment). There is no work that you are aware of to standardise pain assessment and management. Your clinical experience informs your thinking that this as hoc approach is not in the patients' best interests.

Consider how you, as one health care professional, can approach making a service guideline happen.

Top tip: Are there guidelines in the service for other activities? How were these developed? Who else is interested in pain management? View the following website: www.whocancerpain.wisc.edu/eng/14_2/resources.html to support your arguments to introduce pain assessment processes.

Key points

- Pain remains the most common symptom associated with acute, chronic and palliative care
- Pain assessment provides the only way to ensure that methods to manage pain are appropriate and effective
- Pain impacts on an individual's physical and psychological wellbeing and is an important challenge to functional ability
- Pain is a determinant of quality of life
- Pain is a key indicator of effectiveness of services and interventions
- Pain assessment identifies patterns of pain occurrence
- Pain assessment is complex and subjective
- Many pain assessment tools exist – which to choose?

Pain assessment at clinical service level, regardless of setting, should be appropriate, consistent and documented. This means that the assessment can be reviewed, audited and taken into consideration when planning care. The literature clearly emphasises the role of the nurse, as the continued carer, in assessment of pain. It must be recognised that different nurses provide the care at different times as part of a multidisciplinary team. The challenge lies in combining the varied assessments of all contributors to create a cohesive patient plan. While the experience of pain is unique for each individual, a process for monitoring this pain must be developed in a collaborative manner.

Table 6.1 details a sample audit with results which could support the development of a pain assessment strategy.

Table 6.1 A sample audit

Steps	Actions	Sample results
STEP 1	Audit of patient caseload: 1. How many patients go through the service per week? 2. What is the case mix of patients? 3. What is the average length of stay? 4. How many on average have pain?	A general surgical ward with mixed age group performs on average 10 bowel resections and 10 transurethral prostatectomies per week. Average length of stay is five days. Fifty per cent of patients have moderate pain on day two experienced on movement.
STEP 2	Audit of pain experiences: 1. What types of pain? 2. What levels of pain? 3. What is the average age group?	Maximum pain score is 8 on a 1–10 scale and minimum 2. Pain peaks on day one post surgery and generally decreases.
STEP 3	Current methods of pain assessment: 1. Regular? 2. Single or multidimensional scales? 3. Is this documented?	Pain is assessed in an ad hoc manner depending on nurse on duty.
STEP 4	How many health professionals have responsibility for the patient's pain management? 1. Identify who and when the various health professionals interact with the patients. 2. How often are decisions made with regard to pain management?	A surgical patient will on average have at least six different health professionals evaluate and monitor their progress, i.e. nurse(s) × 3 from each shift, doctor(s), anaesthetist, physiotherapist.
STEP 5	Now there is a baseline of pain and the patterns of pain on the clinical area are identified.	

Over to you

What information can you gain from the above audit to support your thinking on developing pain assessment processes for the above surgical ward?

Pain assessment – moving forward

Pain assessment techniques determine the intensity, quality, location and duration of pain; they aid diagnosis and provide a process to evaluate the relative effectiveness of different therapies. The language of pain frequently provides the key to diagnosis and may even suggest the course of therapy (Melzack and Katz, 1994).

Two decades ago, McCaffrey said that the 'assessment of pain and its relief is no simple matter' (McCaffrey 1983). There is little since then to suggest that assessment of pain has become less complex. Indeed there is much research to indicate that pain continues to be under-assessed. Current assessment and the recording of pain by nurses has been shown to be incomplete, inaccurate (often underestimated) and descriptive of location rather than severity of pain (Harrison, 1993; Bowman, 1994; Field, 1996; Idvall *et al.,* 2002; Sloman *et al.*, 2005).

The patient's self report of pain remains the single most reliable indicator of pain and should be recorded in the patient's record as the pain assessment (Acute Pain Management Guideline Panel, 1992; McCaffery and Ferrell, 1994; Moore *et al.*, 2004). It is pleasing to note that new approaches for assessment of pain have emerged and a number of policy and guidance documents give clear direction that assessment of pain in a consistent manner is key to successful pain management.

The nurse's possession of adequate knowledge is critical to effective pain management. The nurse spends the most time with the patient in pain, assesses the patient's pain level, evaluates the information based on assessment and communicates patient progress to the multidisciplinary team. Therefore, nurses, in particular, have a direct responsibility for the provision of measures to relieve pain. Regardless of clinical speciality, pain is a common clinical situation encountered by nurses. Nurses are with patients during their care episodes, when patients report the presence of pain, and monitoring pain control depends on the expertise of the individual nurse. Yet formal pain assessment, such as the use of a pain score or pain chart, does not appear to be regular practice for nurses (Hastings, 1995; Francke, 1996; Taylor, 1997).

Key points Top tips

- The assessment of pain is integral to the planning and implementation of nursing care to relieve pain (Zalon, 1993)
- Assessment of pain should be included in the regular recording of vital signs (Black, 1991)
- Accurate pain assessment is essential for good nursing and medical care, for judging the status and progress of patients, the impact and efficacy of treatments and even for reaching a proper diagnosis (Choiniere *et al.*, 1990)
- The patient's self report is the most reliable indicator of pain

Pain charts and tools have been developed to assist in the accurate measurement and documentation of patients' subjective pain.

Benefits of using such pain assessment charts and tools can be summarised as follows:

- information obtained helps to establish the pattern of pain
- recording when pain intervention occurs and relating the level of pain with the timing and the type of medication/pain intervention provides evaluation of the effects of such intervention
- including the patient in monitoring their pain involves them in their plan of care and supports objective thinking about their pain
- the care provider and the patient can be more specific setting targets for pain relief.

Reflective activity

Assessment of pain requires the recognition that patients' self report of pain is the single most reliable indicator. Consider a time when you found the above statement challenging and felt that your own clinical experience was more relevant.

Difficulties in communicating pain

Pain assessment tools utilised must conform to the communication capabilities of the patient. Some patients will not be able to verbalise their pain or may not be able to describe their pain in the manner requested by the health care professional.

Existing pain charts and tools support the measurement of various aspects of pain. However, not all patients can communicate their pain. One of the biggest challenges in pain assessment is capturing the non-verbal aspects of pain. This is of particular interest to nurses working in clinical areas such as critical care.

Challenges to pain assessment approaches

- There are differences reported in patients' and nurses' assessment of pain.
- Patients depend on health professionals to recognise their pain.
- Some patients are cognitively impaired and therefore cannot express their pain, e.g. confused, unconscious or elderly.
- Patients may have low expectations of pain relief and therefore underestimate pain.

- Patients may not report pain as they may not understand the importance of pain assessment and documentation in reducing pain.
- Pain assessment may not be documented.
- Lack of a consistent approach from the health care team.
- Health professionals consider they know best.
- Lack of a systematic approach to pain assessment.

Evidence base

Key position statements regarding pain assessment and documentation

Timeline

- **2005** National Cancer Institute: *'Failure to assess pain is a critical factor leading to undertreatment. Assessment involves both the clinician and the patient. Assessment should occur: At regular intervals after initiation of treatment, at each new report of pain and at a suitable interval after pharmacologic or nonpharmacologic intervention, e.g., 15 to 30 minutes after parenteral drug therapy and 1 hour after oral administration.'*
- **2003** The Royal College of Anaesthetists and The Pain Society, the British Chapter of the International Association for the Study of Pain: *'Pain and its relief must be assessed and documented on a regular basis. Pain intensity should be regarded as a vital sign and along with the response to treatment and side effects should be recorded as regularly as other vital signs such as pulse or blood pressure.'*
- **2001** The Joint Commission on Accreditation of Healthcare Organisations (JCAHO) implemented pain assessment and management standards and began to assess compliance. *'In the initial assessment, the organisation identifies patients with pain. When pain is identified, the patient can be treated within the organisation or referred for treatment. The scope of treatment is based on the care setting and services provided. A more comprehensive assessment is performed when warranted by the patient's condition. This assessment and a measure of pain intensity and quality (e.g., pain character, frequency, location, duration) appropriate to the patient's age, are recorded in a way that facilitates regular reassessment and follow up according to the criteria developed by the organisation.'*
- **2000** American Pain Society provides a statement on pain assessment and treatment in the managed care environment highlighting timely and effective assessment of pain: *'all patients benefit from timely and effective assessment and treatment by their primary care providers.'*
- **1990** The report of the Royal College of Surgeons and Anaesthetists Working Party on Pain after Surgery: *'postoperative intervention can be assessed and improved only if some form of measurement of effect is made. The report suggests that the routine use of a simple pain assessment system, with treatment based on assessment is essential, if progress is to be made.'*

Advance Nurse Practitioner

Finding an approach to pain assessment that is used consistently to support all the clinical team to make appropriate decisions regarding pain management requires consideration and . . . hard work! Firstly, I would recommend identifying key interested parties such as the anaesthetist, ward nurses, practice development etc. Gather up some evidence examples to share and, if possible, some exemplars of charts/tools from other organisations.

The pain experience

Pain is not a single sensory experience. It has many dimensions and facets. These must be considered and understood before approaches to pain assessment can be decided upon. Pain appears to have three main dimensions: sensory, emotional and an intensity aspect. Clinical pain measures consist of behavioural measurements, observational data, self-reported behaviours and subjective pain reports.

Pain is an individual experience which will be influenced by many factors including culture, coping strategies, fear, anxiety, cause of pain and previous experiences.

The box below details the pain aspects to be considered when assessing pain.

- Location
- Onset
- Duration
- Quality or characteristics (neuropathic or nerve, nociceptive, visceral)
- Aggravating factors
- Interventions/relieving factors
- Associated symptoms, e.g. dizziness, headache, nausea, sweating, palpitations, shortness of breath, vomiting
- Pain and sleep
- Pain caused by physical activity such as coughing, deep breathing or movement

A pain history, as outlined in Table 6.2, can be utilised. Some pain histories will be quite short and others will be longer depending on your patient case mix. You should choose which aspects suit your patients' needs, e.g. a patient having a simple surgical procedure may need questions with regard to pain intensity on a regular timescale while a patient diagnosed with cancer metastasis may require a multidimensional comprehensive pain history.

Table 6.2 Elements to be considered when taking a pain history

Domain	Process
Circumstances associated with pain onset	For example coughing, exercise
Primary site of pain	Consider utilisation of a body chart (see Figure 6.2)
Radiation of pain	Consider utilisation of a body chart (see Figure 6.2)
Character of pain	A multidimensional pain measurement tool as described later in this chapter will support this description, e.g. is pain throbbing, sharp, aching etc.?
Intensity of pain	A single-dimensional pain measurement tool as described later in this chapter will support this description, e.g. use of a visual analogue scale or verbal rating scale. It is important also to assess intensity of pain in certain circumstances such as at rest and on movement. Utilise time periods to support this such as: – at present – during last week Ask questions such as lowest and highest rating experienced by patients
Factors altering pain	Ask what makes it worse and what makes it better, e.g. is the pain aggravated on movement?
Associated symptoms	Does the pain bring on other symptoms such as nausea, dizziness, vomiting?
Temporal factors	Is pain present continuously or otherwise?
Effect of pain on activities	Is pain preventing the patient carrying out their daily activities?
Effect of pain on sleep	Is pain interrupting sleep? It is important to ask whether pain wakes the patient up or if it prevents the patient from sleeping
Medications taken for pain	Is the patient taking any medications for pain at time of history taking? List these medications, how often they are taken and which the patient rates as being helpful. How long has the patient been on these medications? Are there any unwanted side effects from the medication?
Other interventions used for pain	Is the patient using any other interventions such as physiotherapy, acupuncture etc.? (See chapter 8.) Encourage the patient to think about this.
Health professionals consulted for pain treatment	Has the patient been treated for this pain before? Are they currently attending anyone?

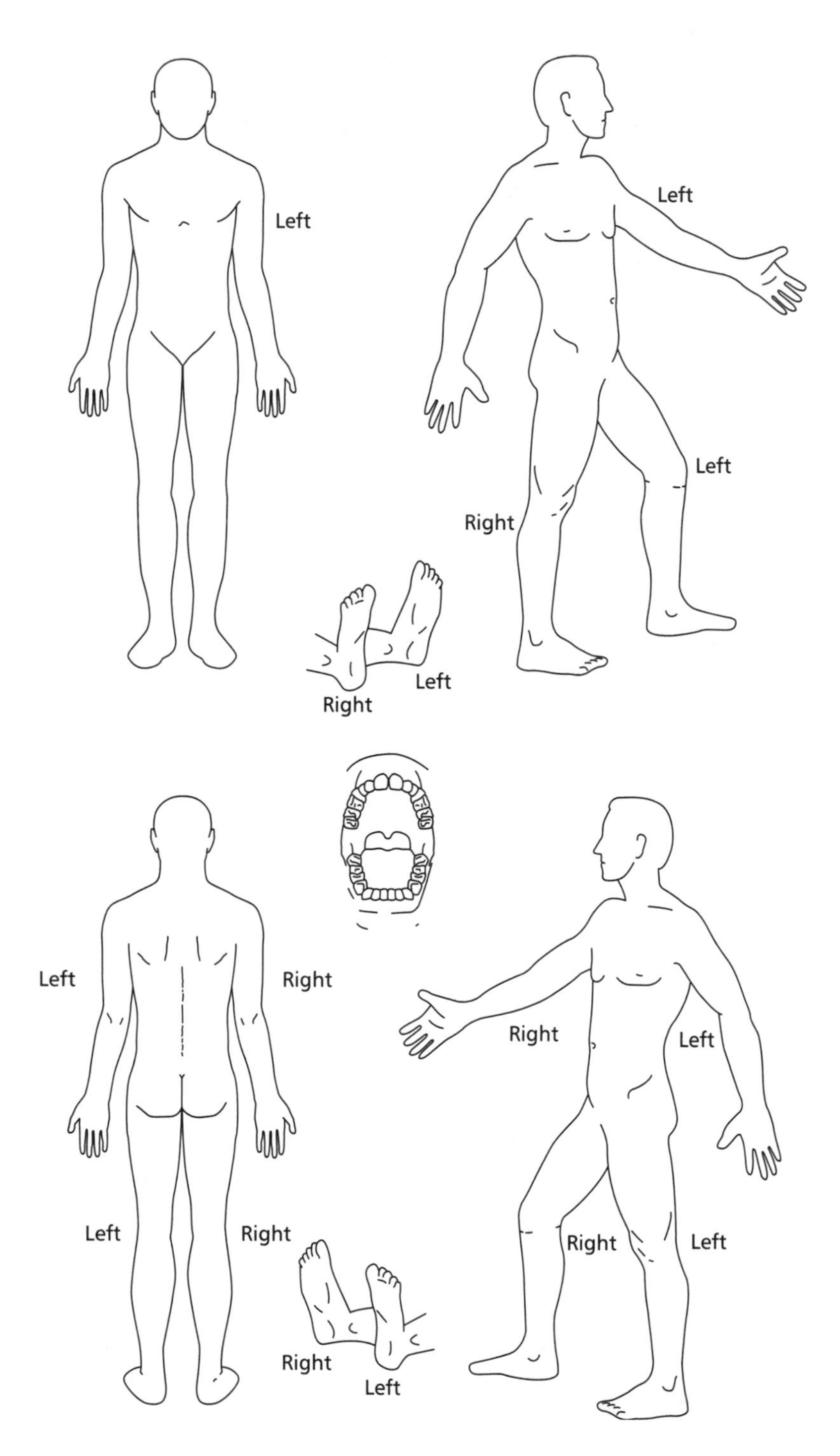

Figure 6.2 Body chart

The box below highlights other information that is useful to gather when obtaining a pain history.

- Expectations of outcome of pain treatment
- Patient's belief concerning the causes of pain (are there cultural issues?)
- Reduction in pain required to resume 'reasonable activities'
- Patient's typical coping response for stress or pain, including presence of anxiety
- Psychiatric disorders (e.g. depression or psychosis)
- Family expectations and beliefs about pain, stress and postoperative course
- Ways the patient describes or shows pain
- Patient's knowledge, expectations and preferences for pain management

The pain history list above has been adapted from the National Health and Medical Research Council, Australia Acute Pain Management Scientific Evidence of 1999 and 2005 (www.nhmrc.gov.au).

Clinical pain measurement tools

The pain history should involve the use of pain assessment tools. There are numerous tools described in the literature and the following qualities should be considered when choosing the tool best suited for your patient case mix – see the box below.

Validity of tool. *This is an expression of the degree to which a measurement measures what it proposes to measure.*
Reliability of tool. *Reliability is defined as determining that the instrument is measuring something in a reproducible and consistent fashion. A reliable instrument is stable, i.e. repetition of that measurement gives the same result.*
Ease of use for both patients and health care professionals. *If the tool is very lengthy or complex to use it may not be suitable for particular patient groups. Depending on the patient case mix there will be different dimensions of pain to capture.*

Pain tools are generally categorised as single and multidimensional with single-dimensional capturing the intensity aspect of pain.

Single-dimensional tools

The three most commonly used single-dimensional tools are the Visual Analogue Scale (VAS), the Numeric Rating Scale (NRS) and the Verbal

Keywords

Pain intensity
This is a measure of the strength of the pain, i.e. how bad the pain is. Pain intensity is usually relative to previous pain experienced. It does not take into account quality of pain, e.g. burning, shooting

Rating Scale (VRS) (see Figure 6.3). **Pain intensity** only is measured. Such tools are suitable for patient groups such as postoperative patients or those with minor injuries. This type of measurement is often included as part of a multidimensional scale.

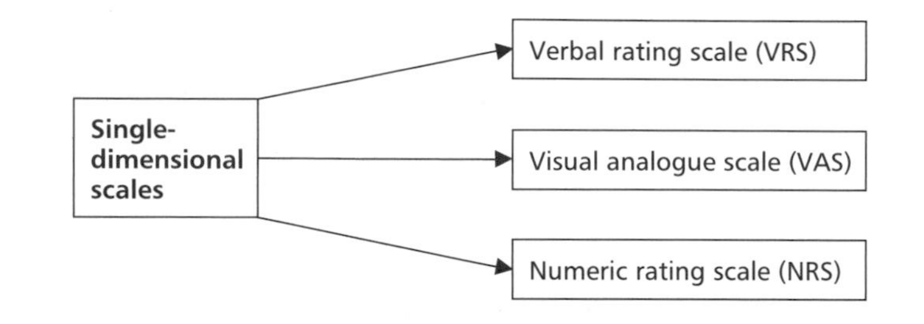

Figure 6.3 Single-dimensional scales

Evidence base

Much of the seminal research on the reliability, validity and sensitivity of these scales was completed in the 1970s and 1980s. However, new research compares these scales with various groups of patients. In 1974 Huskisson described the VAS as sensitive in measuring pain and pain relief. Early comparisons of the VAS and the VRS suggest that the VAS gives a closer assessment of patient experience (Ohnhaus, 1975; Scott and Huskisson, 1976; Sriwatanakul, 1983).

More recently, Cork *et al.* (2004) surveyed 85 chronic pain patients to determine if the simple VRS could be substituted for the VAS to measure pain intensity in chronic pain patients. Pearson correlation coefficient ($r = 0.906$) and p value (< 0.0001) showed excellent correlation between the two, although VRS showed a tendency to be higher than VAS ($p = 0.068$). The authors proposed that the VRS provides a useful alternative to the VAS scores in assessment of chronic pain. Breivik *et al.* (2000) suggest that a four-point VRS was less sensitive than the 100 mm VAS in acute pain but that the 11-point Numeric Rating Scale (NRS) compares well to the VAS and that choice should be based on subjective preferences.

Coll *et al.* (2004), following an integrated literature review, found the VAS to be methodologically sound, conceptually simple, easy to administer and unobtrusive to the patient.

Verbal Rating Scale (VRS) and Verbal/Graphic Rating Scale

See Figures 6.4 and 6.5.

Verbal Rating Scale (VRS)

- No pain
- Mild pain
- Severe pain
- Very severe pain

Figure 6.4 Four-point verbal rating scale

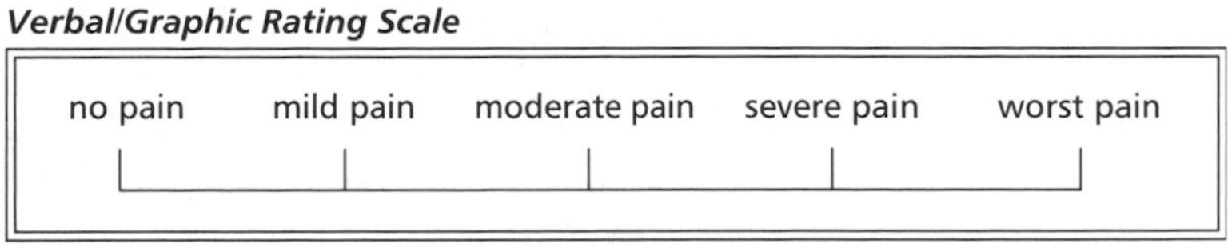

Verbal/Graphic Rating Scale

no pain | mild pain | moderate pain | severe pain | worst pain

Figure 6.5 Five-point verbal/ graphic rating scale

Descriptive rating scales consist of a list of adjectives which describe different levels of pain.These types of scale generally comprise 4–5 word categories. These word categories consist of descriptive pain words. The patient is asked to pick the word which best describes their pain. These words are then given a score. The least intense descriptor is given a score of zero, the next a score of one and so on. The descriptive rating scale can be given in verbal or written form. The patient's intensity score is the number associated with the word they choose as most descriptive of their pain level.

Visual Analogue Scale (VAS)

See Figures 6.6 and 6.7.

Figure 6.6 The VAS for pain severity measurement

Figure 6.7 The VAS for treatment effect

A Visual Analogue Scale (VAS) is a method of providing a simple way of recording subjective estimates of pain. It can be used to measure severity or improvement.

A VAS provides a continuous scale for estimation of the magnitude of pain. It consists of a straight line, the ends of which are defined in terms of the extreme limits of the pain experience. The scale, conventionally a 10 cm long straight line, may be printed either horizontally or vertically. Each end of the scale is marked with labels that indicate the range being considered. Phrases such as 'pain as bad as it could be' and 'no pain' can be used. The patient is asked to place a mark on the line at a point representing the severity of his pain. Measuring the distance of a patient's mark from zero scores the scale. The scale requires only about 30 seconds to complete. A comparison of 5 cm, 10 cm, 15 cm and 25 cm lines suggested that the 10 cm and 15 cm lengths have the least measurement error and that the 5 cm line provides the greatest error (Seymour, 1985). A number of steps are involved in the construction of the VAS.

The sensation or response and the extremes of that response must be clearly defined. End phrases and descriptive words should be short, readily understood and not so extreme that they will never be employed. Definite cut-off points must be made for the line, which should be of a length that can be interpreted as a unit. Although verbal labels define end points of the VAS it is recommended that neither numbers nor verbal labels be used to define intermediate points, as this may cause a clustering of scores around a preferred digit. As the scale is continuous, the restriction of a 3- or 5-point rating scale is overcome. McDowell and Newell (1996) conclude that the VAS is a more sensitive and precise measurement than descriptive pain scales and that, for the VAS, horizontal lines, are preferred to vertical.

Numeric Rating Scale (NRS)

See Figure 6.8.

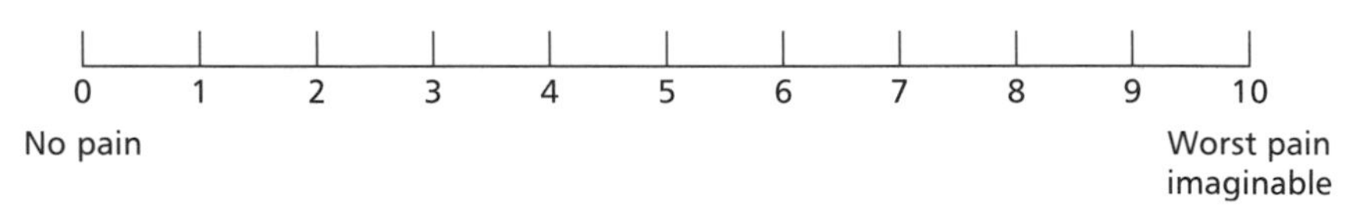

Figure 6.8 Numeric Rating Scale

This is a variation of the VAS and consists of asking the patient to rate their pain on a numerical scale 0–10 with 0 equalling no pain and 10 equalling worst pain imaginable. It has been described as sensitive to change and may be preferred for statistical purposes. The NRS, like the VAS, can be of varying lengths. The 101 cm length appears to have some advantages. It has 101 response categories, which may overcome the limitations in response categories found in other scales.

Pros and cons of single-dimensional scales

Pros:

- Can be used quickly.
- Only minimal instruction to patients is required, i.e. easy to administer.
- Are easily scored.
- Very sick patients are not taxed.
- Have been tested as reliable and valid pain intensity measures.
- Can be incorporated easily into current patient documentation, e.g. as part of vital signs chart.

Cons:

- Most assume that patients are either literate or numerical.
- Pain is treated as a single dimension only.

- *VRS*. With the VRS there is reliance on the use of words, i.e. it is necessary for the patient to translate a feeling into specific words which may not express exactly what the person is experiencing. Words can be ambiguous and the same word does not necessarily mean the same thing to each patient. It limits choices and improvements in pain relief. Effect of analgesic cannot always be measured due to the word restrictions.
- *VAS*. The estimation of pain intensity with the VAS requires an ability to transform a pain experience into a visual display, which involves perceptual judgement and accuracy. This perceptual ability is likely to influence the results, therefore the use of the VAS may not be possible in the elderly, the seriously ill or patients with organic brain disease (Seymour, 1985). It is suggested that as much as 7% of the population would not be able to use it (Kremer, 1981). Factors which may influence reliability and validity are learning, memory and perceptual judgement (Carlsson, 1983; Coll, 2004).
- *NRS*. There may be biases associated with this scale. Some patients may have a preference or an aversion to certain numbers, leading them to consistently choose or avoid those numbers.

Multidimensional measurements of clinical pain intensity

Multidimensional scales involve subscales, which represent different aspects of pain (see Figure 6.9). In an attempt to capture all aspects of pain, multidimensional scales are developing to support assessment of pain for those patients who cannot verbalise their pain experience.

The McGill Pain Questionnaire (MPQ) both long (LF-MPQ) and short (SF-MPQ) and the Non-verbal Pain Scale (NVPS) are described. The McGill Pain Questionnaires are the most widely utilised in both research and clinical practice for multidimensional measurement of pain.

The SF-MPQ was developed to provide a brief assessment. The LF-MPQ and SF-MPQ can be either interviewer administered or self-administered. The LF-MPQ takes 5–10 minutes to administer, and the SF-MPQ 2–5 minutes.

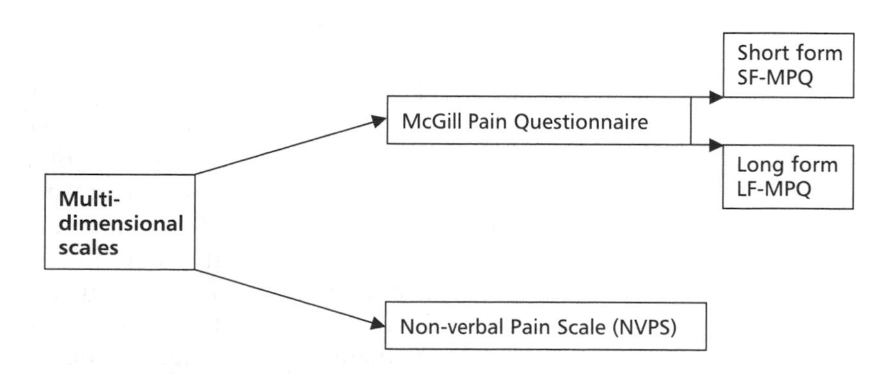

Figure 6.9 Multidimensional scales

Evidence base

The McGill Pain Questionnaire (MPQ) was developed in 1971 following much debate regarding language and meaning of pain (Melzack and Torgeson, 1971). Debate continues and conflicting views may be found with regard to factorial validity and transferability of the scales in the MPQ to other languages. The MPQ has now been translated into Spanish, Danish, Arabic, Chinese, French, German, Italian, Japanese, Norwegian, Polish and Slovak. In 1987, Melzack found the sensory, affective and total scores of the MPQ and SF-MPQ to be significantly correlated (Melzack, 1987). The MPQ has been more widely used in chronic pain than acute pain. It is suggested that acute pain involves less differentiation of sensory, affective and evaluative language dimensions (Reading, 1982).

The SF-MPQ compares well with the long form of the MPQ (Melzack, 1987; Dudgeon *et al.*, 1993). More recently, Grafton *et al.* (2005) studied 57 patients with osteoarthritis and suggested that it is a highly reliable measure of pain

The adapted Non-verbal Pain Scale (NVPS) described here is currently being tested by the University of Rochester Medical Centre.

Long-form McGill Pain Questionnaire (LF-MPQ)

The long McGill Pain Questionnaire (see Figure 6.10) comprises a top sheet to record necessary medical information, line drawings of the body to indicate the distribution of the pain, and word descriptors. The questionnaire consists primarily of three major classes of word descriptors – sensory, affective and evaluative – that are used by patients to specify subjective pain experiences. It contains an intensity scale and other items to determine the properties of the pain experience. The questionnaire was designed to provide quantitative measures of clinical pain that can be treated statistically. The three measures commonly used are: (1) the **p**ain **r**ating **i**ndex (**pri**) based on two types of numerical values that can be assigned to each word descriptor; (2) the number of words chosen; (3) the **p**resent **p**ain **i**ntensity (**ppi**) based on a 1–5 intensity scale, in which each number is associated with the following words: 1 = mild, 2 = discomforting, 3 = distressing, 4 = horrible and 5 = excruciating. The pri is based on values assigned to each pain word chosen, which are added up for a total score.

The MPQ groups pain-related words into three major classes and 16 subclasses. The classes are, firstly, words that describe the sensory qualities of the experience in terms of temporal, spatial, pressure, thermal and other properties; secondly, words that describe affective qualities in terms of tension, fear and autonomic properties that are part of the pain experience; and thirdly, evaluative words that describe the subjective overall intensity of the total pain experience.

Generally, the questionnaire requires 5–10 minutes to complete, but depending on patient profile, can take up to 20 minutes. The MPQ provides quantitative information that can be treated statistically. It is

McGILL PAIN QUESTIONNAIRE

RONALD MELZACK

Patient's Name ______________________ Date ____________ Time ________ am/pm

PRI: S ________ (1–10) A ________ (11–15) E ________ (16) M ________ (17–20) PRI(T) ________ (1–20) PPI ________

1 FLICKERING, QUIVERING, PULSING, THROBBING, BEATING, POUNDING	11 TIRING, EXHAUSTING
2 JUMPING, FLASHING, SHOOTING	12 SICKENING, SUFFOCATING
3 PRICKING, BORING, DRILLING, STABBING, LANCINATING	13 FEARFUL, FRIGHTFUL, TERRIFYING
4 SHARP, CUTTING, LACERATING	14 PUNISHING, GRUELLING, CRUEL, VICIOUS, KILLING
5 PINCHING, PRESSING, GNAWING, CRAMPING, CRUSHING	15 WRETCHED, BLINDING
6 TUGGING, PULLING, WRENCHING	16 ANNOYING, TROUBLESOME, MISERABLE, INTENSE, UNBEARABLE
7 HOT, BURNING, SCALDING, SEARING	17 SPREADING, RADIATING, PENETRATING, PIERCING
8 TINGLING, ITCHY, SMARTING, STINGING	18 TIGHT, NUMB, DRAWING, SQUEEZING, TEARING
9 DULL, SORE, HURTING, ACHING, HEAVY	19 COOL, COLD, FREEZING
10 TENDER, TAUT, RASPING, SPLITTING	20 NAGGING, NAUSEATING, AGONIZING, DREADFUL, TORTURING
	PPI: 0 NO PAIN, 1 MILD, 2 DISCOMFORTING, 3 DISTRESSING, 4 HORRIBLE, 5 EXCRUCIATING

BRIEF	RHYTHMIC	CONTINUOUS
MOMENTARY	PERIODIC	STEADY
TRANSIENT	INTERMITTENT	CONSTANT

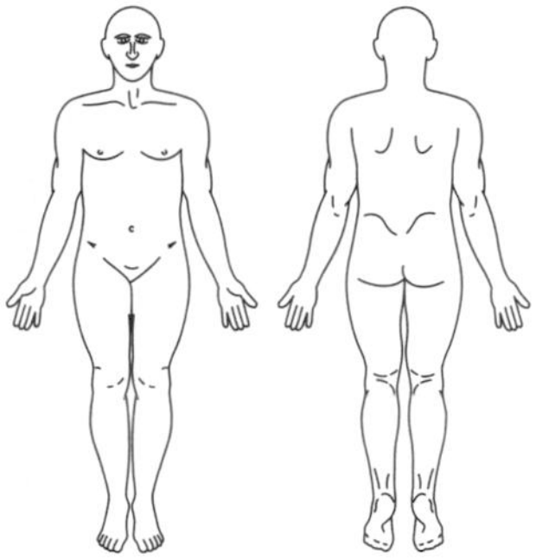

E = EXTERNAL

I = INTERNAL

COMMENTS:

© R. Melzack, 1975

Figure 6.10 Long-form McGill Pain Questionnaire (LF-MPQ). Reproduced with permission from Elsevier. First published in Pain Journal 3: 277–299, Melzack, © 1975 International Association for the Study of Pain.

sufficiently sensitive to detect differences among different methods to relieve pain. It also provides information about the relative effects of a given manipulation on the sensory, affective and evaluative dimensions of pain.

It is important for the interviewer to ensure that the patient understands the meaning of the words as some of the words may be beyond the patient's vocabulary. The length of time to administer the MPQ restricts its widespread use in ward situations where time is a precious commodity.

Short-form McGill Pain Questionnaire (SF-MPQ)

A short-form MPQ (SF-MPQ) has been developed (see Figure 6.11). The main component of the SF-MPQ consists of 15 descriptors (11 sensory and 4 affective), which are rated on an intensity scale as 0 = none, 1 = mild, 2 = moderate or 3 = severe. Three pain scores are derived from the sum of the intensity rank values of the words chosen for sensory, affective and total descriptors. The SF-MPQ also includes the present pain intensity (ppi) index of the standard MPQ and a Visual Analogue Scale.

This may be useful where qualitative information is important as well as pain intensity scores. It takes less time than the long form of the MPQ.

SHORT FORM McGILL PAIN QUESTIONNAIRE

PATIENT'S NAME:_______________________ DATE:________

	NONE	MILD	MODERATE	SEVERE
THROBBING	0)____	1)____	2)____	3)____
SHOOTING	0)____	1)____	2)____	3)____
STABBING	0)____	1)____	2)____	3)____
SHARP	0)____	1)____	2)____	3)____
CRAMPING	0)____	1)____	2)____	3)____
GNAWING	0)____	1)____	2)____	3)____
HOT-BURNING	0)____	1)____	2)____	3)____
ACHING	0)____	1)____	2)____	3)____
HEAVY	0)____	1)____	2)____	3)____
TENDER	0)____	1)____	2)____	3)____
SPLITTING	0)____	1)____	2)____	3)____
TIRING-EXHAUSTING	0)____	1)____	2)____	3)____
SICKENING	0)____	1)____	2)____	3)____
FEARFUL	0)____	1)____	2)____	3)____
PUNISHING-CRUEL	0)____	1)____	2)____	3)____

NO PAIN |________________________________| WORST POSSIBLE PAIN

0	NO PAIN	____
1	MILD	____
2	DISCOMFORTING	____
3	DISTRESSING	____
4	HORRIBLE	____
5	EXCRUCIATING	____

Figure 6.11 Short-form McGill Pain Questionnaire(SF-MPQ). Reproduced with permission from Elsevier. First published in Pain Journal 30 (2): 191–197, Melzack, © 1987 International Association for the Study of Pain.

Adult Non-verbal Pain Scale (ANVPS)

Cognitively impaired patients present unique challenges to pain assessment. Scales which require verbal or numeric skills are not useful and alternative approaches must be used. The ANVPS, for example, takes account of multiple indicators of pain: posture, usual behaviours, self report, facial expression, mood and activities of daily living (see Figure 6.12).

Possible manifestations of pain are presented below. Remember, however, that observation is interpretative and dependent on the health professional assessing the pain behaviours.

Adult non-verbal pain scale University of Rochester Medical Center

Categories	0	1	2
Face	No particular expression or smile	Occasional grimace, tearing, frowning, wrinkled forehead	Frequent grimace, tearing, frowning, wrinkled forehead
Activity (movement)	Lying quietly, normal position	Seeking attention through movement or slow, cautious movement	Restless, excessive activity and/or withdrawal reflexes
Guarding	Lying quietly, no positioning of hands over areas of body.	Splinting areas of the body, tense	Rigid, stiff
Physiology (vital signs)	Stable vital signs	Change in any of the following: * SBP > 20 mm Hg * HR > 20/minute	Change in any of the following: * SBP > 30 mm Hg * HR > 25/minute
Respiratory	Baseline RR/SpO_2 Complaint with ventilator	RR > 10 above baseline or 5% ↓SpO_2 mile asynchrony with ventilator	RR > 20 above baseline or 10% ↓SpO_2 severe asynchrony with ventilator

Abbreviations: HR, heart rate; RR, respiratory rate; SBP, systolic blood pressure; SpO2, pulse oximetry. Instructions: Each of the 5 categories is scored from 0–2, which results in a total score between 0 and 10. Document total score by adding numbers from each of the 5 categories. Scores of 0–2 indicate no pain, 3–6 moderate pain, and 7–10 severe pain. Document assessment every 4 hours on nursing flow-sheet and complete assessment before and after intervention to maximize patient comfort. Sepsis, hypo-volemia, hypoxia need to be excluded before interventions.

Figure 6.12 Adult non-verbal pain scale. Reproduced with permission of the University of Rochester Medical Centre. First published in Dimensions in Critical Care Nursing, 2003.

Behavioural indicators:

- Verbal expression
- Aggression
- Agitation
- Crying
- Facial grimaces
- Restlessness
- Increased confusion
- Change in appetite
- Withdrawal
- Sleep disruption
- Moaning
- Fidgeting
- Guarding or rubbing of body parts

Wegman (2005) describes the adult non-verbal pain scale which is a 10-point scale with five categories that are scored on 0, 1 or 2 point system. This scale is an adaptation of the original NVPS scale (Odhner *et al.*, 2003). Further research was conducted on this revised scale in 2005 (Freeland, 2006). The research involved 111 observations, and internal reliability exceeds minimum expectations for newly developed scales at .79 (Coefficient alpha).

Staff nurse in an intensive-care unit

I find it quite difficult to be confident in my assessment of pain when most of my patients are unconscious and cannot communicate their pain. I have, however, become more self assured in my decisions after increasing my understanding of the physiology of pain on the advice of a more senior nurse. She said that once I understand the pain pathways and the body's physiological responses I would view communication of pain differently to my current approaches. She was right. I am beginning to realise that not all communication of pain is verbal or coherent but that by paying attention to all the non-verbal signs, being more aware of physiological signs such as tachycardia which could indicate pain, and by approaching the patient more holistically I can be more confident in my assessment of pain.

Key points Top tips

- Pain assessment is integral to overall pain management
- Pain assessment should be included as part of overall pain strategy
- Pain assessment approaches utilised should be based on audit of patient need
- Pain tools chosen should be appropriate to patient case mix
- Assessment of pain should occur initially and regularly thereafter
- Pain should be assessed as part of assessment of vital signs
- Pain should be assessed both at rest and during activity
- Pain relief from interventions should be assessed for adequacy

Pain persisting beyond usual course

Sue is a 42-year-old accountant who presented to her general practitioner with pain in her right side (see case study in Chapter 4). She is fit – running 45–60 minutes on the treadmill five days a week. She is slim and eats sensibly. She is studying part-time for a law degree and seems to be very busy with a full life. She has had the pain for some time and is now getting anxious. The general practitioner suspects gallstones and refers Sue to the local hospital. Three weeks later, Sue is reviewed by the general surgeon who elicits a full history and performs a physical examination. Agreeing with the general practitioner, he suspects gallstones and orders an ultrasound to confirm diagnosis. Sue is subsequently diagnosed with gallstones despite being an atypical profile. The ultrasound identifies three rather large stones in the gallbladder and one in the biliary tract. She is scheduled for a laparoscopic cholecystectomy the next month.

When admitting Sue to the ward, the nurse asks details about her pain, eliciting that the pain occurs randomly more likely in the afternoon than the morning. While the pain is generally in the right iliac fossa region Sue suffers some right shoulder pain at times. The pain has been coming and going for about 12 months. Sue describes it as a sharp biting kind of pain which is very acute. She also mentions a residual dull pain that comes and goes more often in the same region.

The nurse is satisfied that the pain is classic for gallstones and asks Sue if the three student nurses on the ward can ask her a few questions about the pain. Sue agrees, happy that a rather straightforward diagnosis is made which can be rectified by what the surgeon has described as a simple procedure.

Sue returns from her surgery which the general surgeon declares is a success. Sue, herself, is relieved to be back in the ward but is rather alarmed with the amount of pain she is suffering. She is on a PCA morphine pump and though a bit nervous of it, is utilising it satisfactorily. Her pain is rated four- to six-hourly on a 10 cm Visual Analogue Scale. After two days, her pump is removed and she is commenced on a low-dose nonsteroidal anti-inflammatory and discharged home the next day.

Sue's partner Michael is rather alarmed at the slow rate of Sue's recovery. Like Sue, he had expected her to be up and about within two days. Sue's pain persists over the next three weeks but she is reassured by her review with the surgeon at three weeks who tells her that everything is getting back to normal. Six weeks later, Sue still gets pain from time to time.

- What additional information would you have elicited from Sue's pain history?
- Was Sue's pain assessed appropriately after her surgery?
- If Sue's pain persists what course of action would you recommend her to take?

Rapid recap

1. What qualities should a pain assessment tool have?
2. What are the three main pain intensity tools used?
3. What can a pain audit contribute when deciding which pain assessment tool to use?

References

Acute Pain Management Guideline Panel (1992). *Acute pain management: operative or medical procedures and trauma: Clinical practice guideline.* Agency for Health Care Policy and Research, Public Health Service, U.S. Department of Health and Human Services, Washington DC.

American Pain Society (2000) *Pain assessment and treatment in the managed care environment. A position statement from the American Pain Society.* American Pain Society, Seattle.

Black A.M.S. (1991) Taking pains to take away pain. *BMJ*, **302**: 1165–1166.

Bowman J.M. (1994) Perception of surgical pain by nurses and patients. *Clinical Nursing Research*, **3**(1): 69–76.

Breivik E.K., Bjornsson G.A., Skovlund E. (2000) A comparison of pain rating scale by sampling from clincial trial data. *Clinical Journal of Pain*, **16**(1): 22–28.

Carlsson A.M. (1983) Assessment of chronic pain: I. Aspects of the reliability and validity of the Visual Analogue Scale. *Pain*, **16**: 87–101.

Choiniere M., Melzack R., Girard N., Rondeau J. and Paquin M.J. (1990) Comparisons between patient's and nurses' assessment of pain and medication efficacy in severe burns injuries. *Pain*, **40**: 143–152.

Coll A.M., Jamal A. and Mead D. (2004) Postoperative pain assessment tools in day surgery: a literature review. *Journal of Advanced Nursing*, **46**(2): 124–133.

Cork R.C., Isaac I., Elsharydah A., Saleemi S., Zavisca F., Alexander L., (2004) A Comparison of the Verbal Rating Scale and the Visual Analog Scale for pain assessment. *The Internet Journal of Anesthesiology*, **8**(1) www.ispub.com/ostia/index.php?xmlFilePath=journals/ija/vol8n1/vrs.xml.

Dudgeon D., Raubertas R.F. and Rosenthal S.N. (1993) The Short-Form McGill Pain Questionnaire in chronic cancer pain. *Journal of Pain and Symptom Management*, **8**(4): 191–195.

Field L. (1996) Are Nurses still underestimating patients' pain postoperatively? *British Journal of Advanced Nursing*, **5**(13): 778–784.

Francke A.L., Garssen B., Abu-Saad H.A. and Grypdonck M. (1996) Qualitative needs assessment prior to a continuing education program. *The Journal of Continuing Education in Nursing*, **27**(1): 34–41.

Freeland N., Odhner M., Wegman D., and Ingersoll G.L. (2006). Refinement of a nonverbal pain assessment scale for nonverbal, critically ill adults. Unpublished research report, University of Rochester Medical Center, Rochester, NY.

Grafton K.V., Foster N.E. and Wright C.C. (2005) Test-retest reliability of the Short-Form McGill Pain Questionnaire: assessment of intraclass correlation coefficients and limits of agreement in patients with osteoarthritis. *Clinical Journal of Pain,* **21**(1): 73–82.

Harrison A. (1993) Comparing nurses' and patients' pain evaluations: A study of hospitalised patients in Kuwait. *Social Science Medicine*, **36**(5): 683–692.

Hastings F. (1995) Introduction of the use of structured pain assessment for post-operative patients in Kenya: implementing change using a research based co-operative approach. *Journal of Clinical Nursing*, **4**: 169–176.

Huskisson E.C. (1974) Measurement of pain. *Lancet*, **9**: 1127–1131.

Idvall E., Hamrin E., Sjostrom B. and Unosson M. (2002) Patient and nurse assessment of quality of care in postoperative pain management. *Quality & Safety in Health Care*, **11**(4): 327–334.

Joint Commission on Accreditation of Healthcare Organisations (2000–2001) *Standards for ambulatory care accreditation policies, standards, intent statements by the Joint Commission on Accreditation of Healthcare Organizations.* Joint Commission on Accreditation of Healthcare Organisations.

Kremer E., Atkinson J.H. and Ignelzi R.J. (1981) Measurement of pain: patient preference does not compound pain measurement. *Pain*, **10**: 241–248.

McCaffrey M. (1983) *Nursing the patient in Pain,* 2nd edn. Harper and Row, London, p. 276.

McCaffrey, M., and Ferrell, B.R. (1994). Nurses' assessment of pain intensity and choice of analgesic dose. *Contemporary Nurse*, **3**(2): 68–74.

McDowell I. and Newell C. (1996) Measuring health: a guide to rating scales and questionnaires, 2nd edn. Oxford University Press, Oxford, pp. 341–351.

Melzack R. (1975) The McGill Pain Questionnaire: major properties and scoring methods. *Pain*, **1:** 277–299.

Melzack R. (1987) The short-form McGill Pain Questionnaire. *Pain*, **30**: 191–197.

Melzack R. and Katz J. (1994) Pain measurement in persons in pain. In: *Textbook of Pain,* 3rd edn. (eds. Melzack R. and Wall P.). London, Churchill Livingstone, p. 337.

Melzack R. and Torgeson W.S. (1971) On the language of pain. *Anaesthesiology*, **34**: 50–59.

Moore A., Edwards J., Barden J. and McQuay H. (2004). Measuring pain. In *Bandolier's Little Book of Pain.* Oxford University Press, Oxford, p. 7.

National Cancer Institute. (www.nci.nih.gov/cancertopics/pdq/supportivecare/pain/HealthProfessional/page2#Section_44). National Health and Medical Research Council, Australia Acute Pain Management Scientific Evidence of 1999 (www.nhmrc.gov.au).

Odhner M., Wegman D., Freeland N., Steinmetz A. and Ingersoll G.L. (2003). Assessing pain control in nonverbal critically ill adults. *Dimensions in Critical Care Nursing*, **22**: 260–267.

Ohnhaus E.E. and Adler R. (1975) Methodological problems in the measurement of pain: a comparison between the verbal rating scale and the visual analogue scale. *Pain*, **1**: 379–384.

Reading A.E. (1982) A comparison of the McGill Pain Questionnaire in chronic and acute pain. *Pain,* **13**, 185–192.

Royal College of Anaesthetists and The Pain Society, the British Chapter of the International Association for the Study of Pain (2003) *Pain Management Services – Good Practice.* The Royal College of Anaesthetists and The Pain Society, London, p. 4.

Royal College of Surgeons and Anaesthetists (1990) *Commission on the Provision of Surgical Services, Report on the Working Party on Pain after Surgery*. Royal College of Surgeons and Anaesthetists, London.

Scott J. and Huskisson E.C. (1976) Graphic representation of pain. *Pain*, **2**: 175–184.

Seymour R.A., Simpson J.M., Charlton J.E. and Phillips M.E. (1985) An evaluation of length and end-phrase of Visual Analogue Scales in dental pain. *Pain*, **21**: 177–185.

Sloman R., Rosen G., Rom M. and Shir Y. (2005) Nurses' assessment of pain in surgical patients. *Journal of Advanced Nursing*, **52**(2): 125–132.

Sriwatanakul K., Kelvie W., Lasagna L., Calimlin J.F., Weis O.F. and Mehta G. (1983) Studies with different types of visual analog scales for measurement of pain. *Clinical. Pharmacological Therapeutics,* Aug: 234–239.

Taylor H. (1997) Pain scoring as a formal pain assessment tool. *Nursing Standard*, **11**(35): 40–42.

Wegman D.A. (2005) Tool for pain assessment. *Critical Care Nurse*, **25**(1):14.

Zalon M.L. (1993) Nurse's assessment of postoperative patients' pain. *Pain*, **54**: 329–334.

7 Pain interventions

Learning outcomes

By the end of this chapter you should be able to:

★ Understand the types of pain interventions available and the benefits of combining approaches to pain management

★ Understand the main pharmacological approaches to pain management

★ Identify the location of modulation of pain on the pain pathway for various pain interventions.

Introduction

There are many interventions used to manage pain, from strong opioids to transcutaneous electrical nerve stimulation (TENS) machines to acupuncture. It is important for you as a health professional to be knowledgeable about these interventions, their actions and interactions. Clinical services often rely on the way they have managed pain for many years and are slow to introduce new methods of pain relief. However, the evidence for pain interventions is constantly changing and services should take the time to review and audit the levels and effectiveness of their pain intervention strategies. The choice of intervention to manage pain should be based on accurate, appropriate assessment of the patient's pain and knowledge of the physiology of pain. This chapter details the more common pain interventions highlighting where on the pain pathway they are effective. Different analgesic schemes to support maximisation of medication usage are described.

Pain interventions

Chapter 4 described the physiology of pain in detail and highlighted the importance of understanding where on the pain pathway different interventions modulate pain.

Reflective activity

Consider the last pain intervention you performed. Was it administering an opioid to a postoperative patient? Consider where on the pain pathway this intervention modulated the patient's pain.

Health care professionals can treat pain more effectively and efficiently if they identify the type of pain and then select the most appropriate intervention. Thus, the pain assessment process is critical to successful interventions. Chapter 6 detailed the various types of pain assessment and in which clinical situations they are most effective. Patients may be suffering from more than one type of pain and, therefore, may need combination interventions to manage their pain experience. The choice of pain interventions, like any aspect of patient care, should be based on the best available evidence. As technology, skills and knowledge grow, nurses and health care professionals need to develop systems to ensure they are accessing, interpreting and utilising the best available evidence. Throughout this book reference has been made to evidence-based practice and pain resources available. These should be accessed regularly in order to ensure the chosen pain intervention is appropriate, up to date and evidence based. In particular, medications and their dosages change and new medicines come on the market. This chapter details medications, their use and side-effects in order to give you a better understanding of approaches to managing pain and their mechanisms. However, given the dynamic nature of medications and the individual responses of patients to interventions, you should always refer to your most up-to-date source of information before choosing or administering medications.

The extent of pain interventions available is broader than just medications, for example there is physiotherapy, cognitive behavioural therapy, TENs, relaxation etc. This means that health professionals need to engage in a systematic approach to assessment and monitoring of a patient's pain to ensure that the patient is receiving the most effective intervention or combination of interventions. Figure 7.1 details the pain assessment and monitoring cycle.

Table 7.1 will remind you of the various types and causes of pain.

Table 7.1 Types and causes of pain

	Cutaneous pain	Somatic pain	Visceral pain	Neuropathic pain
Cause	Produced by stimulation of the pain receptors in the skin, e.g. superficial burn.	Produced by stimulation of pain receptors in the deep structures, i.e. muscles, bones, joints, tendons, ligaments.	Produced by stimulation of pain receptors in the viscera, e.g. stomach, kidney, gallbladder, urinary bladder, intestines.	Radiating or specific, e.g. trigeminal neuralgia, limb amputation.
Description	Pin prick, stabbing, or sharp. Can be accurately localised due to the large number of receptors in the skin.	Unlike cutaneous pain, deep somatic pain is dull, intense and prolonged. It is usually associated with autonomic stimulation, e.g. sweating, vomiting and changes in heart rate.	Is poorly localised and is often referred or radiates to other sites.	Electric, tingling or shooting in nature. It may be continuous or paroxysmal in presentation.

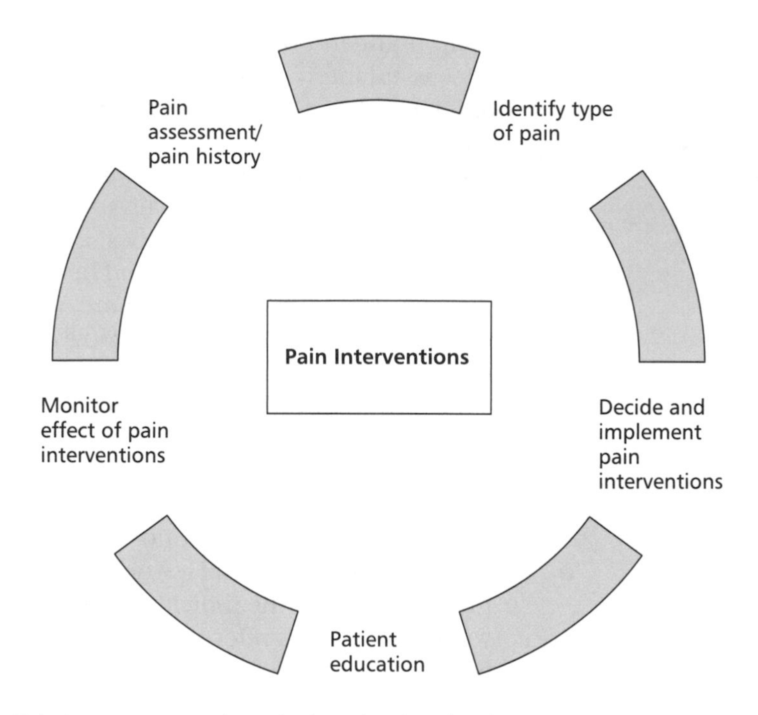

Figure 7.1 Assessment and monitoring of patients' pain and pain interventions

Reflective activity

Consider the last two pain interventions you performed. What type of pain did the patients have? How did they describe their pain? How did you assess the pain? Looking back on how you assessed the pain, would you do anything different?

Table 7.2 details the more common pain interventions and the part of the pain pathway at which these are effective. See Figure 4.1 page 48 to remind you of the pain pathways.

Medications for pain management

Medications are the mainstay of pain management. The three main classes of medication effective for pain management are opioids, nonsteriodal anti-inflammatories (NSAIDs) and anticonvulsants. Each is described below. Acetaminophen (paracetamol) is also described.

Table 7.2 Common pain interventions

Nervous system	Specific area	Aim of modulation	Interventions
CNS	Brain and spinal cord	The inhibition of the release of chemical transmitters at the various synapses within the dorsal horn of the spinal cord and brain as a result of the uptake of naturally occurring morphine-like substances by specialised receptors on relevant neurons. These receptors are known as mu (μ), delta (δ) and kappa (κ) receptors.	Opioid drugs will bind to these specialised receptors and reduce pain perception.
CNS	Brain	To inhibit prostaglandin synthesis.	Acetaminophen (paracetamol)
CNS	Spinal cord	To prevent transmission of pain signals through the dorsal horn in the spinal cord in order that they cannot travel to the brain (Gate Control Theory).	TENs Heat and cold
CNS	Spinal epidural space	To block or modulate pain signals in the spinal epidural space preventing their transmission to the brain.	Nerve blocks. Both epidural local anaesthetics and epidural opioids can produce analgesia.
PNS	Nerve fibre	To prevent the transmission of pain signals along the nerve cell. Nerve conduction depends on sodium ion channels in the axon.	Local anaesthetic agents such as lignocaine are absorbed across the membrane of the neurone and block the sodium channels. This means that the electrical impulse cannot be transmitted.
PNS	Nociceptor	To prevent transmission of the pain signal at the site of pain. Pain at the site is exacerbated by the presence of certain chemicals, such as prostaglandins or bradykinin.	NSAIDs act at the site of pain reducing the presence of prostaglandins.

Acetaminophen (paracetamol)

Paracetamol or acetaminophen acts by inhibition of prostaglandin synthesis (production of potent, hormone-like substances that mediate a wide range of physiological functions) in the CNS. It is an analgesic and antipyretic. It has no anti-inflammatory effects. Side-effects are minimal once taken in recommended doses. However, those with liver or renal impairment are advised to avoid it.

Opioids

Opiates originally came from the seed capsule of the opium poppy (*Papaver somniferum*). Opioids are derivatives of opium and include naturally occurring opium derivatives, partially synthetic derivatives of morphine and synthetic compounds (Irving and Wallace, 1997). Opium derivatives have had a mixed history over time, and are used in many formulations for perception-altering purposes. Opioids are potent analgesics. They have been both used and abused, challenging governments to develop regulations and legislation to support a balance for opioid use. Thus, while opioids form the cornerstone of the pharmacological armoury for the treatment of pain (Ferrante, 1996) and are often the drugs of choice in the first-line management of acute pain, opioids are controlled substances and subject to regulation. Common opioids in use are morphine, diamorphine, pethidine, methadone, hydromorphone, oxycodone, fentanyl and buprenorphine (McQuay, 1999).

In 1973, it was first discovered that opioids act by binding to specific sites in the CNS (Park and Fulton, 1991). Mu (μ), sigma (σ) and kappa (κ) opioid receptors are found throughout the nervous system and produce analgesia. When activated they reduce the excitability of nociceptors. The potency or intensity of the analgesic effect is dependent upon, firstly, access to the receptor and, secondly, binding affinity at the site (Ferrante, 1996).

Although opioid compounds are active in the PNS, it is in the CNS that they primarily produce analgesia by inhibiting nociceptive transmission. McQuay (1999) states that there is little difference between the various opioids in speed of onset and duration of effect. Faster onset and longer effect are achieved by changing the route of administration or formulation. The onset time by IV route is approximately two minutes and by IM route 20 minutes (the more lipophilic the drug the faster the onset). According to Collins *et al.* (1998), morphine reaches maximum concentration at different times depending on formulation:

- Immediate-release oral formulations – one hour
- Controlled-release oral formulations – three hours
- Once-daily formulations – nine hours.

Opioids produce a number of side-effects which require monitoring and management. Table 7.3 highlights the key side-effects, clinical assessment and management techniques required as detailed by Fine and Portenoy (2004).

The Royal College of Surgeons and Anaesthetists report (1990) recommends that in order to improve the efficacy of opioid administration, the dose and frequency should be adjusted in response to the needs of each patient. Patients should have an individualised treatment regimen. Efficacy and side-effects should be recorded regularly and treatment modified accordingly.

In 2001, the American Academy of Pain Medicine, the American Pain Society and the American Society of Addiction Medicine agreed a consensus document on definitions related to the use of opioids and the treatment of pain. The consensus document states that most specialists agree that patients treated with prolonged opioid therapy usually do develop physical dependence and sometimes develop tolerance but do not usually develop addictive disorders. The actual risk is, however,

Table 7.3 Opioids: key side-effects, clinical assessment and management techniques (Source: Fine and Portenoy, 2004)

Side effect	Description	Clinical assessment	Management
Constipation	Most common and persistent side-effect is bowel dysmotility leading to constipation.	Depends on time, course of its development and medical setting. Is the relationship with the drug clear? Other contributing factors such as inactivity and dehydration may be present. History taking and physical examination should be completed.	Tolerance to opioids develops very slowly and a large proportion of patients require laxative therapy while taking opioids. For some patients dietary modification is adequate. Fluid intake should be increased. Mobility should be encouraged.
Nausea and vomiting	May occur after taking an opioid; however, tolerance usually develops rapidly.	Assess for contributing factors such as constipation or other medications.	Administration of an anti-emetic at the time of nausea. Routine prophylactic administration of an anti-emetic agent is not typically indicated.
Somnolence and cognitive impairment	Somnolence (mental clouding) typically wanes over a period of days or weeks. Can range from mild to severe. Cognitive impairment can range from slight inattention to extreme confusion. Persistence is atypical.	Assess for contributing factors.	Address any contributing causes. If analgesia is adequate it may be possible to reduce dose. Fine and Portenoy (2004) suggest that if these are not effective there is a large body of clinical experience in the use of psychostimulants to treat opioid-induced cognitive impairment.
Pruritus	May occur with any opioid and may be caused by the opioid mediated release of histamine from mast cells.	Assess extent and duration of itch.	Pharmacological management with trial use of an antihistamine.
Respiratory depression	Rarely a problem when used according to accepted guidelines. Tolerance usually develops quickly, allowing escalation of the dose incrementally. Combining opioids with other medications such as hypnotics or benzodiazepines should be monitored closely.	Close monitoring of respiration and sedation scores.	Naloxone should be administered for symptomatic respiratory depression.

unknown and probably varies with genetic predisposition among other factors. They recommended the following definitions of **addiction**, **physical dependence** and **tolerance** and their use (see Keywords).

Chapter 10 provides more detail on the use of opioids and chronic non-malignant pain and Chapter 11 describes the use of opioids for cancer pain.

Keywords

Addiction
Addiction is a primary, chronic, neurobiologic disease, with genetic, psychosocial and environmental factors influencing its development and manifestations. It is characterised by behaviours that include one or more of the following: impaired control over drug use, compulsive use, continued use despite harm, and craving

Physical dependence
This is a state of adaptation that is manifested by a drug-class specific withdrawal syndrome that can be produced by abrupt cessation, rapid dose reduction, decreasing blood level of the drug, and/or administration of an antagonist

Tolerance
This is a state of adaptation in which exposure to a drug induces changes that result in a diminution of one or more of the drug's effects over time

Nonsteroidal anti-inflammatories (NSAIDs)

Nonsteroidal anti-inflammatory drugs (NSAIDs) are among the most widely used medications on both a prescription basis and over the counter (Abramson and Weaver, 2005). NSAIDs have anti-inflammatory, analgesic and antipyretic effects and inhibit thrombocyte aggregation. NSAIDs act by decreasing prostoglandin production by inhibiting the production of the enzyme cyclo-oxygenase.

A major limiting factor is the risk of gastrointestinal toxicity. Recent studies indicate that this risk has declined 67% since 1992 (Fries *et al.*, 2004). This is as a result of the use of lower doses of NSAIDs, use of gastroprotective agents such as proton pump inhibitors (PPIs), and the introduction of the selective cyclo-oxygenase-2 (COX-2) inhibitors.

Possible cardiovascular risks are being debated with regard to COX-2 inhibitors. The FDA advisory committee (USA) voted that all COX-2 inhibitors that have been approved for use in the USA significantly increase the risk of cardiovascular events (American College of Rheumatology, 2005). NSAIDs should be used with caution with asthma, impaired renal function and those on antiplatelet regimes including aspirin.

Systematic reviews have found no important differences in effect between different NSAIDs or doses, but have found differences in toxicity related to increased doses and possibly due to the nature of the NSAID itself (Gotzche, 2000).

Anticonvulsants

Anticonvulsants are a group of medicines commonly used for treating epilepsy, and they are also effective for treating pain. Nurmikko *et al.* (1998) state that anticonvulsant drugs have an established role in the treatment of chronic neuropathic pain, especially when patients complain of shooting sensations.

Reflective activity

Consider the profile of your patients (e.g. surgical, older person, renal etc.). What types of analgesics do you use in your clinical area? Take a sample of five patient medication charts. It is likely you use all of the above described medications. Discuss with a pharmacist and nurse specialist the use and actions of the analgesics you use regularly.

Routes of analgesia

It is possible to administer analgesia via many routes but some are more effective than others. Pharmokinetics describes the uptake, distribution and elimination of the drug. There is wide variation between individuals and, sometimes, within the same individual at different times. Patient characteristics which may influence analgesic pharmacokinetic variability include age (the elderly have a diminished volume of distribution), hepatic disease, renal disease, acid base balance, hypothermia, hypothyroidism and concurrent drug administration. Figure 7.2 details the main methods of administration of analgesia.

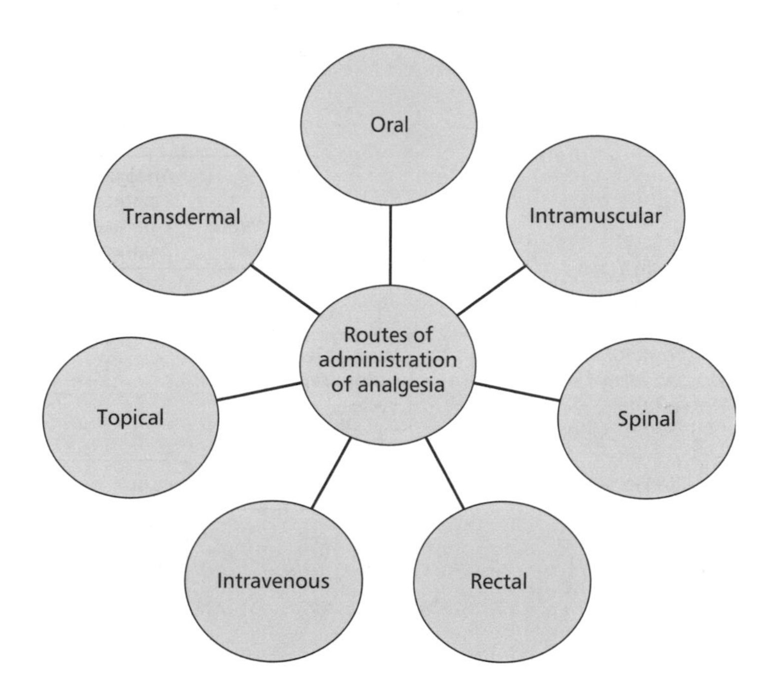

Figure 7.2 Main methods of administration of analgesia

Evidence base

'Epidural analgesia is highly effective for controlling acute pain after surgery or trauma to the chest, abdomen, pelvis or lower limbs. It can, however, cause serious, potentially life threatening complications and its safe, effective management requires a co-ordinated multidisciplinary approach.' This is the introduction to the document *Good practice in the management of continuous epidural analgesia in the hospital setting* produced by the Royal College of Anaesthetists, The Royal College of Nursing, The Association of Anaesthetists of Great Britain and Ireland, The British Pain Society and The European Society of Regional Anaesthesia and Pain Therapy (2004). Go to the British Pain Society website (www.britishpainsociety.org) to access this document. Review the recommendations for use of epidural analgesia.

Analgesic schemes

It is evident that there is often benefit from combining different analgesics as part of a regime for a patient. This is more effective if it is planned in the form of a patient protocol using analgesic schemes.

Moore *et al.* (2003) describe the three-pot system: paracetamol, paracetamol and opioid combination and NSAIDs. The authors state that the system is based on best evidence and uses cheapest available analgesics. Further research is underway. There are two systems, one for those who can take NSAIDs and one for those who cannot (see Figures 7.3 and 7.4).

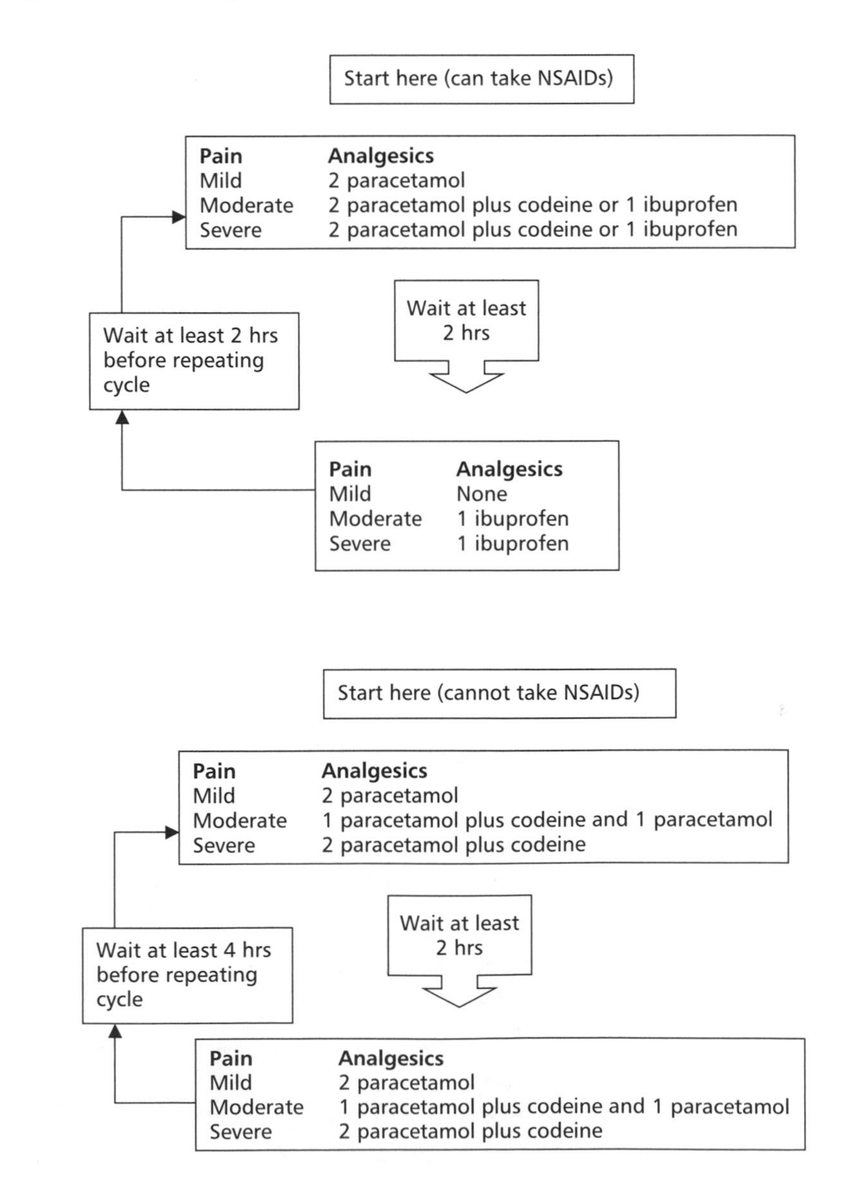

Figure 7.3 A simple scheme (three pot) for acute and chronic pain relief using paracetamol/opioid combination drugs for patients who can take NSAIDS (Moore et al. *2003, p. 432). Reproduced with permission from* Bandolier's Little Book of Pain: An Evidence Based Guide to Treatments *by* Moore et al.*, 2003 Oxford University Press.*

Figure 7.4 A simple scheme (three pot) for acute and chronic pain relief using paracetamol/opioid combination drugs for patients who cannot take NSAIDS (Moore et al. *2003, p. 433). Reproduced with permission from* Bandolier's Little Book of Pain: An Evidence Based Guide to Treatments *by* Moore et al.*, 2003 Oxford University Press.*

The WHO analgesic ladder was a major landmark in the management of pain (see Figure 7.5). It was first produced by the WHO in 1986. The WHO (2006) website states 'If a pain occurs, there should be a prompt oral administration of drugs in the following order: non-opioids (aspirin or paracetamol); then, as necessary, mild opioids (codeine); or the strong opioids such as morphine, until the patient is free of pain. To maintain freedom from pain, drugs should be given "by the clock", that is every 3–6 hours, rather than "on demand". This three-step approach of administering the right drug in the right dose at the right time is inexpensive and 80–90% effective.' Moore *et al.* (2003) suggest that while the WHO analgesic ladder can relieve pain for approximately 80% of patients, pain management must be optimal, which is not always the case in practice. They suggest using the ladder going upwards for chronic pain, starting with non-opioids and in reverse starting with strong opioid opioids +/– non-opioid for acute pain.

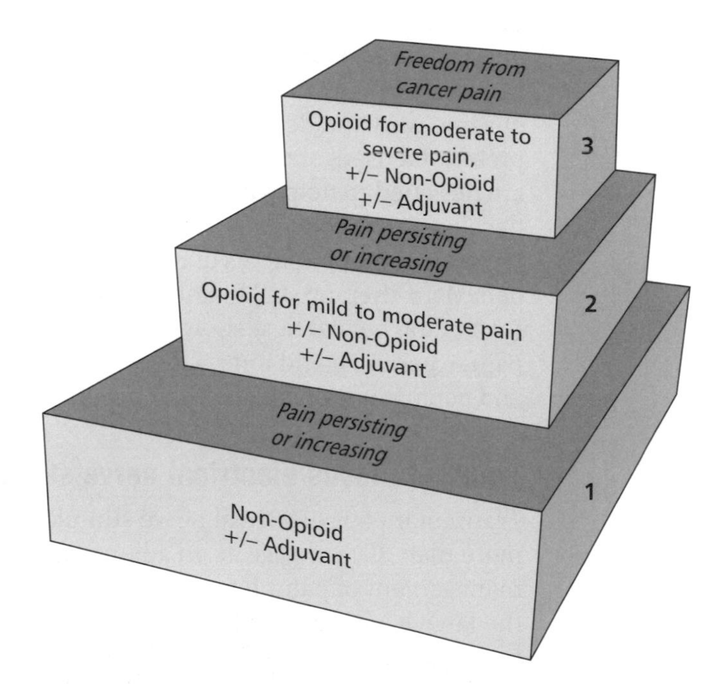

Figure 7.5 WHO's pain relief ladder. World Health Organization Analgesic Ladder (WHO, 1986)

Clinical Pharmacist

I work as a clinical pharmacist supporting five medical wards, one of which is an oncology unit. I audit medication charts weekly and order stocks for the ward areas. What concerns me most is the reactive prescribing for medications in general and specifically for pain. In order to address this, I have met with the clinical nurse manager and consultant and presented my findings. As a result, we are devising an analgesic protocol to try to make analgesic prescription and administration more consistent and planned. We will combine this with short education sessions and information material.

Non-pharmacological approaches

It is appropriate to consider and utilise non-pharmacological pain management options to support patients through their pain experience, particularly those patients with mild or persistent chronic pain. Reference is made here to cognitive behavioural therapy, TENs, heat and cold, and acupuncture.

Cognitive-behavioural therapy

Given that pain is a personal and unique experience for the individual, the emotive and contextual aspects of pain will influence the pain experience. For those with chronic pain extending over long periods of time it is logical to consider complementary approaches to support the individual to cope with both the psychological and functional facets of that pain. Turk (2003) suggests that cognitive-behavioural (CB) approaches should be viewed as important complements to more traditional, pharmacological, physical and surgical interventions for chronic pain and describes CB approaches as: 'characterised by being present focused, active, time-limited and structured. CB interventions are designed to help patients identify maladaptive patterns and acquire, develop and practice more adaptive ways of responding'. A systematic review and meta-analysis of randomised controlled trials of cognitive behaviour therapy (CBT) and behaviour therapy for chronic pain (excluding headaches) showed that CBT was effective in reducing the pain experience and improving positive behaviour expression, appraisal and coping in individuals with chronic pain (Morley *et al.,* 1999).

Transcutaneous electrical nerve stimulation (TENS)

Transcutaneous electrical nerve stimulation (TENS) was introduced more than 30 years ago as an adjunct to the pharmacological management of pain. It involves application of electrical stimulation to the large low-threshold fibres through the skin. It is based on the principles of the Gate Control Theory (see Chapter 4). The effectiveness of TENS remains under debate. In 1996, Carroll *et al.*, following systematic review, judged that of 15 out of 17 randomised controlled trials (RCTs) there was no benefit achieved with TENS compared to placebo for postoperative patients.

A Cochrane systematic review found limited and inconsistent evidence to support the use of TENS as an isolated intervention in the management of chronic lower back pain. Only two RCTs met the criteria for inclusion (Khadilkar *et al.* 2005). Moore *et al.* (2003), on review of the evidence for TENS, state that the clinical bottom line is that there is no conclusive evidence that TENS is effective for either postoperative or chronic pain.

Heat and cold

The effectiveness of heat and cold is described anecdotally. However, research in the form of well-designed trials is needed to assess the effectiveness of both for relieving pain. Where used, heat and cold should be applied with caution following assessment of the patient to ensure that there are no contraindications.

Heat stimulates the thermoreceptors in the skin and deeper tissues. Theoretically, this can help reduce pain by closing the gating system in the spinal cord (see Chapter 4 for details on Gate-Control Theory). Heat is used extensively for muscle spasm, menstrual pain and joint stiffness. It can be applied in many ways, for example hot baths, heating pads, hot water bottles or heat packs. Chandler *et al.* (2002) caution that use of heat therapy in hospitals is often limited due to safety issues with heating devices.

Cold will cause vasoconstriction and reduce swelling and should be applied using ice packs or compresses. Cold should not be applied directly to the skin. Atraksinen *et al.* (2004) describe a randomised, blinded controlled trial assessing whether cold gel reduced pain and disability in minor soft-tissue injurics (n = 74). The active cold gel with menthol and ethanol showed decreased pain at rest and movement compared with the placebo group.

Acupuncture

Acupuncture involves inserting fine needles into specific points in the surface of the body. The needles are often rotated to provide stimulus. Moore *et al.* (2003) state that while acupuncture is widely used, there is little evidence that it is effective. They highlight the difficulties with the quality of the studies particularly in relation to the correctness of the acupuncture procedure, appropriateness of control interventions and blinding of participants. They state, on review of available trials, that the clinical bottom line is that acupuncture for back pain is ineffective, that there is no convincing evidence for acupuncture for neck and back pain and that there is no evidence of it being more effective in arthritis. A strong case is made for high-quality, large long-term trials.

One more recent randomised controlled trial (Berman *et al.* 2004) concluded that acupuncture appeared to provide improvement in function and pain relief as an adjunctive therapy for osteoarthritis of the knee when compared with credible sham acupuncture and education control groups (570 patients over 26 weeks).

See Chapter 8 for further discussion of alternative and complementary therapies.

Key points Top tips

- There are many pain interventions available, from medications to TENs, acupuncture and cognitive behavioural therapy
- Knowledge of the physiology of pain is critical to deciding on the most appropriate pain intervention
- There is benefit in combining different analgesics as part of a regime for a patient
- Non-pharmacological approaches support pain management; however, there is limited evidence for effectiveness of some of these such as heat and cold

Patterns of analgesic use

Jennifer is a 50-year-old journalist who has had a hysterectomy. She is two days post surgery but is mobilising very slowly and appears to be in some pain. Margaret is the nurse on duty. She is an experienced surgical nurse and sits down with Jennifer to review her progress. Jennifer was on a PCA morphine pump for 24 hours, which, on review of her pain scores, had been providing adequate pain relief.

Margaret begins by asking Jennifer how her pain has been, what her current pain score on VAS is now and whether she is able to walk or take deep breaths. Jennifer's pain score is 3.5 at rest but 7 on mobilisation. Margaret realises that pain scores had only been taken at rest, not during movement, which explained why Jennifer's PCA pump was discontinued.

Margaret reviewed Jennifer's medication chart and is concerned when she sees that she is prescribed paracetamol, two different NSAIDs and morphine 10 mg intramuscularly (IM) all on a pre re nata (PRN) basis. It seems that she has been seen by three different doctors in the past 24 hours for her pain and each has prescribed a different medication. For a summary of what has been administered since discontinuation of the PCA pump see Table 7.4.

- What other information should Margaret seek before deciding on the next steps?
- Why is it so important to assess pain on movement as well as pain at rest for surgical patients?
- If you were to rewrite Jennifer's medication plan, what would you do?

Table 7.4 Medication Chart

Time	Medication
9 a.m.	PCA discontinued
12 midday	NSAID (no. 1) PR
6 p.m.	Paracetamol
11 p.m.	Morphine 10 mg IM
8 a.m.	NSAID (no. 2) PO
2 p.m.	Margaret is reviewing pain medications and deciding on next steps

Rapid recap

1 Name two side-effects of opioids.
2 What are the principles of the WHO analgesic ladder?
3 What is the medication of choice for chronic neuropathic pain?
4 What are the major effects of NSAIDs?
5 What are the limiting factors for use of NSAIDs?

References

Abramson S.B. and Weaver A.L. (2005) Current state of therapy for pain and inflammation. *Arthritis Research & Therapy*, **7**(suppl 4): S1–S6.

American Academy of Pain Medicine, the American Pain society and the American Society of Addiction Medicine (2001). Definitions related to the use of opioids and the treatment of pain. A consensus document from the American Academy of Pain Medicine, the American Pain Society, and the American Society of Addiction Medicine. Available at http://painmed.org/productpub/statements/.

American College of Rheumatology, The safety of COX-2 inhibitors: deliberations from February 16–18, 2005, FDA meeting. http://www.rheumatology.org/publications/hotline/0305NSAIDs.asp.

Atraksinen O.V., Kyrklund N., Latvala K., Kouri J.P., Gronblad M. and Kolari P. (2004) Cold gel reduced pain and disability in minor soft-tissue injury. *Journal of Bone and Joint Surgery*, **86**(5): 1101.

Berman B.M., Lao L., Langenberg P., Lin Lee W., Gilpin A.M.K. and Hochberg M.C. (2004) Effectiveness of acupuncture as adjunctive therapy in osteoarthritis of the knee. *Annals of Internal Medicine*, **141**(12): 901–910.

Carroll D., Tramer M., McQuay H. and Moore A. (1996) Randomisation is important in studies with pain outcomes: systematic review of transcutaneous electrical nerve stimulation in acute postoperative pain. *British Journal of Anaesthesia*, **77**: 798–803.

Chandler A., Preece J. and Lister S. (2002) Using heat therapy for pain management. *Nursing Standard*, **17**(9): 40–42.

Collins S.L., Faura C.C., Moore R.A. and McQuay H.J. (1998) Peak plasma concentrations after oral morphine – a systematic review. *Journal of Pain and Symptom Management*, **16**: 388–402.

Ferrante F.M. (1996) Principles of Opioid Pharmacotherapy: Practical Implications of Basic Mechanisms. *Journal of Pain and Symptom Management*, **11**(5): 265–273.

Fine P.G. and Portenoy R.K. (2004) *A Clinical Guide to Opioid Analgesia.* McGraw-Hill Inc, USA.

Fries J.F., Murtagh K.N., Bennerr M., Zatarain E., Lingala B. and Bruce B. (2004) The rise and decline of nonsteroidal anti-inflammatory drug-associated gastropathy in rheumatoid arthritis. *Arthritis Rheum*, **50**: 2433–2440.

Gotzche P.C. (2000) Non-steroidal anti-inflammatory drugs. *BMJ*, **320**: 1058–1061.

Irving GA and Wallace MS (1997) Opioid Pharmacology. In *Pain Management for the Practising Physician.* Churchill Livingstone, Edinburgh, pp.17–30.

Khadilkar A., Milne S., Brosseau L., Robinson V., Saginur M., Shea B. *et al.* (2005). Transcutaneous electrical nerve stimulation (TENS) for chronic low-back pain. *The*

Cochrane Database of Systematic Reviews. Issue 3. Art. No.: CD003008. DOI: 10.1002/14651858.pub2.

McQuay H. (1999) Opioids in pain management. *The Lancet*, **353**: 2229–2232.

Moore A., Edwards J., Barden J. and McQuay H. (2003) *Bandolier's Little Book of Pain*. Oxford University Press, Oxford.

Morley S., Eccleston C. and Williams A. (1999) Systematic review and meta-analysis of randomised controlled trials of cognitive behaviour therapy for chronic pain in adults, excluding headache. *Pain*, **80**: 1–13.

Nurmikko T.J., Nash T.P. and Wiles J.R. (1998) Recent advances: control of chronic pain. *BMJ*, **317**: 1438–1441.

Park G. and Fulton B. (1991) *The Management of Acute Pain.* Oxford Medical Publications, Oxford.

Royal College of Nursing, The Association of Anaesthetists of Great Britain and Ireland, The British Pain Society and The European Society of Regional Anaesthesia and Pain Therapy (2004) *Good practice in management of continuous epidural analgesia in the hospital setting*. Available at www.britishpainsociety.org/pub_pub.html.

Royal College of Surgeons and Anaesthetists (1990) *Commission on the Provision of Surgical Services, Report of the Working Party on Pain after Surgery*. Royal College of Surgeons and Anaesthetists, London.

Turk D.C. (2003) Cognitive-behavioural approach to the treatment of chronic pain patients. *Regional Anaesthesia and Pain Medicine*, **28**(6): 573–579.

World Health Organization (2006) Pain Ladder website. Available at: www.who.int/cancer/palliative/painladder/en/index.html.

8 The role of complementary and alternative therapies in pain management

Lynne Wigens

Learning outcomes

By the end of this chapter you should be able to:

- ★ Define and discuss a range of complementary and alternative medicine
- ★ Appreciate the influence and widespread use of complementary and alternative therapies in the field of pain care
- ★ Examine a range of evidence relating to complementary and alternative medicine
- ★ Discuss the practitioner–patient relationship and its impact on pain management.

Defining complementary alternative medicine (CAM)

Complementary and alternative medicine (CAM) has been defined as health care practices and products outside the realm of conventional medicine, which are yet to be validated using scientific methods (Eisenberg, 2002). No book on pain management could ignore the role of complementary and alternative therapies. Within this chapter a range of therapies are reviewed, with particular reference to the evidence base for practice. The ways that health care practitioners may integrate CAM is discussed, and it is concluded that there is a need for a strong practitioner–patient relationship, whether the care delivered is conventional or complementary/ alternative.

The term 'complementary medicine' tends to be used for therapies that have been widely integrated with conventional medicine, e.g. aromatherapy and massage combined with palliative care analgesics.

Alternative medicine is used as a collective term for treatments that are used in place of conventional medicine, e.g. accessing osteopathy rather than seeking conventional medical management of lower back pain. Traditional medicine refers to health practices, approaches, knowledge and beliefs incorporating plant, animal and

I am receiving palliative care treatment for my cancer and I go to the hospice for day treatment, and as part of this I have 30-minute sessions of aromatherapy and massage. They use lavender and rose oils in a grapeseed oil base. As the treatment begins I start to feel more relaxed. Over time the aches and pains across my body seem to reduce. The massage is so relaxing, even though only the lightest touch and stroking is used. It seems to get the tension out from my neck and shoulders and I feel heavenly for quite a while afterwards.

mineral-based medicines, spiritual therapies, manual techniques and exercises, applied singularly or in combination to treat, diagnose and prevent illnesses or maintain wellbeing (World Health Organization (WHO), 2003).

Different complementary therapies have very different philosophies and practices, but most share a common view of health and healing, emphasising wellness that is perceived to stem from a balance between the mind, body and its environment. Complementary therapies concentrate on the whole person in a '**holistic**', individualised approach, often calling for the patient to actively change their lifestyle or behaviours. Complementary therapists believe that this self-healing is the basis of all healing. These factors may be major contributing factors to the increasing popularity of complementary therapies in the management of pain.

Keywords

Holistic care

This is concerned with the interrelationships of body, mind and spirit in an ever-changing environment (Dossey, 2000). It means that care has to focus on the whole person rather than on one specific problem or disease.

Underpinning concepts/beliefs about CAM therapies

- Illness is a disharmony or deviation from health, affecting the person as part of the larger environment/universal system
- Health is a balance of opposing forces, and treatment is about strengthening healing forces
- Holistic in nature, strengthening the wellbeing of the whole person
- Less high-technology interventions or diagnostics are used
- The patient should be active in their cure

Reflective activity

What experience have you had with CAM? Have you received complementary or alternative therapies yourself, or treated others?

Think about your current view of CAM and its potential use in pain management.

Use of complementary alternative medicine (CAM)

Complementary medicine has been used by patients for their pain and other health care needs for many years (Lewith *et al.*, 2002). The UK annual expenditure on CAM in 2004/05 is likely to be in the region of £4.5 billion (Smallwood, 2005, p. 22). Within the NHS confederation leading-edge report on the future of acute care (Black, 2005) the complementary and alternative medicine market is viewed as being undervalued. A successful approach to the future should include the non-NHS 'health economy' which is much bigger than some health care practitioners realise (Black, 2005). It is suggested that the current UK market for CAM could be in excess of £5 billion. Within one small market town in 2004 there were 40 CAM practitioners (double that of the locality's general practitioners (GPs), and this is almost completely funded from discretionary payments by individuals (Black, 2005).

Over to you

Take a look at at least two of these websites and identify what they indicate about the current status of CAM within the UK.

- The British Holistic Medical Association
  www.bhma-sec.dircon.co.uk
- Institute for Complementary Medicine
  www.icmedicine.co.uk
- Complementary Medical Association
  www.the-cma.org.uk
- The British Complementary Medicine Association
  www.bcma.co.uk

There is clearly a large proportion of the UK population choosing to access CAM and in many instances funding their own treatments. In Vallerand *et al.*'s (2005) study with 723 participants about their self-treatment choices they found that participants reported using the following pain treatments:

- 75% were taking non-opoid analgesics
- 11% were taking adjuvant analgesics
- 29% were taking herbal products and supplements
- 68% were using non-pharmacological modalities, e.g. osteopathy
- 28% had not informed their primary care practitioners of their self-treatment choices.

Reflective activity

Are you surprised by Vallerand *et al.*'s (2005) findings?

From your own experience as a giver and receiver of health care do you think similar findings would be found if the study was replicated in your locality?

It has been suggested that the rise in the take up of CAM is due to the failure of conventional medicine to respond to patients as individual human beings (Graham, 1999). Motivators that have been identified as leading to patients accessing CAM include:

- the need for chronic symptom relief
- perceived effectiveness
- safety (fewer side-effects) and non-invasiveness
- emphasis on the whole person 'high touch/ low tech'
- sense of patient control over the treatment process, and being 'heard'
- good therapeutic relationship and support
- accessibility
- dissatisfaction with conventional medicine (Smallwood, 2005).

There has been an increase in acceptance of Eastern and other healing practices and methods that reflect the interconnectedness between mind and body, and this has occurred at a time when there has been a growth in self-help and support programmes. The number of adults that use some form of complementary or alternative therapy is high, with many people obtaining their own therapies through health-food stores and the internet, as well as consulting therapists. In order to provide optimal and comprehensive care of patients with pain, health care professionals need to be aware of the patient's use of CAM. Caring is not fully realised without an understanding of the personal meaning in a patient's pain.

Any list of what is considered a CAM changes continually as some become integrated into conventional medicine and additional alternative therapies become more widely known. Table 8.1, therefore, can only give a brief description of some complementary and alternative therapies most commonly used to reduce pain and is not an exhaustive list.

CAM therapies can be classified into five main groups:

1. Alternative medical systems: Based on Eastern medical philosophies such as traditional Chinese medicine and Ayurveda (ancient Indian subcontinent medicine).
2. Mind–body interventions: These use the mind's capacity to affect bodily functions and problems, e.g. meditation.
3. Physiological treatments: These often use substances found in nature, such as herbs. An example of this could be dietary supplements.
4. Manipulation of the body therapies: These are based on the manipulation of one or more body parts, e.g. chiropractic.
5. Energy therapies: These involve the use of energy fields, e.g. Reiki.

Table 8.1 CAM therapies and treatments in common use

Acupuncture	Chinese medical treatment that involves inserting fine needles into the skin at specific points in the body. It seems to relieve pain by diverting or changing the painful sensations that are sent to the brain from damaged tissues and also by stimulating 'natural painkillers' (endorphins, encephalins and serotonin).
Alexander technique	The technique teaches people to prevent unwanted and harmful habits, such as muscle tension and poor posture, by improving an individual's body alignment.
Aromatherapy and massage	Plant oils are massaged into the skin, inhaled or used in the bath, which have a variety of properties. Examples of those that are used to alter pain perception include peppermint and lavender. Massage involves a manual technique in which a rhythmic movement using a variety of strokes, kneading or tapping is undertaken to move the muscles and soft tissue of the body. Massage reduces anxiety and stress levels and relieves muscular tension and fatigue.
Chiropractic	Chiropractors manipulate the joints to try to restore alignment, as problems in alignment are viewed as causing a range of problems in other organs. Chiropractic works towards the vertebrae in the spine being in the correct position, and if there are problems with the alignment of the vertebrae this is thought to cause a range of problems in other organs.
Dietary supplements	These supplements are often bought from chemists, health-food shops or in 'added nutritional foods'. One example is Omega 3 essential fatty acids which are naturally found in oily fish, and thought to be beneficial for people with inflammatory forms of arthritis
Healing	Healing links with a range of belief systems, which may be religious, spiritual, social or cultural, e.g. faith healing. The healer assesses an individual's 'energy field' and then tries to pass energy to the body by gentle touch or by sweeping their hands over the body. Distance healing tries to achieve this through thought, meditation or prayer.
Herbal medicine	Whole plant derived medicines that assist in mobilising the healing mechanisms within the body. Herbal medicine has been present throughout history. One example is Phytodolor which is used to reduce rheumatic pain
Hypnotherapy	The person allows the hypnotherapist to bring them into a daydream or trance-like state in which they can relax and drift away, making use of the unconscious mind.
Homeopathy	Problems are treated with low-dose preparations that induce similar symptoms. Homeopathy is based on the principle that 'like can be cured with like'. Homoeopathy usually also requires a change in lifestyle, which could include a change of diet, more relaxation or exercise, to complement the treatment.
Meditation	Generally involves eliciting a relaxation state, centred on breathing which allows the person to focus attention freely from one thought to the next.
Osteopathy	Osteopathy uses manipulation of the bones and other parts of the musculoskeletal system to diagnose and treat. Osteopaths believe that manipulation of the muscles and joints helps the body to combat illness and heal itself.
Reflexology	Reflexology is a treatment where varying degrees of pressure are applied to the reflex areas of the feet, and sometimes hands, to promote health and wellbeing. It is based on the belief that every part of the body is connected by 'pathways' which terminate in the soles of the feet, palms of the hands, ears, tongue and head.
Reiki	Reiki energy is perceived to flow from the practitioner who lightly touches the patient in certain positions with their hands. Hand positions are static, there is no manipulation of the tissues. and this is done to tap into healing energy (termed *ki*).
Relaxation	This includes techniques such as progressive muscle relaxation (to increase body awareness and decrease perception of pain), guided imagery (using the imagination to reframe the pain and increase a sense of control), biofeedback (learning to use relaxation to achieve a measurable end point) and distraction.
Shiatsu	The term 'Shiatsu' means 'finger pressure', and various parts of the practitioner's body (fingers, thumbs, palms, forearms, even feet and knees) are used to apply pressure to the patient's body. This can be targeted at general areas or specific points, and this is done in conjunction with stretches, joint rotations and joint manipulation to give a 'holistic' treatment that works on the internal energy of the patient.

The evidence base for CAM therapies with pain management

The evidence base for CAM therapies continues to be an area for debate. The knowledge claims have been particularly contested, as much of the past research into CAM has been with limited sample sizes and with flawed methodologies. The main criticisms of CAM are outlined here.

Critique of alternative therapies

- Focuses on the individual rather than on overall public health issues, so can be seen as individual responsibility-based and 'blaming'
- Encourages a cultural nostalgia for an imagined past society/wholeness
- Calls for changed consciousness and lifestyle that is difficult to enact
- Evidence base for many of the therapies is inconclusive
- Only accessible by higher-income population
- May stop patients accessing conventional treatments that could assist

There continues to be a questioning of CAM by some conventional health care practitioners. The findings from the following article within the evidence-base activity is discussed on the General Osteopathic Council's website www.osteopathy.org.uk.

Evidence base

Canter *et al.* (2005) found that the evidence does not demonstrate that spinal manipulation is an effective intervention for any condition. Given the possibility of adverse effects, this review did not suggest that spinal manipulation was a treatment to be recommended.

Take a look at this systematic review article and then look at the response that osteopaths made to this.

The General Osteopathic Council's response to this paper was to state that spinal manipulation was only one element of treatment, which also includes guidance on lifestyle, diet and exercise. They also state that chronic pain is rarely a single problem and that this is influenced by psychological and social factors and requires treatment that is tailored to individual needs. Another paper could also be looked at that suggests that there is a statistically significant reduction in lower back pain with osteopathic manipulation (Licciardone *et al.*, 2005).

Despite the difficulties in determining an agreed evidence base for CAM treatments, the WHO has suggested that the scientific evidence does presently support some CAM therapies. Scientific evidence from randomised clinical trials is strong for many uses of acupuncture, some herbal medicines and for some of the manual therapies (WHO, 2003). Further research is needed to ascertain the efficacy and safety of several other practices and medicinal plants (WHO, 2003). Acupuncture has been proven effective in relieving postoperative pain, nausea during pregnancy, nausea and vomiting resulting from chemotherapy, and dental pain, with extremely low side-effects. It can also alleviate anxiety, panic disorders and insomnia (WHO, 2003).

An example of the evidence associated with acupuncture is a randomised controlled trial on the effectiveness of acupuncture pain relief for patients with osteoarthritis of the knee, in comparison to sham acupuncture (Berman *et al.*, 2004). This study found that acupuncture seems to improve function and pain relief when compared to sham acupuncture or patient education control groups. While some scientific evidence, such as this, exists regarding CAM therapies, there continue to be difficulties with researching CAM that may not be assisted by using the same forms of research methodologies.

Keywords

Placebo

A placebo is a sham treatment without biological activity, which in many cases induces biological and/or psychological effects in people. There are thought to be two main reasons why this can occur: a) as a conditioned response, b) due to patient expectations. Endorphins can be released which mediate the placebo effect, mimicking the pain relief anticipated by the treatment

Key difficulties in developing clinical guidelines based on evidence are the **placebo** effect and minimising bias, and this has led to an emphasis on randomised controlled trials.

The effectiveness of a placebo pain reliever varies as a function of its believed effectiveness, e.g. a morphine placebo, will be believed to be more potent than an aspirin placebo. It is argued that, as many CAM approaches attempt to induce self-healing by changing the context and meaning of illness, some could view this as the placebo effect of the treatment, and this interacts with the non-placebo elements of the therapy in complex ways (Moerman, 2000). The placebo effect could even be reconceptualised as a legitimate part of treatment (Lewith *et al.*, 2002).

Randomised controlled trials try to 'bracket out' human factors such as the ability to self-heal, which is viewed as important in CAM. The point of CAM therapies are that they are individualised, and this makes them difficult to be properly judged through randomised controlled trials. It is argued that what is needed is 'real', long-term studies, which requires considerable funding, but there is a lack of incentive to fund research in CAM, as research funding routes for treatments are often from pharmaceutical companies.

It could be argued that researching CAM treatments for pain purely using the accepted 'gold standard' of randomised controlled trials is not the appropriate way to judge the efficacy of a CAM treatment. As these therapies are holistic they may be better researched from a naturalistic approach. The objective of naturalistic research is to understand and describe human experience, as expressed by those who participate in the experience.

GP/Complementary Therapist

My experience as a GP who is also a complementary therapist has led me to believe that complementary therapies can augment conventional medicine in ways that add considerable value. However, there seems to be a huge gap between what we as practitioners and patients say about the effectiveness of treatments, and what researchers tell us.

Keywords

The Quality Adjusted Life Year (QALY)
This has been created to determine the outcomes from treatments and other health-influencing activities by combining the quantity and quality of life. The basic idea of a QALY is that it takes one year of perfect health-life expectancy to be worth 1, but regards one year of less-than-perfect life expectancy as less than 1. QALYs can therefore provide an indication of the benefits gained from a variety of therapies in terms of quality of life as well as survival for the patient. Quality of life is an important component of pain management

The high levels of CAM usage suggest that many people hold the belief that the benefits of these therapies outweigh their costs. However, the majority of CAM remain to be evaluated in terms of cost benefits, apart from certain CAM therapies such as acupuncture and some manipulative therapies (Herman *et al.*, 2005). Within the UK cost effectiveness studies show that spinal manipulation and acupuncture can represent an additional cost to usual care, but that estimates of cost per **Quality Adjusted Life Year** (QALY) compare favourably to conventional treatments (Canter *et al.*, 2005).

Spinal manipulation is viewed as a cost-effective addition to 'best care' for back pain in general practice, judged by participant's average QALY (UK BEAM, 2004). Manipulation alone probably gives better value for money than manipulation followed by exercise (UK BEAM, 2004). This one example of a cost benefit study illustrates the complexity of the evidence-base movement for CAM therapies. Not only is there a call for studies that support efficacy, but there is also a requirement for cost-effectiveness to be determined if these treatments and therapies are to be integrated into NHS and health service practice.

Integrating CAM into conventional health care working

Some CAM practitioners have been seeking acceptance and integration into conventional health service working. There has been a drive for the professionalisation of CAM therapists that has involved promotion of the cause from a political, educational and clinical perspective through the following strategies:

- accreditation of educational qualifications and the profession
- alignment to science and evidence-based practice
- contained knowledge claims, differentiating the CAM therapy from other treatments

- boundary construction to ensure that others do not encroach on this area of practice.

For instance the General Osteopathic Council's website www.osteopathy.org.uk places on its front page the statement that the role of the General Osteopathic Council as a regulatory body was placed in statute in the Osteopaths Act of 1993. The other information mirrors the regulations expected of other conventional health care professions. There is a register of practitioners who are the only ones able to call themselves an osteopath. Courses to qualify as an osteopath are a four- to five-year honours degree with underpinning clinical training, and there is a requirement for continuous professional development (CPD) and protection of the public from poor clinical practice. The internet site also indicates that the uptake of this treatment is on the increase, with over 700 million osteopath consultations in the UK in 2005. As integration of CAM progresses within the NHS, and as CAM practitioners also seek professionalisation, there is a risk that barriers could be increased between CAM and patients. The practitioner–patient relationship is addressed in the next section of this chapter.

The Smallwood Report (2005), commissioned by the Prince of Wales, examined the 'Big five' CAM areas of osteopathy, chiropractic, acupuncture, homeopathy and herbal medicine, which are the areas most frequently referred to by doctors. The Department of Health recommends that only those therapies that are statutorily regulated or have robust self-regulation should be available through public funding. As well as looking at the available evidence regarding efficacy, there was also an examination of the potential cost effectiveness of these treatments. For instance back pain accounts for 200 million days lost from work each year at an estimated cost of £11 billion (Smallwood, 2005). The report looked at ways to integrate CAM into NHS treatment, so that it would be possible to increase accessibility by low-income families.

The Smallwood report (2005) found that the most effective CAM therapies (acupuncture for postoperative pain and nausea, and for lower back pain and migraine; manipulation therapies for acute and chronic lower back pain and arthritis) correspond to areas where there is an existing 'effectiveness gap' in NHS services. The typical effectiveness reported for both CAM and conventional practices is in the order of 70–80% (Lewith *et al.*, 2002). It was suggested that where CAM treatments were available this could reduce costs by reducing the number of medical consultations, prescribed medications and pain clinics (Smallwood, 2005 p.12).

At the moment there are a range of methods for the delivery of CAM within the UK. These include:

- private CAM practitioner
- cam service delivered by a GP or nurse
- on-site multidisciplinary team that incorporates CAM practitioners

- specialist CAM centres which are part of the NHS
- specialist CAM centres contracted by the NHS
- off-site CAM practitioners contracted by the NHS
- hospital-based services using CAM
- palliative care services providing CAM. (Adapted from Smallwood, 2005.)

Many nurses, midwives and other health professionals have integrated CAM into their practice or are aware of patients also accessing this independently. The Nursing and Midwifery Council (NMC) offer the following guidance: 'If a registrant is caring for a patient/client who wishes to continue their homeopathic or alternative therapies, then this decision must be respected. If administering the alternative treatments, the registrant must have undertaken an approved course and be deemed competent to offer this treatment. CAM should be discussed with the multidisciplinary team.' (NMC, 2006).

The NMC Code of Professional conduct: Standards for conduct, performance and ethics (2004) also states that the practitioner should ensure that the use of complementary or alternative therapies is safe and in the interests of patients and clients, and discussed with the interprofessional team as part of the therapeutic process. The patient must always have granted consent for CAM use.

Surveys indicate that nurses are mainly using a narrow range of therapies, e.g. aromatherapy and massage, reflexology, therapeutic touch, relaxation and visualisation, acupuncture, hypnotherapy, shiatsu, homeopathy and herbalism (Rankin Box, 1997).

Example of a text for a visualisation session for a patient with pain

- Sit comfortably and relax as much as possible
- Concentrate on your breathing, allow it to become slow, gentle and rhythmic
- Become aware of your pain and picture it as being in the shape of a circle
- Direct your breathing into this circle, allowing it to become deeper
- As you do this allow the circle to become larger and larger, and as the circle increases in size the pain begins to decrease
- Continue to increase the size of the circle and feel the pain decrease until it has gone away

Lewandowski *et al.* (2005) examined how verbal descriptors of pain changed with the use of guided imagery techniques. Participants in the treatment group who received guided imagery over a four-day consecutive period moved their description of pain away from 'pain is never ending' prominent in both the control group and the pre-treatment group towards a description of pain as 'changeable' (Lewandowski *et al.*, 2005).

One measure of the integration of CAM into NHS practice is in evidence in the development of competencies for practice for a range of complementary therapies.

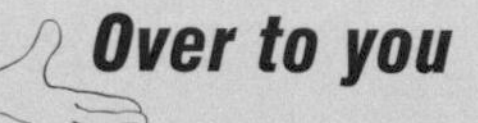

Over to you

Take a look at the Skills for Health website and the range of therapies with competency frameworks already developed. www.skillsforhealth.org.uk/frameworks.

Find out what complementary therapies are used for pain care in your current practice area.

NHS Trusts should have local guidance for health care professionals who wish to undertake complementary therapies so that the suitability of this can be assessed. This integration of CAM into nursing could be seen as just another caring task for nurses or as nurses re-engaging with the caring component of the role. Whatever the viewpoint, nurses need to practice with an appropriate degree of critical self-appraisal, and to risk assess any pain intervention against the possible benefits.

Evidence base

Take a look at the following two articles:

Chandler A. and Lister S. (2002) Using heat therapy for pain management. *Nursing Standard* 17(9): 40–42.

This article looks at some of the safety issues in using heat therapy within a hospital setting.

Snyder M. and Wieland J. (2003) Complementary and alternative therapies: what is their place in the management of chronic pain? *Nursing Clinics of North America* 38(3): 495–508

This article looks at the therapies nurses might consider when planning care for patients with chronic pain, particularly those which can promote self-care, e.g. massage and imagery. They advocate for nurses to weigh the risks and benefits before suggesting a therapy to promote comfort, and to evaluate its effectiveness.

Complementary therapies are increasingly available in the NHS and are beginning to compete with conventional medicines for scarce resources

(Zollman and Vickers, 1999). Given the known association between low quality studies and positive outcomes, further high-quality research is needed covering the range of CAM approaches. Better treatment outcomes are likely to occur when the psychological contributors as well as the physiological factors involved in pain are addressed. The appeal of CAM depends on its philosophical assumptions as well as its practical effects, and its acceptance and growth in popularity means that practitioners working in the field of pain care need to understand these therapies and to integrate these where deemed appropriate.

Case study

Accessing alternative and complementory therapies

Jennie, a 41-year-old pharmacist living in London, started to suffer with lower back pain when she was expecting her third child. She was told that the loosening of the ligaments caused this, which is a common problem during pregnancy. Six months after she gave birth to her third son the situation had not improved, and she was finding it hard to get out of bed, walk or to get down the stairs. Her GP had referred her to a physiotherapist but the lower back pain continued. She was taking strong analgesics, prescribed by her doctor, and decided to attend an Alexander Technique class. Despite this, she continued to complain about her pain, and felt that the drugs she was taking only served to mask her symptoms rather than deal with an underlying problem. Her GP decided to refer her to a local homeopathic hospital. Her first appointment was for 45 minutes, where she was questioned about her lifestyle, diet and stress and given a tailor-made homeopathic treatment. The doctors who treated her were also conventionally trained. She then attended for follow-ups (lasting about 20 minutes) whilst taking two homeopathic pills twice a day 'to tighten the ligaments' until the pain eased. Over the course of the next few months her pain eased and she ceased treatment.

Although back pain is the most common reason patients use CAM, patients' knowledge about these therapies is limited (Sherman *et al.*, 2004).

Jennie paid for her Alexander Technique classes but received her homeopathic treatment through NHS funding. There is an expressed wish by some patients for the more established forms of alternative therapy to be more widely available on the NHS.

- What issues have been highlighted by the case study in relation to this?

The practitioner–patient relationship

Although both conventional medicine and complementary therapies emphasise the quality of the relationship between the practitioner and patient, the variability of the practitioner–patient relationship remains apparent. In fact some CAM practitioners use their therapy as a springboard for an attack on conventional approaches (Vickers, 1998). It is helpful, therefore, to look at the nurse–patient relationship as this will be a component of pain care management.

When Watson (1996) discusses a 'caring science' she includes art and humanities in this, acknowledging the connections between individuals, others, the community and the wider world. She particularly advocates the many ways of knowing combining clinical and empirical enquiry with other forms such as personal, intuitive and spiritual knowing. Watson first started to develop her 'Theory of Human Caring' during the 1970s, and was trying to make explicit the values, knowledge and practice of human caring which nurses bring to the health setting. This is particularly pertinent in the pain care field, as the management of pain is often not a measure of 'cure'.

An adapted version of Watson's (1996) 10 key caring processes are listed here.

1. Delivering practice within a humanistic, altruistic and loving context.
2. Being authentically present, instilling faith and hope.
3. Cultivating sensitivity to self and to others.
4. Developing and sustaining a helping-trusting relationship.
5. Being present and supportive of the expression of positive and negative feelings.
6. Using self creatively, and all ways of knowing, as part of the caring process.
7. Engaging in genuine teaching-learning experiences.
8. Creating a healing environment (physical/psychological/social) including dignity and respect.
9. Assisting with basic needs.
10. Attending to the spiritual dimensions of life and death.

Just from reading these 10 factors it becomes clear that Watson looks at nursing as at the intersection between the personal and the professional, seeing communication as central to a **therapeutic relationship**.

Keywords

Therapeutic relationship
This requires the practitioner to respect and have genuine interest in the person, to show emotional warmth, tolerance and non-judgemental acceptance of the patient. It also calls for the practitioner to use 'self' in nursing interactions, whilst maintaining awareness of their limitations and adherence to ethical codes. The patient also plays a part in a therapeutic relationship by trusting and co-operating with the intervention and being motivated to understand the treatment/ care

In a similar manner to complementary and alternative therapy practitioners, the therapeutic relationship moves away from a 'modern', technical view of nursing towards a focus on the uniqueness of each situation and the wholeness of the person, regardless of illness and/or pain. The nurses' own life experiences are seen as integral to their patient care. As Watson states:

> I emphasize that it is possible to read, study, learn about, even teach and research the caring theory; however, to truly 'get it', one has to personally experience it; thus the model is both an invitation and an opportunity to interact with the ideas, experiment with and grow within the philosophy, and living it out in one's personal/ professional life.
> (Watson, 1996 p. 161)

Reflective activity

Think about your worst experience of pain and how this affected your thoughts about pain and pain care management.

Then go on to think about a patient care episode where you were pleased with the pain care that was achieved. What were the factors about the pain care that made this a pleasant incident to reflect on?

The patient–practitioner relationship and the explanatory frameworks provided by CAM are perceived as important components of the therapeutic process (Cartwright and Torr, 2005). Strong therapeutic relationships impact on patient satisfaction, treatment concordance and compliance and treatment outcomes (Leach, 2005). CAM treatments are viewed as serving a variety of functions beyond the relief of symptoms such as pain; these include increased energy and relaxation, the facilitation of coping, enhancing self- and others' awareness (Cartwright and Torr, 2005).

Key points

- Alternative and complementary therapies take a holistic, individualised approach to a patient, and believe that this self-healing is the basis of all healing
- CAM treatments are widely used within the UK, with over £4.5 billion spent on this annually
- CAM therapies can be classified into five main groups: alternative medical systems, mind–body interventions, physiological treatments, manipulation of the body therapies, and energy therapies
- The evidence base for CAM therapies continues to be an area for debate, although the scientific evidence from randomised clinical trials is strong for many uses of acupuncture, some herbal medicines and for some of the manual therapies
- CAM therapists are becoming more regulated and there is increasing professionalisation
- If a health care professional uses CAM within their practice they must have undertaken an approved course and be deemed competent to offer this treatment. CAM should be discussed with the patient and consent must be given, and the multidisciplinary team should be involved
- The patient–practitioner relationship is perceived as an important component of the therapeutic process, and has a part to play in pain care management.

Rapid recap

1. Outline the main differences between complementary medicine and alternative medicine.
2. List four underpinning concepts held by CAM practitioners.
3. What CAM treatments currently have the strongest empirical evidence base for their use in pain care?
4. What is a placebo effect?
5. What would a health care practitioner need to do if they wanted to integrate and deliver complementary therapies within their care setting?
6. Define a therapeutic practitioner–patient relationship.

References

Berman B., Lao L., Langenberg P., Lee W., Gilpin A. and Hochberg M. (2004) Effectiveness of acupuncture as adjunctive therapy in osteoarthritis of the knee. *Annals of Internal Medicine*, **141**: 901–910.

Black A. (2005*) The future of Acute Care*. A personal view commissioned by the NHS Confederation. NHS Confederation, London.

Canter P., Coon J. and Ernst E. (2005) Cost effectiveness of complementary treatments in the United Kingdom: systematic review. *British Medical Journal*, **331**: 880–881.

Cartwright T. and Torr R. (2005) Making sense of illness: the experiences of users of complementary medicine. *Journal of Health Psychology*, **10**(4): 559–572.

Dossey B (2000) *Holistic nursing: A handbook for practice,* 3rd edn. Aspen, New York.

Eisenberg D. (2002) Complementary and integrative medical therapies. *Current trends and future trends*. Harvard Medical School, Boston.

Graham H. (1999) *Complementary therapies in context. The psychology of healing*. Jessica Kingsley Publishers, London.

Herman P., Craig B. and Caspi O. (2005) Is complementary alternative medicine (CAM) cost-effective? A systematic review. *Biomed Central Complementary Alternative Medicine*, **5**: 11.

Leach M. (2005) Rapport: a key to treatment success. *Complementary therapies in Clinical Practice*, **11**(4): 262–265.

Lewandowski W., Good M. and Draucker C. (2005) Changes in the meaning of pain with the use of guided imagery. *Pain Management Nursing*, **6**(2): 58–67.

Lewith G., Jonas W. and Walach H. (eds) (2002) *Clinical Research in Complementary Therapies. Principles, Problems and Solutions*. Churchill Livingstone, Edinburgh.

Licciardone J., Brimhall A. and King L. (2005) Osteopathic manipulative treatment for low back pain: a systematic review and meta-analysis of randomized controlled trials. *Biomedical Central Musculoskeletal disorders*, **6**: 43.

Moerman D. (2000) Cultural variations in the placebo effect: Ulcers, anxiety and blood pressure. *Medical Anthropology Quarterly*, **14**: 15–72.

Nursing and Midwifery Council (2004) *Code of Professional conduct: Standards for conduct, performance and ethics.* NMC, London.

Nursing and Midwifery Council (2006) *A–Z Advice sheet. Complementary, Alternative therapies and Homeopathy.* NMC, London.

Rankin Box D. (1997) Therapies in Practice: a survey assessing nurses' use of complementary therapies. *Complementary Therapies in Nursing and Midwifery*, **3**: 92–99.

Sherman K., Cherkin D., Connelly M., Erro J., Savetsky J., Davis R., *et al.* (2004) Complementary and alternative medical therapies for chronic low back pain: What treatments are patients willing to try? *Biomedical Central Complementary Alternative Medicine*, **4**(1): 9.

Smallwood Report (2005) *The role of complementary and alternative medicine in the NHS.* An investigation into the potential contribution of mainstream complementary therapies to health care in the UK. Freshminds, London.

UK BEAM trial team (2004) United Kingdom back pain exercise and manipulation randomised trial: cost effectiveness of physical treatments for back pain in primary care. *British Medical Journal*, **329**(7479): 1381.

Vallerand A., Fouladbakhsh J., Templin T. (2005) Patients' choices for the self-treatment of pain. *Applied Nursing Research*, **18**(2): 90–96.

Vickers A. (ed)(1998) *Examining Complementary Medicine 'The Skeptical Holist'.* Stanley Thornes, Cheltenham.

Watson J (1996) Watson's theory of transpersonal caring. In *Blueprint for use of nursing models: Education, research, practice and administration*. Walker P. and Neuman B. (eds.) NLN Press, New York, pp. 141–184.

World Health Organization. (2003) *Traditional medicine*. Fact sheet no. 134. WHO, Geneva.

Zollman C. and Vickers A. (1999) ABC of complementary medicine. What is complementary medicine? *British Medical Journal*, **319**: 693–696.

9 Management of acute pain

Learning outcomes

By the end of this chapter you should be able to:

- ★ Describe various forms of acute pain
- ★ Detail methods of managing acute post-operative pain
- ★ Examine the importance of acute pain management.

Introduction

This chapter will help you to consider the broader context of acute pain. Methods of managing acute pain are discussed. Traditionally, acute pain is described as pain that subsides as healing takes place, i.e. it has a predictable end and it is of brief duration, usually less than six months. An example of acute pain could be postoperative pain that occurs following surgery. There are, however, many examples of acute pain – from pain due to trauma, pain due to disease processes such as pancreatitis, to chest pain (see Figure 9.1). Within this chapter surgical urgent care and medical acute-pain management will be discussed.

Good practice in acute-pain management

The Institute for Clinical Systems Improvement (ICSI) (2004) suggests that acute pain is not a diagnosis, it is a symptom. It usually

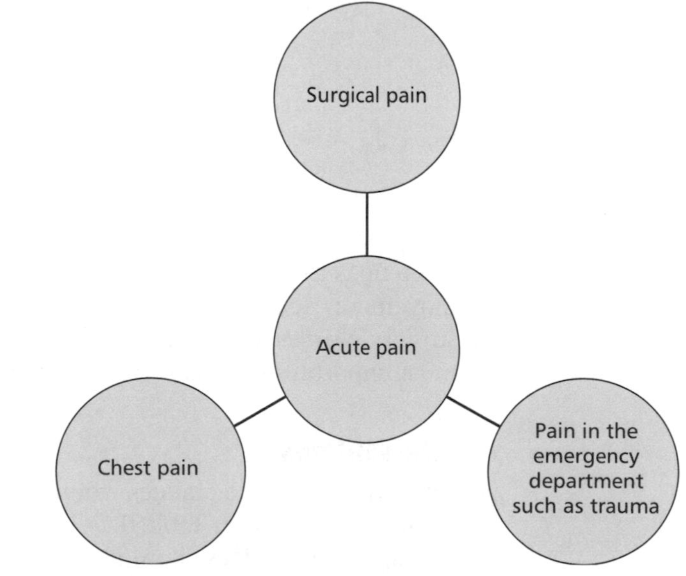

Figure 9.1 Common acute-pain presentations

occurs following surgery or an acute trauma. However, the cause is not always clear. ICSI recommends that a general history, pain history, clinical examination, diagnostic studies and specialty consultation should be conducted as appropriate.

Particular elements of acute pain assessment are outlined within Chapter 6. Figure 9.2 summarises the important issues for acute-pain assessment.

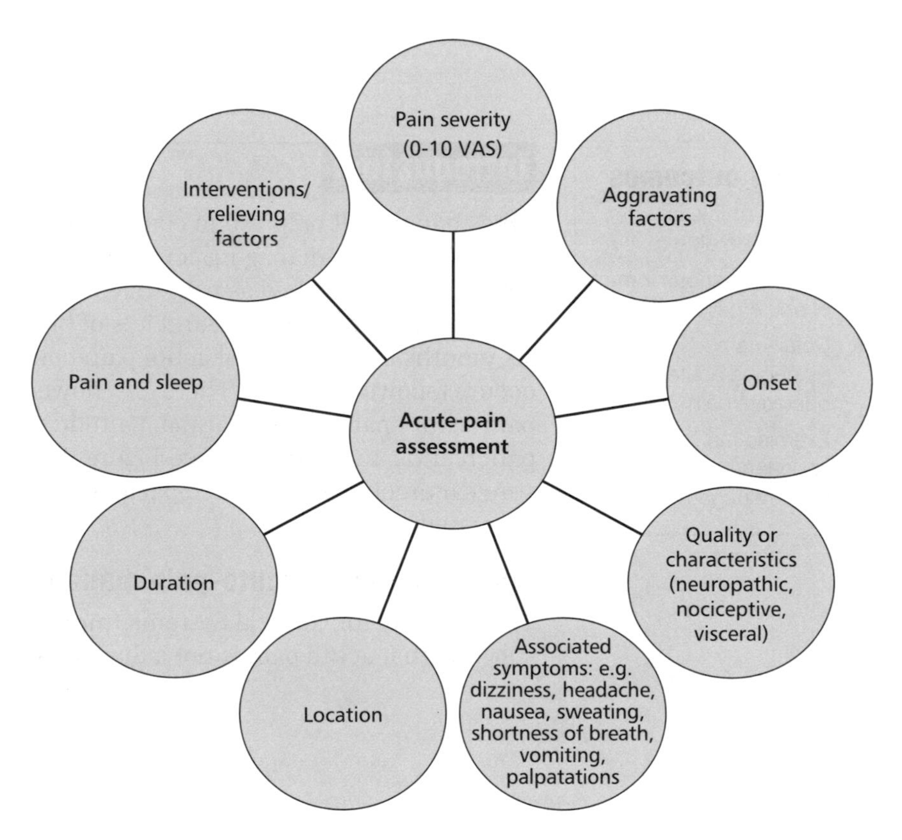

Figure 9.2 Assessment of acute pain

A top tip is to remember that pain assessment is dynamic, e.g. on rest, movement, deep breathing and coughing especially after trauma and surgery. Nurses nowadays decide and document which activity is the most important to assess in regards to pain.

Interventions

There are extensive choices when treating acute pain and examples of these are outlined in Table 9.1. The decisions in relation to acute-pain management choices should be based on sound decision making and the best available evidence.

Table 9.1 Modalities for treatment of acute pain

	Examples
Pharmacological preparations	Opioids NSAIDs Acetaminophen (paracetamol)
Operative procedures	Surgery
Physical modalities	Physiotherapy Manipulation Elevation Heat and cold TENs
Regional anaesthesia	Epidural infusions of local anaesthetics +/– opioids Nerve blocks +/– opioids
Complementary modalities	Acupuncture, relaxation, massage

Surgical pain

Acute pain is an expected outcome of any surgical procedure and the extent of surgery will not necessarily be a reliable indicator of the likely level of pain to be suffered. It is the most common form of acute pain. Postoperative pain studies demonstrate that patients suffer moderate to severe pain after their surgery (Apfelbaum *et al.*, 2003; Mac Lellan, 2004; Coll and Ameen, 2006). Dolan *et al.* (2002) conducted a review of published data considering three types of analgesic techniques and concluded that over the period 1973–1999 there has been a highly significant reduction ($p < .0001$) of the incidence of moderate to severe pain at rest of 1.9 (1.1–2.7%) per year. Table 9.2 summarises the incidence of overall moderate to severe and severe pain based on

Table 9.2 Incidence of overall moderate to severe and severe pain based on method of analgesic administration (Source: Dolan *et al.*, 2002)

Confidence intervals in brackets	Intramusclar (IM)	PCA	Epidural	Overall
Moderate to severe pain	67.2 (58.1–76.2)	35.8 (31.4–40.2)	20.9 (17.8–24)	29.7 (26.4–33)
Severe pain	29.1 (18.8–39.4)	10.4 (8.0–12.8)	7.8 (6.1–9.5)	10.9 (8.4–13.4)

method of analgesic administration using pooled data from 165 papers, 20 000 patients for the time period. Mean and 95% confidence interval are reported.

> **Over to you**
>
> Consider Table 9.2. What does it mean? Is intramuscular (IM) administered analgesia the least effective? Go to your library and get a full copy of the article: Dolan S.J., Cashman J.N. and Bland J.M. (2002). Effectiveness of acute postoperative pain management: I. Evidence from published data. *British Journal of Anaesthesia*. 89(3): 409–423. Three levels of evidence are identified for the papers used in the study. What does this mean?

Various physiological responses are associated with acute pain (Jurf and Nirschl, 1993). The stress response initiated by the autonomic nervous system causes increased respiratory rate, increased cardiac output, elevated blood pressure, increased skeletal muscle tone, vasospasm of peripheral vessels and smooth-muscle relaxation. Pain can cause emotions such as anxiety and fear, which also stimulate the autonomic nervous system. These physiological consequences of acute pain, which are present with post surgical pain, if allowed to reach a certain intensity, can lead to a number of complications.

Undertreated severe pain is associated with a number of harmful effects including decreased lung volumes, myocardial ischaemia, decreased gastric and bowel motility and anxiety (Macintyre and Ready, 2002). High pain scores have been associated with an increased incidence of nausea. The psychological effects of pain include both anxiety and fear.

The objectives of postoperative pain control should be to minimise discomfort, facilitate recovery and avoid treatment related side-effects. Management of surgical pain is complicated and challenging because of large variations in pain experience, analgesic requirements and the many techniques available to treat postoperative pain (Lynch *et al.*, 1997). The choice of the most appropriate method to manage pain will be governed by the nature of the surgery, the intensity and expected duration of the pain, the availability of medication and expertise, and patient factors such as illness, age, contraindications and psychological state. Patients vary greatly in their responses to surgery and procedures as well as their responses to pain and interventions and their personal preferences.

As surgery becomes less invasive and a high proportion of surgical procedures become day surgery cases, the forms of pain care interventions will need to maintain the same levels of effectiveness, whilst being appropriate for patient self-administration.

Pharmacological intervention is the mainstay in the management of acute-pain management, although non-pharmacological approaches provide good adjunctive therapy in many cases (Day, 1997). An integrated approach to pain management is recommended that includes:

- cognitive-behavioural interventions such as relaxation, distraction and imagery: these can be taught preoperatively and can reduce pain, anxiety and the amount of drugs needed for pain control
- systematic administration of nonsteroidal anti-inflammatory drugs (NSAIDs) or round-the-clock administration of opioids
- patient-controlled analgesia (PCA) which usually means self-medication with intravenous doses of an opioid; this can include other classes of drugs administered orally or by other routes
- epidural analgesia, usually an opioid and/or local anaesthetic injected intermittently or infused continuously
- intermittent or continuous local neural blockade (examples of the former include intercostal nerve blockade with local anaesthetic or cryoprobe; the latter includes infusion of local anaesthetic though an interpleural catheter)
- transcutaneous electrical nerve stimulation.

Pharmacological management of postoperative pain

The use of medications remains the mainstay of management of acute pain. It is evident, however, that patients continue to suffer from moderate to severe pain while in hospitals.

Pharmacological management of mild to moderate postoperative pain should begin (unless there is a contraindication) with paracetamol and/or an NSAID. Opiates remain the mainstay of systemic analgesia for the treatment of moderate to severe pain (Australian and New Zealand College of Anaesthetists and Faculty of Pain Medicine, 2005). The concurrent use of paracetamol, opioids and NSAIDs often provides more effective analgesia than either class alone, while reducing side-effects. Refer to chapter 7 for more detail on these medications, their use and side effects.

Studies support the concept that current prescribing patterns are not supporting comprehensive pain management.

Evidence base

Timeline

- **2003** Manias following prospective audit of 100 patients found that while almost all patients received some form of infusion, the use of 'as required' analgesics varied from one-third to over two-thirds of patients during the postoperative period.
- **2001** Schafheutle *et al*. surveyed nurses who stated that analgesic regimens were 'inadequate', 'inappropriate' or 'ineffective'. This included insufficient frequency of dosing or inadequate flexibility in the choice of analgesics.
- **1997** Boer *et al.* found that the prescribed daily dose of morphine was only received by 4.2% of patients.
- **1997** Mac Lellan reported that 97% of patients were prescribed more than one analgesic. Mean number of doses of analgesia administered daily varied from 1.4 to 3.2. Mean amount of analgesia administered varied from 4% to 41% of the maximum possible for the first five days post surgery.
- **1994** Oates reported that those patients with moderate and severe pain received only 36% of their prescribed analgesics.
- **1993** Juhl found that 91% of patients had analgesia prescribed pro re nata (PRN) and that only 4% of patients had analgesics prescribed for regular use. He found that on average patients received 70% of the maximally prescribed dose of analgesic during the first 24 hours and an average of 43% during the following day.
- **1992** Closs examined patterns of analgesic provision and found that the number of doses given peaked at two points during the 24-hour cycle. The highest number of doses were give between 8 a.m. and 12 noon and 8 p.m. and 12 midnight. Fewer doses were given at night between midnight and 4 a.m. In the study, pain was found to be the most common form of night-time sleep disturbance, with analgesics helping more patients to get back to sleep than any other intervention. Almost 50% of patients said that their pain was worse at night. Analgesic provision at night, therefore, did not appear to be explicitly related to need. Of those prescribed intermittent opioids (n = 79) they received 23 ± 2% of their theoretical maximum dose.

Pain in the accident and emergency department and emergency assessment unit

Pain is the single most common reason for presentation to emergency departments (Campbell *et al.* 2004). Cordell *et al.* (2002), reviewing charts over one week (n = 1665), found a prevalence of 52.2% of pain, identifying it as the chief complaint for visits to emergency departments (United States). A Canadian survey of 525 patients showed that abdominal, chest and musculoskeletal pain were the most common diagnoses and that more than two-thirds of patients had non-traumatic pain (Todd *et al.*, 2004). Patients in this survey had high levels of pain; however, only half the patients received any analgesic and these analgesics were administered two hours after presentation to the emergency department. Patients themselves, however, reported high levels of satisfaction and 88% of those untreated for pain did not request analgesics.

When approaching the management of pain in the emergency department it is important to remember that the basic principles of pain management as detailed in Chapter 7 apply. Campbell *et al.* (2004) demonstrate that by implementing a pain protocol in the emergency department a significant improvement in early pain management and patient satisfaction can be achieved. Regular documented pain assessment is of particular importance in the emergency department (see chapter 6) as the patient's pain level may alter very quickly, indicating a deteriorating condition. Cohen *et al.* (2004) suggest that improved pain management in trauma patients leads to not only increased comfort but has been shown to reduce morbidity and improve long-term outcomes. The authors' approach to the management of pain in trauma is outlined in Table 9.3.

Table 9.3 Modalities for treatment of trauma pain (Source: Cohen *et al.*, 2004)

	Examples
Pharmacological preparations	• Acetaminophen (paracetamol) • NSAIDs • Opioids • Ketamine (blocks NMDA receptors – see chapter 4) • Local anaesthetics • Tricyclic antidepressants (neuropathic pain) • Anticonvulsants (neuropathic pain) • Clonadine (analgesic and sedative properties) • Benzodiazepines (lacks analgesic properties but promotes sedation and muscle relaxation) • Entonox • Corticosteroids
Physical modalities	Physiotherapy Manipulation TENs Heat and cold Immobilisation
Regional anaesthesia	Regional blocks Epidural infusions
Complementary modalities	Acupuncture, TENs, hypnosis, psychological interventions

A top tip to bear in mind is: although the acute pain interventions listed in the table are used within urgent-care settings, nursing advice and patient care also impacts on the total pain experience of the patient, e.g. elevation of a limb that has been plastered following a fracture.

In addition to the usual effects and side effects of pain interventions, there are special considerations for many of the traditional approaches to pain management for the emergency department. These are detailed below (Dolan, 2000).

Morphine:

- is usually administered intravenously in the emergency department.
- causes sedation
- should not be given to shocked patients
- if used in coronary thrombosis may further slow pulse and lower blood pressure.

NSAIDs:

- are particularly effective in relieving pain with inflammation and thus are useful for musculoskeletal disorders and trauma to peripheral tissues.

Non-pharmacological management:

- information to reduce anxiety
- immobilisation and elevation
- warm and cold compresses
- distraction.

Entonox is a gaseous mixture of 50% oxygen and 50% nitrous oxide (Butcher, 2004). Common examples of its use are with soft tissue injury such as wounds, abscesses, burns, traumas to the abdomen, chest and limb fractures. Entonox is contraindicated in pneumothorax and certain head injuries where there is an altered state of consciousness.

The use of local anaesthesia can be particularly effective for pain in the emergency department (Edwards, 2000). Local infiltration into an area surrounding a wound site is ideal for the suturing and extensive cleaning of minor wounds and the drainage of superficial abscesses. Peripheral nerve blocks such as ring block will achieve anaesthesia of a digit. Intravenous regional anaesthesia (Bier's block) can be used for anaesthesia below the elbow, for example, for manipulation of a Colles fracture. NSAIDS are particularly effective in musculoskeletal conditions, and specifically in limb trauma involving major joints (knee etc.).

Chest pain

Chest pain is one the common acute-pain problems that patients present with to the health services. Research indicates that chest pain can account for 2–4% of all new attendances at emergency departments per year in the UK (Herren and Mackway-Jones, 2001). There are many different causes of chest pain, some of which are life threatening, requiring medical interventions (Tough, 2004) (see Table 9.4).

Table 9.4 Causes of chest pain

Cardiovascular	Myocardial infarction Unstable angina
Pulmonary	Pleurisy Pulmonary embolism Pneumothorax Pneumonia
Musculoskeletal	Costocondritis Trauma
Gastrointestinal	Reflux Gastric ulcers Gallstones Pancreatitis
Non-organic	Anxiety

Management of chest pain is complex and begins with a detailed assessment of the pain. The cause of the pain should be established without delay.

McAvoy (2000) and Efre (2004) refer to acronyms such as OLD CART to assist in this assessment. OLD CART is described in Table 9.5.

Table 9.5 OLD CART chest pain assessment

Onset	When did the pain start? Are you having the pain now?
Location	Where is the pain located?
Duration	How long have you been having this pain? Does it come and go or is it consistent?
Characteristics	What does it feel like? Pressure, tightness, heaviness? Describe how it feels.
Accompanying symptoms	What other signs and symptoms accompany the pain? Nausea? Sweating? Dizziness or light-headedness?
Radiation	Does the pain travel to any other part of your body– your left or right arm, neck, jaw, back?
Treatment	What makes the pain better or worse? If you've had this kind of pain before, what relieved it?

In addition to determining previous medical history, risk factors, medication history and social history the points described in Figure 9.3 should assist in establishing the cause of pain (Tough, 2004).

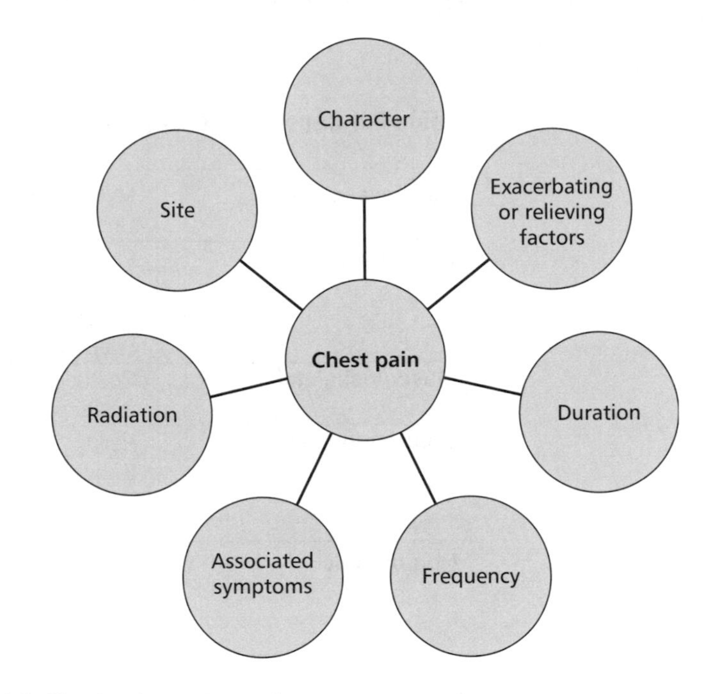

Figure 9.3 Chest pain assessment

Detail of further diagnostic tests and treatment for chest pain to relieve symptoms, limit myocardial damage and risk of cardiac event are beyond the scope of this book. Indeed, much immediate treatment is now protocol and guideline driven with the use of various chest pain assessment services such as chest pain assessment units, rapid access chest pain clinics as well as emergency departments and coronary care units (Aroney *et al.*, 2003; Fox, 2001). One of the medications of choice to relieve cardiovascular chest pain based on appropriate assessment is morphine administered intravenously. However, it should be noted that for **acute coronary syndrome** effective and early pain relief remains a clinical priority and no one agent offers the ideal solution to controlling the pain.

Keywords

Acute coronary syndrome

Describes a spectrum of illness from ischaemic heart disease to unstable angina to myocardial infarction

Nitrates have an important role and Opie and Gersh (2001) recommend that at onset of the angina the patient should rest in the sitting position and take nitroglycerin sublingually (0.3–0.6 mg) every five minutes until the pain is relieved. Castle (2003) suggests that the early use of opiates, betablockers and nitrates, as well as reassuring patients, have vital roles to play in providing effective analgesia for acute coronary syndrome.

However, the pain may not be cardiac in origin, for example the pain may be musculoskeletal in origin and if so an NSAID may the most appropriate medication.

Triage Nurse

I started working as a triage nurse in the accident and emergency department about two years ago. One of the most difficult pains to manage is chest pain because it can be anything from a heart attack to a strained muscle. In order to be confident in my triage decision, I always do a comprehensive pain assessment.

Good practice in acute-pain management should be part of a continuous cycle as outlined in Figure 9.4.

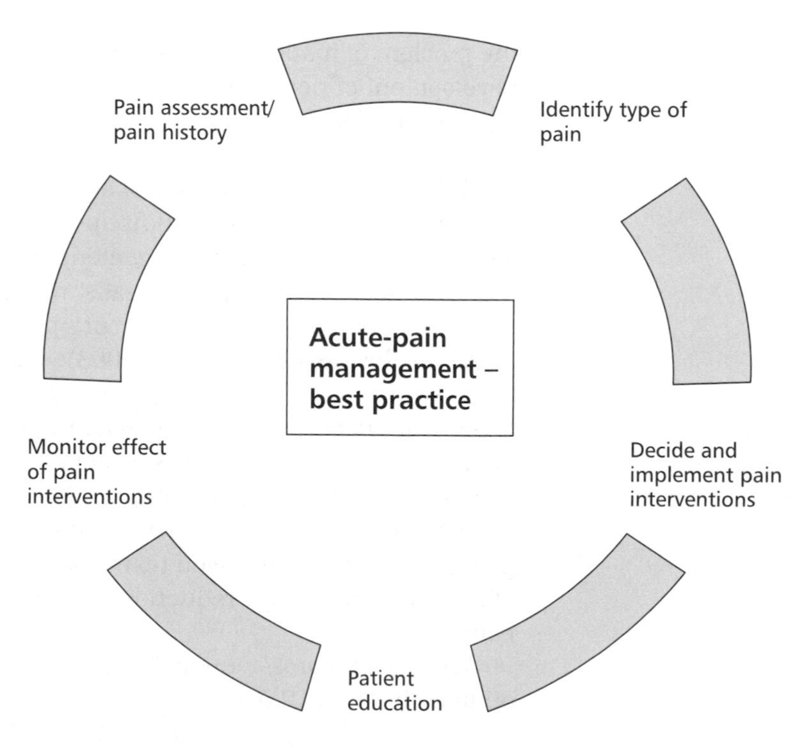

Figure 9.4 Good practice in acute-pain management

Reflective activity

Consider the last patient you managed with acute pain. How did you assess his/her pain? How were decisions regarding pain interventions for that acute pain made? Was the effect of the pain interventions monitored? For example, if the person was on an opioid did it relieve the pain? Could the patient mobilise? Did the patient have any side-effects, e.g. constipation?

Acute-pain services

Formalising services for acute pain has become a trend within the health service ever since Ready *et al.* in 1988 drew attention to the potential role of an acute-pain service by developing an anaesthesiology-based postoperative pain management service. They said: 'just as chronic pain management has become a special area of medical practice, treatment of acute pain deserves a similar commitment by practitioners with special expertise'.

The evidence as detailed in Chapter 3 suggests that the solution to the problem of inadequate postoperative pain relief lies not so much in development of new techniques but in the development of a formal organisation for better use of existing techniques (Rawal and Berggren, 1994).

Following systematic review of acute pain teams, McDonnell *et al.* (2003) describe them as multidisciplinary teams who assume day-to-day responsibility for the management of postoperative pain. Team members generally include specialist nurses, anaesthetists and pharmacists. The prevalence of acute-pain teams in the UK is high, as evidenced by McDonnell *et al.* (2003). Their survey highlights the fact that 84% of acute English hospitals performing inpatient surgery for adult patients have an acute pain team (McDonnell *et al.*, 2003). Surveys report that 42% of US hospitals (Warfield, 1995) have acute-pain management programmes and 53% of Canadian hospitals (Zimmermann and Stewart 1993). Many benefits are ascribed to acute-pain services and acute-pain teams, such as a more consistent standard of postoperative care, a reduction in pain scores and increased safety for patients.

Acute pain teams/acute pain services encompass the qualities outlined in Figure 9.5.

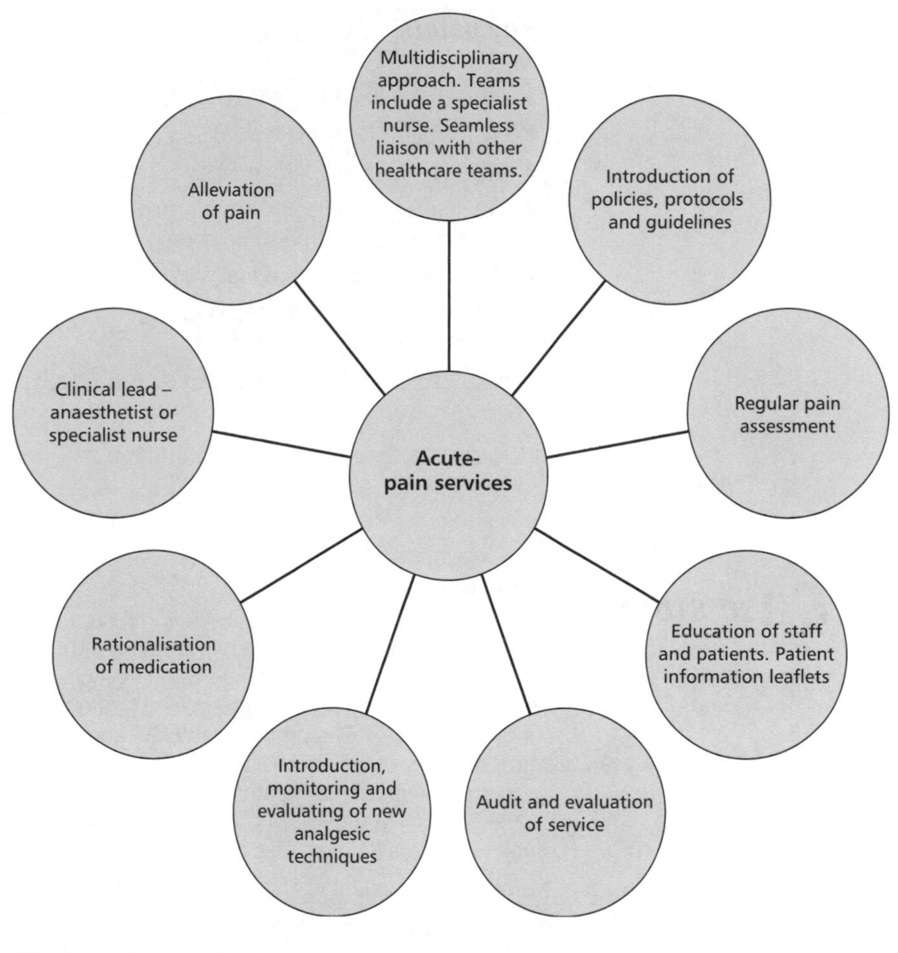

Figure 9.5 Acute-pain services

Impact of technology on acute-pain management

Technology has revolutionised modes of analgesic administration allowing for smaller, more regular amounts of analgesia to be administered in a safe and efficient manner. Patients have been enabled to take more control over pain relief. It is suggested that refinements in systemic opioid administration and epidural analgesia have by far the greatest impact on the quality of acute-pain care (Carr *et al.*, 2005). Patient-controlled analgesia (PCA) using traditional intravenous opioids, the newer concept of fentanyl hyrdrochloride patient-controlled transdermal system (PCTS) and epidural analgesia are reviewed in Chapter 3.

Key points Top tips

- The prevalence of acute pain is high
- Acute pain should be considered a symptom not a diagnosis
- Evidence-based approaches should be used to manage acute pain
- There are extensive choices for managing acute-pain, from pharmaceutical preparations to complementary modalities
- Medications remain the mainstay of management of acute severe pain
- Assessment and monitoring of the effect of the pain intervention is the key to good acute-pain management

Chest pain

Mark is a 45-year-old man who owns and runs a corner shop. He is married with three children all under the age of six. He is sitting at home watching TV one Saturday night and suddenly experiences a sharp pain in his chest and neck. His wife Sarah drives him to the local emergency department. On arrival, the nurse assesses his pain as 7 out of 10 on a VAS. She performs an ECG and takes bloods. She begins to take a detailed history of his pain as outlined in Table 9.6.

In addition to assessing Mark's pain the nurse begins a history, asking Mark about his medical history, family history of illness, any medications that he is on and any other risk factors for heart disease.

- Why is it important in Mark's case history to assess more than his pain?
- How do you think Mark's pain should be managed?
- How would you explain Mark's pain to him and his wife Sarah?

Table 9.6 Mark's OLD CART chest pain assessment

Onset	Pain is present on attending the emergency department. Mark's pain began suddenly two hours ago.
Location	The pain is in his chest and radiating up his neck.
Duration	The pain is constant and has been present for about two hours.
Characteristics	Mark describes the pain as 'pressure'
Accompanying symptoms	He is sweating slightly and feels somewhat nauseous. He thinks this may be due to feeling very anxious.
Radiation	The pain is radiating up his neck.
Treatment	Mark has never had this kind of pain before and is not on medication for his heart.

Rapid recap

1. Define acute pain.
2. Name and give examples of three modalities for reducing acute pain.
3. What special considerations are there with the use of morphine in the accident and emergency department?
4. When assessing chest pain what does the acronym OLDCART stand for?
5. What are the main qualities of an acute pain service?

References

Acute Pain Management Guideline Panel. (1992) *Acute Pain Management: Operative or Medical Procedures and Trauma. Clinical Practice Guideline*. Rockville, MD: AHCPR Pub. No. 92–0032. Agency for Health Care Policy and Research, Public Health Service, US Department of Health and Human Services.

Apfelbaum J.L., Chen C., Mehta S.S. and Gan T.J. (2003) Postoperative pain experience: results from a national study suggest postoperative pain continues to be undermanaged. *Anesthesia and Analgesia*, **97**: 534–540.

Aroney C.N., Dunlevie H.L. and Bett J.H.N. (2003) Use of an accelerated chest pain assessment protocol in patients at intermediate risk of adverse cardiac events. *Medical Journal of Australia*, **178**(8): 370–374.

Australian and New Zealand College of Anaesthetists and Faculty of Pain Medicine (2005) Acute pain management: scientific evidence. Available at www.anzca.edu.au/publications/acutepain.htm.

Boer C., Treebus A.N., Zuurmond W.W.A. and de Lange J.J. (1997) Compliance in administration of prescribed analgesics. *Anaesthesia*, **52**: 1177–1181.

Butcher D. (2004) Pharmacological techniques for managing acute pain in emergency departments. *Emergency Nurse*, **12**(1): 26–35.

Campbell P., Dennie M., Dougherty K., Iwaskiw O. and Rollo K. (2004) Implementation of an ED protocol for pain management at triage at a busy level 1 trauma centre. *Journal of Emergency Nursing*, **30**(5): 431–437.

Carr D.B., Reines D., Schaffer J., Polomano R.C. and Lande S. (2005) The impact of technology on the analgesic gap and quality of acute pain management. *Regional Anesthesia and Pain Medicine*, **30**(3): 286–291.

Castle N. (2003) Effective relief of acute coronary syndrome. *Emergency Nurse*, **10**(9): 15–19.

Closs S.J. (1992) Patients' night-time pain, analgesic provision and sleep after surgery. *International Journal of Nursing Studies*, **29**(4): 381–392.

Cohen S.P., Crhisto P.J. and Moroz L. (2004) Pain management in trauma patients. *American Journal Physical Medicine Rehabilitation*, **83**: 142–161.

Coll A.M. and Ameen J. (2006) Profiles of pain after day surgery: patients' experiences of three different operation types. *Journal of Advanced Nursing*, **53**(2): 178–187.

Cordell W.H., Keene K.K., Giles B.K., Jones J.B., Jones J.H. and Brizendine E.J. (2002) The high prevalence of pain in emergency care medical care. *American Journal of Emergency Medicine*, **20**(3): 165–169.

Day R. (1997) A pharmacological approach to acute pain. *Professional Nurse Supplement*, **13**(1): S9–S12.

Dolan B. (2000) Physiology for A&E Practice in *Accident and Emergency Theory into Practice.* Balliere Tindell, London.

Dolan S.J., Cashman J.N. and Bland J.M. (2002) Effectiveness of acute postoperative pain management: I. Evidence from published data. *British Journal of Anaesthesia*, **89**(3): 409–423.

Edwards B. (2000) Physiology for A&E Practice in *Accident and Emergency Theory into Practice.* Balliere Tindell, London.

Efre A.J. (2004) Gender bias in acute myocardial infarction. *Nurse Practitioner*, **29**(11): 42–55.

Fox K.F. (2001) Chest pain assessment services: the next steps. *Quarterly Journal of Medicine*, **94**: 717–718.

Herren K.R., and Mackway-Jones K. (2001) Emergency management of cardiac chest pain: a Review. *Emergency Medical Journal*, **18**: 6–10.

Institute for Clinical Systems Improvement (ICSI) (2004) Assessment and management of acute pain. ICSI, London.

Juhl I.U. (1993) Postoperative pain relief, from the patients' and the nurses' point of view. *Acta Anaesthesiology Scandinavia*, **37**: 404–409.

Jurf J.B. and Nirschl A.L. (1993). Acute postoperative pain management: A comprehensive review and update. *Critical Care Nursing Quarterly*, **16**(1): 8–25.

Lynch E.P., Lazor M.A., Gellis J.E., Orav J., Goldman L. and Marcantonio E.R. (1997) Patient experience of pain after elective noncardiac surgery. *Anaesthesia and Analgesia*, **85**: 117–23.

Manias E. (2003) Medication trends and documentation of pain following surgery. *Nursing and Health Sciences*, **5**: 85–94.

McAvoy J.A. (2000) Cardiac pain: discover. *Nursing*, **30**(3)34–40.

McDonnell A., Nicholl J. and Reid S.M. (2003) Acute pain teams in England: current provision and their role in postoperative pain management. *Journal of Clinical Nursing*, **12**: 387–393.

Macintyre P.E. and Ready L.B. (2002) *Acute Pain Management, A Practical Guide*, 2nd edn. W.B. Saunders, Elsevier Science, Philadelphia.

Mac Lellan K. (1997) A Chart Audit reviewing the prescription and administration trends of analgesia and the documentation of pain, after surgery. *Journal of Advanced Nursing*, **26**: 345–350.

Mac Lellan K. (2004) Postoperative pain: strategy for improving patient experiences *Journal of Advanced Nursing*, **46**(2): 179–185.

Oates J.D.L., Snowden S.L. and Jayson D.W.H. (1994) Failure of pain relief after surgery. *Anaesthesia*, **49**: 755–758.

Opie L.H. and Gersh B.J. (2001) *Drugs for the heart*, 5th edn. W.B. Saunders, Elsevier Science, Philadelphia.

Pain Management Guideline Panel (1992) Clinicians Quick Reference Guide to Postoperative Pain Management in Adults. *Journal of Pain and Symptom and Management,* **7**(4): 214–228.

Rawal N. and Berggren L. (1994) Organisation of acute pain services: a low cost model. *Pain*, **57**: 117–123.

Ready B.L., Oden R., Chadwick H.S., Benedetti C., Rooke G.A., Caplan R. *et al.*(1988) Development of an anaestiology-based postoperative pain management service. *Anaesthesiology*, **68**(1): 100–106.

Schafheutle E.I., Cantrill J.A. and Noyce P.R. (2001) Why is pain management suboptimal on surgical wards? *Journal of Advanced Nursing*, **33**(6): 728–737.

Todd K.H., Sloan E.P., Chen C., Eder S. and Wanstad K. (2004) Survey of pain etiology, management practices and patient satisfaction in two urban emergency departments. *Journal of the Canadian Association of Emergency Physicians,* **4**(4): 252–256.

Tough J. (2004) Assessment and treatment of chest pain. *Nursing Standard*, **18**(37): 45–53.

Warfield C.A. and Kahn C.H. (1995) Acute pain management programs in U.S. hospitals and experiences and attitudes among U.S. adults. *Anaesthesiology*, **83**: 1090–1094.

Zimmermann D.L. and Stewart J. (1993) Postoperative pain management and acute pain service activity in Canada. *Canadian Journal of Anaesthesia*, **40**(6): 568–575.

10 Management of chronic pain

Learning outcomes

By the end of this chapter you should be able to:

★ Describe various types of chronic pain

★ Detail methods of managing chronic pain

★ Detail special considerations for the role of opioids and chronic non-malignant pain

★ Examine the importance of chronic-pain care management.

Introduction

This chapter will help you to consider chronic-pain care management. The range of interventions for chronic-pain management will be detailed. Chronic pain tends to be considered as pain without apparent biological value that has persisted beyond the normal tissue healing time (usually taken to be three months). It is linked to maladaptive repsonses that have adverse psychological consequences, and often the cause of the pain cannot be identified. Figure 10.1 identifies some common conditions associated with chronic pain. Many other conditions are, however, associated with chronic pain such as cardiovascular disease, gastrointestinal conditions (e.g. ulcers, liver disease), endocrinological conditions (e.g. diabetes), cerebrovascular conditions (e.g. stroke, multiple sclerosis) (Sprangers *et al.*, 2000). Chapter 11 explores cancer pain management and palliative care.

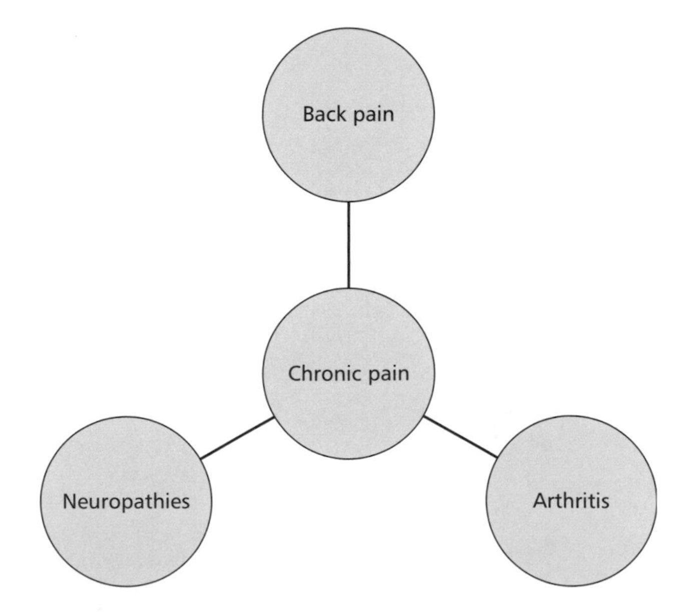

Figure 10.1 Common chronic-pain conditions

Prevalence and effect of chronic pain

Macfarlane (2005) reviewed the epidemiology of chronic pain over the past 50 years. He concluded that the burden of musculoskeletal pain and the prevalence of pain may have increased during the past half century. Early studies on the aetiology of chronic pain concentrated on mechanical injury factors while more recent studies have emphasised the even greater influences of individual psychological factors and the social environment, according to Macfarlane.

Evidence base

Prevalence of chronic pain

Numerous studies chronicle the prevalence of chronic pain. The following timeline identifies key large population studies from the 1990s to the present date.

Timeline

- **2005** Dobson reports from his research that approximately a fifth of adults in Europe have moderate to severe chronic pain and in many cases their symptoms are inadequately managed.
- **2005** Rustoen reports age differences in pain prevalence rates for age groups (younger, middle-aged and older) with prevalence rates of 19.2%, 27.5% and 31.2% respectively (n = 1912, Norwegian study).
- **2005** Veillette *et al.* identifies that between 18% and 29% of Canadians experience chronic pain.
- **2002** Papageorgious *et al.* conducted a seven-year study to document the natural course of chronic widespread pain in a general population (n = 1386). Prevalence at year one was 11% and 10% at year seven. The authors conclude that the proportion of subjects changing from chronic widespread pain to no pain was very low over a seven-year period and suggested that, once established, pain is likely to persist or recur.
- **2000** Palmer *et al.*, report on two prevalence surveys of back pain at an interval of 10 years. Over a 10-year interval the one-year prevalence of back pain, standarised for the age and sex distribution, rose from 36.4% to 49.1%.
- **1999** Elliot *et al.*, in a UK survey of 29 general practices (5036 patients, 72% response rate), report that 50.4% of patients in the community self-report chronic pain. Back pain and arthritis were the most common complaints.
- **1998** American Geriatrics Society (AGS) Panel find that 25–50% of those living in the community suffer significant pain problems
- **1995** Ferrell reports 62% of nursing-home residents have pain.

Chapter 2 identified some of the challenges of reviewing studies that measure prevalence of chronic pain due to variations in population sampled, methods used to collect data and criteria used to define chronic pain. Standard definitions for chronic pain are not available.

Chronic pain has a significant effect on the lives of those experiencing that pain. It also impacts on their friends and their families. Chronic pain affects patients' mood, social relationships and quality of life. Breivik (2004), quoted in a WHO media release, suggests that chronic pain is a disease in its own right.

Chronic pain in the workforce affects individuals who experience a variety of conditions resulting in increased financial burden to their employers and increased use of health care resources. Given the economic impact of chronic pain, it has been suggested that employers and managed care organisations should evaluate the potential benefits in productivity resulting from workplace initiatives such as ergonomic modifications, rest breaks or pain management programmes (Pizzi *et al.*, 2005).

Davidson and Jhangri (2005) report on the impact of chronic pain in haemodialysis patients concluding that chronic pain in these patients is associated with depression and insomnia and may predispose patients to consider withdrawal of dialysis. There is increasing evidence that psychological disorders such as depression (Tsai, 2005) often coexist with or are exacerbated by chronic pain.

A population-based survey of 1953 patients assessed whether chronic pain predicts future psychological distress (McBeth *et al.*, 2002). This was shown to be the case; however, the authors concluded that it is the interaction between chronic widespread pain and physical and psychological co-morbidities that predicts future distress. Unsurprisingly, with ongoing pain, increased pain intensity is associated with decreased quality of life (Laursen *et al.*, 2005).

Good practice in chronic-pain management

Traditionally, pain is treated as a symptom which is appropriate in patients with an acute injury or disease. However, patients are often left with persistent pain once acute injury or disease has subsided. With chronic pain a general history, pain history, clinical examination, diagnostic studies and specialty consult should be conducted as appropriate. Turk and McCarberg (2005) support the concept of an interdisciplinary treatment approach for chronic pain that focuses on self-management and functional restoration. Clinicians have extensive choices when treating chronic pain as outlined in Table 10.1 (Turk, 2002).

Table 10.1 Options for treating chronic pain

	Examples
Pharmacological preparations	Opioids Nonsteroidal anti-inflammatories (NSAIDs) Anticonvulsants Tricyclic antidepressants NMDA antagonists Topical preparations
Operative procedures	For example, surgery for osteoarthritic knee
Physical modalities	Ultrasound Transcutaneous electrical nerve stimulation (TENs) Diathermy Physiotherapy
Regional anaesthesia	For example, epidural infusions of local anaesthetics +/– opioids Nerve blocks +/– opioids Denervation
Neuroaugmentation modalities	Spinal column stimulators (SCSs) Implantable drug delivery systems (IDDSs)
Comprehensive pain rehabilitation programme	Interdisciplinary pain centres Functional restoration programmes
Complementary modalities	Acupuncture, cognitive behavioural therapy, guided imagery, relaxation

Reflective activity

Consider the last patient you managed with chronic pain. How did you assess his/her pain? How were decisions regarding pain interventions for that chronic pain made? Was the effect of the pain interventions monitored? For example, if the person was attending a pain clinic did he/she keep a pain diary? Was the extent of functional status monitored over time to see if, for example, the patient could mobilise better with a particular medication regime? Were any adverse effects experienced?

Management of chronic pain will involve the following:

- pain assessment
- a pain management plan
- progress evaluated at regular intervals
- monitoring of quality of life.

Pain assessment and ongoing monitoring of interventions for pain are critical. Pain assessment processes for chronic pain are described in Chapter 6. Figure 10.2 summarises the important issues for chronic-pain assessment.

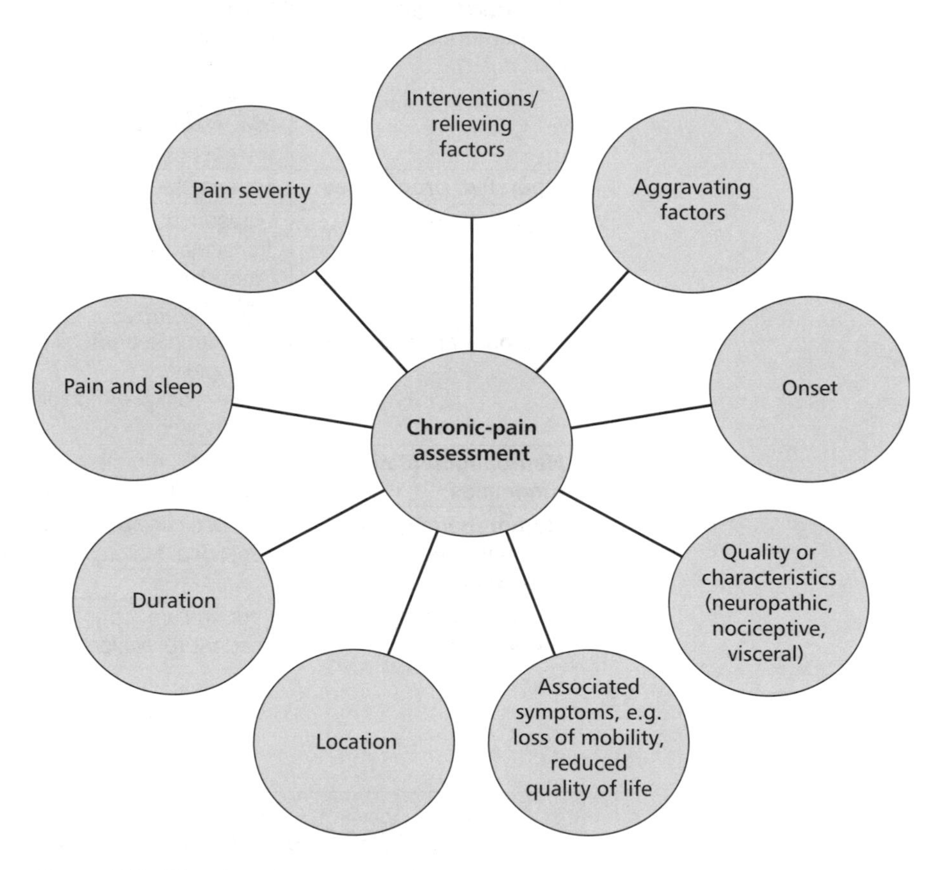

Figure 10.2 Chronic-pain assessment

Long-term use of pharmacological preparations for chronic pain

The long-term use of pharmacological preparations for pain management expose patients to more medication side-effects. Chapter 7 describes each of the medication classes in detail. Patients on long-term medications should be monitored regularly particularly for hepatic and renal effects. In addition, nonsteroidal anti-inflammatories (NSAIDs) have a high risk of gastric toxicity and caution should be utilised for long-term use. The long-term use of opioids in particular is the subject of much debate with special considerations for use in chronic non-malignant pain. These are described below.

Opioids in chronic pain

The goal of management of chronic non-cancer pain is to keep the patient functional, both physically and mentally, with improved quality of life. Relief of pain may be an essential factor in this and opioids are only one aspect of the overall rehabilitative strategy for the patient. The actions and side-effects of opioids are described in Chapter 7. There are, however, special considerations when a person is on long-term opioids with chronic non-malignant pain. Opioids are not effective in every patient with pain and randomised clinical trials (RCTs) indicate that no criteria have been identified that predict a good response to opioids in any particular condition (Kalso, 2005). Devulder *et al.* (2005) evaluated 11 studies of long-term treatment with opioids in patients with chronic non-malignant pain and subsequent quality of life (n = 2877). They suggest that long-term treatment with opioids can lead to significant improvement in functional outcomes including quality of life. However, the authors caution that there remains a need for future long-term rigorous research.

Kalso *et al.* (2003) provide recommendations for using opioids in chronic non-cancer pain. Their conclusions are:

- the management of pain should be directed by the underlying cause of pain
- opioid treatment should be considered for both continuous neuropathic and nociceptive pain if other reasonable therapies fail to provide adequate analgesia within a reasonable timeframe
- the aim of opioid treatment is to relieve pain and improve the patient's quality of life. Both of these should be assessed during a trial period
- the prescribing physician should be familiar with the patient's psychosocial status
- the use of sustained-release opioids administered at regular intervals is recommended
- treatment should be monitored
- a contract setting out the patient's rights and responsibilities may help emphasise the importance of patient involvement
- opioid treatment should not be considered a lifelong treatment.

Misconceptions regarding the use of opioids have made clinicians reluctant to prescribe these agents. Fear of side-effects such as addiction, misuse, tolerance and respiratory depression have contributed as barriers to the effective treatment of chronic pain. In 2001, the American Academy of Pain Medicine, the American Pain Society and the American Society of Addiction Medicine agreed a consensus document on the use of opioids and the treatment of pain and provided definitions of addiction, physical dependence and tolerance. The consensus document states that most specialists agree that patients treated with prolonged opioid therapy usually develop physical dependence and sometimes develop tolerance, but do not usually develop addictive disorders. The actual risk is, however, unknown and probably varies with genetic predisposition among other factors.

The American Academy of Pain Medicine and the American Pain Society in 1997 provided a consensus statement on the use of opioids for the treatment of chronic pain. They reiterate that pain is one of the most common reasons that people consult a physician but that, frequently, it is inadequately treated. They devised the following principles of good practice to guide the prescribing of opioids:

- **Evaluation of the patient**: pain history, assessment of impact of pain on the patient, a directed physical examination, a review of previous diagnostic studies, a review of previous interventions, a drug history and an assessment of coexisting conditions.
- **Treatment plan**: tailored to both the individual and the presenting problem. Consideration should be given to different treatment modalities. If a trial of opioids is selected, the physician should ensure that the patient is informed of the risks and benefits of opioid use and conditions under which they will be prescribed.
- **Consultation**: a consultation with a specialist or with a psychologist may be warranted.
- **Periodic review of treatment efficacy**: this should occur periodically to assess the functional status of the patient, continued analgesia, opioid side-effects, quality of life and indication of medication misuse. Periodic re-examination is warranted to assess the nature of the pain complaint and to ensure that opioid therapy is still indicated. Attention should be given to the possibility of a decrease in global function or quality of life as a result of opioid use.
- **Documentation**: this is essential for supporting the evaluation, the reason for opioid prescribing, the overall pain management treatment plan, any consultation received and periodic review of the status of the patient.

The British Pain Society (2005) together with the Royal College of Anaesthetists, the Royal College of General Practitioners and the Royal College of Psychiatrists prepared a consensus statement *Recommendations for the appropriate use of opioids for persistent non-cancer pain*.

Over to you

Go to the British Pain Society website (www.britishpainsociety.org) and download the recommendations for the appropriate use of opioids for persistent non-cancer pain. Are any of the recommendations in practice in your clinical area? How would you go about creating a local guideline to support best practice in the use of opioids for chronic pain?

Back pain

Back pain is one of the most common types of chronic pain reported. In 1999, Elliot *et al.,* in a UK survey of 29 general practices (5036 patients, 72% response rate) report that 50.4% of patients self-report chronic pain. Back pain and arthritis were the most common complaints. Palmer *et al.* (2000) report on two prevalence surveys of back pain at an interval of 10 years. Measurements included low back pain and low back pain which made it impossible to put on hosiery. Over a 10-year interval, the one-year prevalence of back pain, standarised for the age and sex distribution, rose by 12.7%.

Guzman *et al.* (2001) conducted a systematic review of 10 trials using multidisciplinary rehabilitation for chronic low back pain.
The authors suggest that there was strong evidence that intensive multidisciplinary biopsychosocial rehabilitation with functional restoration improves function when compared with inpatient or outpatient non-multidisciplinary treatments and moderate evidence that such rehabilitation reduces pain. Ospina and Harstall (2003) support this through analysis of the literature concluding that the evidence for the effectiveness of multidisciplinary pain programmes is strong for chronic low back pain.

The Swedish Council on Technology Assessment in Health Care (2000) following a review of the scientific literature (2000 studies) summarised that:

- back pain is common and not usually harmful – patients should stay active; however, it is important to identify rare cases where back pain has a specific cause
- consequences of back pain may be more problematic than the pain itself
- preventative measures are known but not practiced
- societal costs of back pain are three times higher than the total cost of all types of cancer
- more research is needed.

In addition the Swedish Council identify the following evidence in relation to medications for conservative treatments for *reducing* low back pain (see Table 10.2).

Table 10.2 Evidence base for conservative treatments for reducing low back pain (Source: Swedish Council, 2000)

Pain type	Treatment	Level of evidence
Acute and subacute low back pain *(up to 3 or 12 weeks)*	Muscle relaxants (e.g. benzodiazepines) and NSAIDs	Strong
Acute low back pain	Paracetamol	Moderate
Chronic low back pain	Muscle relaxants, NSAIDs	Limited
Acute low back pain	Anti-depressants	None

Beliefs and attitudes

The influence of attitudes and beliefs are increasingly accepted as an important role in the improvement of functioning with back pain. An Australian study evaluated the effectiveness of a population-based, state-wide public health intervention designed to alter beliefs about back pain, influence medical management and reduce disability and costs of compensation (Buchbinder *et al.*, 2001). The authors concluded that such a strategy improves population and general practitioner beliefs about back pain and seems to influence medical management and reduce disability and workers' compensation costs related to back pain.

Brief pain-management techniques

Hay *et al.* (2005), following a randomised clinical trial, suggest that brief pain management techniques delivered by appropriately trained clinicians offer an alternative to physiotherapy incorporating manual therapy and could provide a more efficient first-line approach for management of non-specific subacute low back pain in primary care. The brief techniques involve assessment of psychosocial risk factors with an emphasis on return to normal activity through functional goal setting, with educational strategies to overcome psychosocial barriers to recovery.

Transcutaneous electrical nerve stimulation (TENS)

Transcutaneous electrical nerve stimulation (TENS) was introduced more than 30 years ago as an adjunct to the pharmacological management of pain. However, despite its widespread use, the usefulness of TENS for chronic low back pain is still controversial. In a Cochrane systematic review, Khadilkar *et al.* conclude that there is limited and inconsistent evidence to support the use of TENS as an isolated intervention in the management of chronic lower back pain. Only two RCTs met the criteria for inclusion (Khadilkar *et al.*, 2005).

Injection therapy

Injections with anaesthetics and/or steroids is one of the treatment modalities used in patients with chronic low back pain. Nelemans *et al.* (1999) looked at the evidence in a Cochrane systematic review with 21 RCTs. The authors conclude that convincing evidence on the effects of injection therapies for low back pain is lacking and that there is a need for more well-designed explanatory trials in this field.

Behavioural treatment for chronic low back pain

Behavioural treatment is primarily focused on reducing disability through modification of environmental contingencies and cognitive processes. Ostelo *et al.* (2005), in a Cochrane systematic review of 21 studies, could not conclude from the review whether clinicians should refer patients with chronic low back pain to behavioural treatment programmes or to active conservative treatment.

> **Over to you**
>
> If your patient had chronic back pain and you were giving them an information session to support an improved quality of life, what sort of issues would you discuss with them?

Neuropathic pain

The IASP (1994) define neuropathic pain as pain initiated or caused by primary lesion or dysfunction of the nervous system. An example of neuropathic pain is post-herpetic neuralgia which is a serious complication of herpes zoster (shingles). Herpes zoster is an acute infection that occurs when the latent varicella zoster virus (VZV) is reactivated. Risk factors include immunodeficiency and age.

Abnormal sensory symptoms and signs are associated with neuropathic pain and are outlined in Table 10.3.

Anticonvulsants are a group of medicines commonly used for treating epilepsy but which are also effective for treating pain. Nurmikko *et al.* (1998) state that anticonvulsant drugs have an established role in the treatment of chronic neuropathic pain, especially when patients complain of shooting sensations.

Wiffen *et al.* (2005), following a Cochrane systematic review, conclude that although anticonvulsants are used widely in chronic-pain, surprisingly few trials show analgesic effectiveness (migraine and headache were excluded) (23 trials were included). The authors state that in chronic-pain syndromes other than trigeminal neuralgia, anticonvulsants should be withheld until other interventions have been tried. While gabapentin is increasingly being used for neuropathic pain,

Table 10.3 Abnormal sensory symptoms (Source: IASP, 2004)

Allodynia	Pain due to a stimulus which does not normally provoke pain.
Dysthesias	An unpleasant abnormal sensation, whether spontaneous or evoked.
Hyperalgesia	An increased response to a stimulus which is normally painful.
Heperpathia	A painful syndrome characterised by an abnormally painful reaction to a stimulus, especially a repetitive stimulus, as well as an increased threshold.
Paresthsias	An abnormal sensation, whether spontaneous or evoked.

the evidence would suggest that it is not superior to carbamazepine. In their synopsis, the authors state that neuropathic pain responds well to anticonvulsants and approximately two-thirds of patients who take either carbimazole or gabapentin can be expected to achieve good pain relief.

Arthritis

The term 'arthritis' is a generic term that refers to more than 100 conditions with osteoarthritis and rheumatoid arthritis (RA) being the most common (American Pain Society, 2002). These conditions can cause pain, stiffness and swelling of the joints. These diseases may also affect supporting structures such as muscles, bones, tendons and ligaments and have a systemic effect. Most forms of arthritis cause chronic pain which can range from mild to severe and can last days, months or years. Arthritis sufferers will also suffer acute pain which by its nature is temporary and diminishes over time. Rheumatoid arthritis affects approximately 1% of the US population with peak incidence between the ages of 40 and 60 (Romano, 2006). In patients with rheumatoid arthritis, pain is the most common reason for seeking medical care (Anderson, 2001).

Arthritis pain has many causes and may be due to inflammation of the synovial membrane, tendons, ligaments and muscle strain. A pain management plan for arthritis will need to be developed following a general history, pain history, clinical examination, diagnostic studies and specialty consultation as appropriate. An example of a specific pain scale developed for rheumatoid arthritis (RA) is RAPS (Rheumatoid Arthritis Pain Scale) (Anderson, 2001). This scale has four subscales:

- physiological – clinical manifestations of RA
- affective – moment by moment unpleasantness, distress and annoyance that closely co-vary with the intensity of the painful sensation
- sensory-discriminative – intensity, duration, location and quality of pain sensation
- cognitive – secondary stage of pain-related affect based on cognitive process.

The scale contains statements such as 'I have morning stiffness of one hour or more' and 'Pain interferes with my sleep' and the patient scores from 0 (always) to 6 (never) how he/she rates each question on the scale.

Management of arthritis pain will involve the patient and their health care team developing long-term plans to manage both episodes of acute pain and chronic pain over time, possibly a lifetime. Pain episodes due to inflammation are often unpredictable.

Like any chronic pain, arthritis will affect the patient's quality of life and patients and their families will need support to cope with a lifelong disease. In addition to management of direct pain, a holistic approach relevant to the patient's lifestyle will need to be developed. This will involve supporting the patient to maintain a healthy diet, develop good sleep patterns, exercise as appropriate, be informed of disease pattern and treatments, and ongoing monitoring of quality of life.

The American Pain Society (2002) in a set of clinical guidelines for treating arthritis pain recommend that:

- All treatment for arthritis should begin with a comprehensive assessment of pain and function.
- For mild to moderate pain, acetaminophen is the choice because of its mild adverse effects, over-the-counter availability and low cost.
- For moderate to severe pain from both osteoarthritis and RA, COX-2 NSAIDs are the medications of choice.
- Opioids are recommended for treating severe pain for which COX-2 drugs or non-specific NSAIDs do not provide substantial relief.
- Unless there are medical contraindications most individuals with arthritis including the obese and elderly should be referred for surgical treatment when drug therapy is ineffective and function is severely impaired to prevent minimal physical activity. It is advised that surgery be recommended before the onset of severe deformity and advanced muscular deterioration.
- An ideal weight is maintained.
- Referrals are made for physical and occupational therapy to evaluate and reduce impairments in range of motion, strength, flexibility and endurance.

Table 10.4 identifies some of the options for pain relief for arthritis suffers which will be part of an overall patient management plan. Methods of pain relief chosen will depend on the type of arthritis and the disease progression. Methods chosen should be based on sound pain assessment and adhere to good pain management practice for chronic pain. Further detail of diagnostic tests and treatment options are beyond the scope of this book.

Best practice in chronic-pain management should be part of a continuous cycle as outlined in Figure 10.3.

Table 10.4 Options for pain relief in arthritis

Short-term relief	*Medications*	Paracetamol (Zhang *et al*. 2004 on review of 10 trials suggest that paracetamol reduces osteoarthritis pain more than placebo but does not affect functioning or stiffness) NSAIDs
	Heat and cold	
	Joint protection	Splints and braces to allow joints to rest and protect them from injury
	TENs	
	Physiotherapy	
	Occupational therapy	
Long-term relief	*Medications*	– Biological response modifiers (new drugs for treatment of RA to reduce inflammation) – NSAIDs – Disease modifying anti-rheumatic drugs (DMARDs). – Corticosteroids – Opioids – Intra-articular injections
	Weight reduction	If necessary to reduce stress on weight-bearing joints
	Exercise	As appropriate
	Surgery	An example is knee replacement
	Complementary therapies	For example relaxation, meditation.
	Physical therapy	
	Occupational therapy	

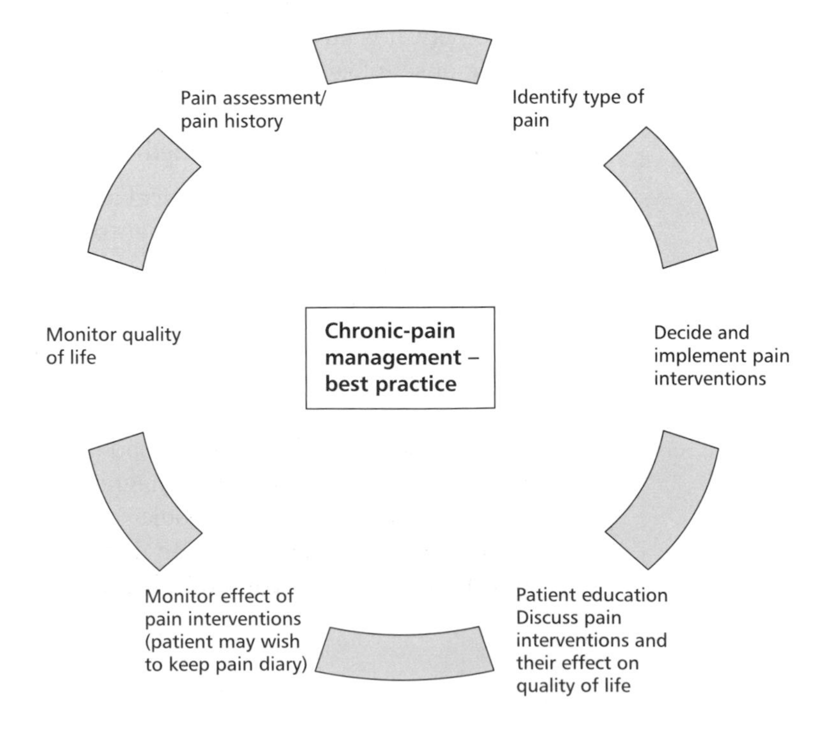

Figure 10.3 Best practice in chronic-pain management

Chronic-pain services

Many patients with chronic pain continue to have symptoms and may develop psychological distress, overall disability and increased dependency on family members and social services even though there are many advances in pain relieving techniques. Pain management programmes have been developed to enhance the patient's physical performance and help them cope more effectively with their pain (Nurmikko *et al.*, 1998).

The literature describes various methods to help patients cope with chronic pain. Van Korff (1998) describes an RCT in a primary care setting in the USA where an educational back pain self-management group intervention led by trained lay people reduced back pain worries and activity limitations and increased self-care confidence at six months. This was achieved in the absence of reducing pain intensity. An internet-delivered cognitive-behavioural intervention with telephone support shows some improved coping skills such as feeling of control over pain and ability to decrease pain in patients with chronic low back pain (Buhrman *et al.*, 2005).

Flor *et al.* (1992) performed a meta-analysis evaluating the efficacy of multidisciplinary treatments for chronic pain and suggested that patients treated in multidisciplinary pain clinics showed improvements in pain and functioning compared to conventional treatments. The analysis showed that patients treated in a multidisciplinary pain clinic are almost twice as likely to return to work than the untreated or unimodally treated patients.

Good practice in chronic-pain management suggests the provision of core services for chronic pain in all district general hospitals and most specialist hospitals (The Royal College of Anaesthetists and The Pain Society, 2003).

Veillette *et al.* (2005) report on a Canadian survey of the availability of hospital-based anaesthesia departments with chronic non-cancer pain services. The survey indicates that 73% of anaesthesia departments offer chronic non-cancer pain services with 26% providing some form of multidisciplinary assessment and treatment. However, 4500 patients were waiting for their first appointment to see a pain consultant, with 67% of these waiting for nine months or more.

Chapter 3 provides more detail on chronic-pain services but it should be borne in mind that there is limited published empirical work on the prevalence or effectiveness of chronic-pain services. It emerges, however, that chronic-pain services should include the elements detailed in Figure 10.4.

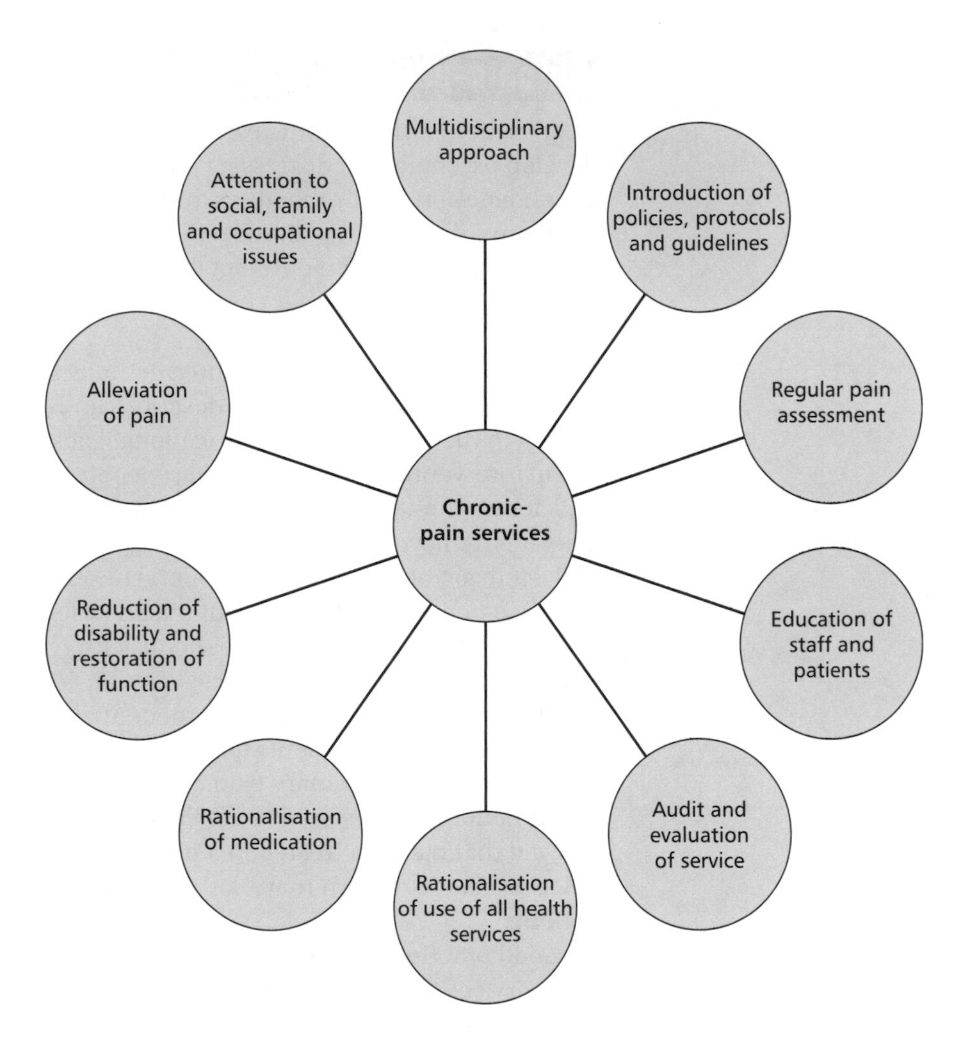

Figure 10.4 Chronic-pain services

Key points Top tips

- The prevalence of chronic pain is high
- Evidence-based approaches should be used to manage chronic pain
- There are extensive choices for managing chronic pain from pharmaceutical preparations to complementary modalities.
- Specific guidance applies to the use of opioids in chronic non-malignant pain
- Assessment and monitoring of the effect of the pain intervention is the key to good chronic-pain management

Chronic back pain

Catriona is a 30-year-old bank clerk. She has had chronic back pain for eight years following an accident at work where a roof collapsed on top of her. She was trapped under the roof and continues to get nightmares about being trapped under the roof. She attends her general practitioner every two months. At this stage she says she is very fed up as she feels that she no longer has the quality of life that her friends have. For example, when her friends start talking about going on holidays and ask her to come along she feels she is preventing them from going on any adventure or active holidays. Catriona has been on every medication from NSAIDs to morphine. She has tried physiotherapy, TENs and swimming. Her doctor thinks that she might benefit from attending a pain clinic in the local hospital.

- How has Catriona's pain affected her quality of life?
- What benefits would Catriona gain from attending the local pain clinic?
- What sort of pain assessment would help monitor Catriona's pain?

RRRRRRapid recap

1. Define chronic pain.
2. What are the main qualities of a chronic-pain service?
3. Name and give examples of three modalities for reducing chronic pain.
4. Identify four principles of good practice to guide the prescribing of opioids for chronic non-cancer pain.
5. How strong is the evidence for the use of TENS with low back pain?

References

American Academy of Pain Medicine, the American Pain Society and the American Society of Addiction Medicine (2001) *Definitions related to the use of opioids and the treatment of pain.* A consensus document from the American Academy of Pain Medicine, the American Pain Society, and the American Society of Addiction Medicine, www.painmed.org/.

American Academy of Pain Medicine and the American Pain society (1997) *The Use of Opioids for the Treatment of Chronic Pain.* A consensus statement from the American Academy of Pain Medicine and the American Pain society. Available at www.painmed.org/productpub/statements/

American Geriatrics Society (AGS) Panel on Chronic Pain in Older Persons (1998) The management of chronic pain in older persons: AGS Panel on Chronic Pain in Older Persons. *Journal of the American Geriatrics Society*, **46**: 635–651.

American Pain Society (2002) APS guideline for the management of pain in osteoarthritis, rheumatoid arthritis and juvenile chronic arthritis. American Pain Society, www.ampainsoc.org.

Anderson D.L. (2001) Development of an instrument to measure pain in rheumatoid arthritis: rheumatoid arthritis pain scale (RAPS). *Arthritis Care & Research,* **45**: 317–323.

Breivik (2004) World Health Organization supports global effort to relieve chronic pain. Geneva: WHO/70/2004.

British Pain Society (2005) Recommendations for the appropriate use of opioids for persistent non-cancer pain. A consensus statement prepared on behalf of the Pain Society, the Royal College of Anaesthetists, the Royal College of General Practitioners and the Royal College of Psychiatrists. www.britishpainsociety.org.

Buchbinder R., Jolley D. and Wyatt M. (2001) Population based intervention to change back pain beliefs and disability: three-part evaluation. *BMJ*, **322**: 1516–1520.

Buhrman M., Faltenhag S., Strom L., Andersson G. and Polly D.W. (2005) An internet-delivered cognitive-behavioural intervention with telephone support improved some coping skills in patients with chronic low back pain. *Journal of Bone and Joint Surgery*, **87**(5): 1169.

Davidson S.N. and Jhangri G.S. (2005) The impact of chronic pain on depression, sleep, and the desire to withdraw from dialysis in haemodialysis patients. *Journal of Pain and Symptom Management*, **30**(5): 465–473.

Devulder J., Richarz U. and Nataraja S.H. (2005) Impact of long-term use of opioids on quality of life in patients with chronic, non-malignant pain. *Current Medical Research and Opinions*, **21**(10): 1555–1568.

Dobson R. (2005) Chronic pain is poorly managed. *BMJ*, **331**: 476.

Elliot A.M., Smith B.H., Penny K.I., Cairns Smith W. and Chambers W.A.(1999) Epidemiology of chronic pain in the community. *Lancet*, **354**: 1248–1252.

Ferrell B.A. (1995) Pain evaluation and management in a nursing home. *Ann Intern Med*, **123**: 681–687.

Flor H., Fydrich T. and Turk D.C. (1992) Efficacy of multidisciplinary pain treatment centres: a meta-analytic review. *Pain*, **49**: 221–230.

Guzman J., Esmail R., Karjalainen K., Malmivaara A., Irvin E. and Bombardier C. (2001) Multidisciplinary rehabilitation for chronic low back pain: systematic review. *BMJ*, **322**: 1511–1516.

Hay E.M., Mullis R., Vohara K., Watson P., Sim J., Minns Lowe C. *et al*. (2005) Comparison of physical treatments versus a brief pain-management programme for back pain in primary care: a randomised clinical trial in physiotherapy practice. *The Lancet*, **365**(9476): 2024–2030.

IASP Task Force on Taxonomy (1994) Classification of chronic pain, 2nd edn (eds. Merskey H. and Bogduk N.). IASP Press, Seattle.

Kalso E. (2005) Opioids for persistent non-cancer pain. *BMJ*, **330**: 156–157.

Kalso E., Allen I., Dellemijn P.L.I., Faura C.C., Ilias W.K., Jenson T.S. *et al*. (2003) Recommendations for using opioids in chronic non-cancer pain. *Europena Journal of Pain*, **7**: 381–386.

Khadilkar A., Milne S., Brosseau L., Robinson V., Saginur M., Shea B. *et al*. (2005) Transcutaneous electrical nerve stimulation (TENS) for chronic low-back pain. *The Cochrane Database of Systematic Reviews.* Issue 3. Art No.: CD003008. DOI: 10.1002/14651858.pub2.

Laursen B.S., Bajaj P., Olesen A.S., Delmar C. and Arendt-Nielsen L. (2005) Health related quality of life and quantitative pain measurement in females with chronic non-malignant pain. *European Journal of Pain*, **9**: 267–275.

McBeth J., Macfarlane G.J. and Silman A.J. (2002) Does chronic pain predict future psychological distress? *Pain*, **96**: 239–245.

Macfarlane G.J. (2005) Looking back: developments in our understanding of the occurrence, aetiology and prognosis of chronic pain 1954–2004. *Rheumatology*, Suppl. **4**: iv23–iv26.

Nelemans P.J., de Bie R.A., de Vet H.C.W. and Sturmans F. (1999) Injection therapy for subacute and benign low-back pain. *The Cochrane Database of Systematic Reviews.* Issue 4. Art. No.: CD001824. DOI: 10.1002/14651858.CD001824.

Nurmikko T.J., Nash T.P. and Wiles J.R. (1998) Recent advances: control of chronic pain. *BMJ*, **317**: 1438–1441.

Ospina M. and Harstall C. (2003) Multidisciplinary pain programmes for chronic pain: Evidence from systematic reviews. Alberta Heritage Foundation for Medical Research – Health Technology Assessment. HTA 30 (Series A): 1–48.

Ostelo R.W.J.G., van Tulder M.W., Vlaeyen J.W.S., Linton S.J., Assendelft W.J.J. (2005) Behavioural treatment for chronic low-back pain. *The Cochrane Database of Systematic Reviews*, Issue 1. Art. No.: CD002014. DOI: 10.1002/14651858. CD002014.pub2.

Palmer K.T., Walsh K., Bendall H., Cooper C. and Coggon D. (2000) Back pain in Britain: Comparison of two prevalence surveys at an interval of 10 years. *BMJ*, **320**: 1577–1578.

Papangeorgious A.C., Silman A.J. and Macfarlane G.J. (2002) Chronic widespread pain in the population: a seven-year follow up study. *Annals of Rheumatology*, **61**: 1071–1074.

Pizzi L.T., Carter C.T., Howell J.B., Vallow S.M., Crawford A.G. and Frank E.D. (2005) Work loss, health care utilization and costs among US employees with chronic pain. *Disease Management and Health Outcomes*, **13**(3): 201–208.

Romano T. (2006) Rheumatologic Pain. In *Weiner's Pain Management A Practical Guide for Clinicians,* 7th edn.(eds. Boswell M.V. and Cole B.E.) American Academy of Pain Management. CRC Press, Taylor and Francis Group.

Royal College of Anaesthetists and The Pain Society, The British Chapter of the International Association for the Study of Pain (2003) *Pain Management Services – Good Practice*. Royal College of Anaesthetists and The Pain Society, London.

Rustoen T., Wahl A.K., Hanestad B.R., Lerdal A., Paul S. and Miaskowski C. (2005) Age and the experience of chronic pain. Differences in health and quality of life among younger, middle-aged and older adults. *Clinical Journal of Pain*, **21**: 513–523.

Sprangers M.A.G., de Regt E.B., Andries F., van Agt H.M.E., Biljl R.V., de Boer J.B. *et al.* (2000) Which chronic conditions are associated with better or poorer quality of life? *Journal of Clinical Epidemiology*, **53**: 895–907.

Swedish Council on Technology Assessment in Health Care (2000) *Back pain, neck pain: an evidence based review*. Swedish Council on Technology Assessment in Health Care. www.sbu.se.

Tsai P.F. (2005) Predictors of Distress and Depression in Elders with Arthritic Pain. *Journal of Advanced Nursing*, **51**(2): 158–165.

Turk D.C. (2002) Clinical effectiveness and cost-effectiveness of treatments for patients with chronic pain. *The Clinical Journal of Pain*, **18**: 355–365.

Turk D.C. and McCarberg B. (2005) Non-pharmacological treatments for chronic pain. A disease management context. *Disease Management and Health Outcomes*, **13**(1): 19–30.

Van Korff M., Moore J.E., Lorig K. *et al.* (1998) Back pain self management groups led by lay people increased self care confidence and reduced activity limitations at 6 months. *Spine*, **23**: 2608–2615.

Veillette Y., Dion D., Altier N. and Choiniere M. (2005) The treatment of chronic pain in Quebec: a study of hospital-based services offered within anesthesia departments. *Canadian Journal of Anesthesia*, **52**(6): 600–606.

Wiffen P., Collins S., McQuay H., Carroll D., Jadad A. and Moore A. (2005) Anticonvulsant drugs for acute and chronic pain. *The Cochrane Database of Systematic Reviews.* Issue 3. Art. No.:CD001133. DOI:10.1002/14651858.CD001133. pub2.

Zhang W., Jones A. and Doherty M. (2004) Review: paracetamol reduces pain in osteoarthritis but is less effective than NSAIDs. *Evidence Based Nursing*, **63**: 901–907.

11
Pain management in palliative care

Deborah Hayden

Learning outcomes

By the end of this chapter you should be able to:

★ Appreciate the history of hospice and palliative care

★ Recognise the need for palliative care

★ Articulate the concept of palliative care with particular reference to total pain

★ Apply the principles of palliative care in order to manage pain in your own work area utilising a holistic and individual approach in order to contribute towards alleviating suffering

★ Perform pain assessment whilst mindful of the psychological, social and spiritual factors that will often impact the person's experience of pain

★ Refer to standard analgesic guidelines to direct safe and effective pharmacological interventions.

Introduction

> 'The right to die with dignity with as little pain and as comfortably as possible is as important as the right to life.'
>
> (Archbishop Desmond Tutu, 2005)

This chapter strives to enhance health care professionals' understanding of the meaning, assessment and management of pain in order to provide true holistic care to adults requiring palliative care. Understandably, people frequently fear the involvement of hospice or palliative care teams in their care. This is often due to their lack of knowledge of what palliative care entails. It is therefore imperative that all health care professionals understand the concept and philosophy of hospice and palliative care in order to alleviate unfounded fears and myths, and contribute positively to people's understanding of the benefits of modern palliative care involvement.

History of hospice care and palliative care

The meaning of the word hospice has changed over the centuries. Historically, hospices were institutions whose primary purpose was to provide nursing and spiritual care to the destitute and frail, with particular focus on the dying aspect of care. The shift to 'modern' palliative care is largely due to the work of Dame Cicely Saunders, who was the woman responsible for establishing the discipline and the culture of palliative care. During her lifetime she worked as a nurse, medical social worker, volunteer and eventually as a physician. Witnessing the unnecessary suffering of others, she strived to improve the care of dying patients.

In response to her distress, Saunders founded the first modern hospice, St. Christopher's Hospice in London, in 1967. By creating St. Christopher's Hospice as a 'place' to care for the dying, she modelled a health care system where personhood and dignity were preserved, pain was managed and family-centred care was provided. Hospice care is therefore a term that is often used to describe a place of care and a philosophy of care offered to people when the disease process is at an advanced stage. The term may be applied in a range of care settings. Saunder's clinical work has transformed society's perspective on the treatment of suffering, and led to the development and expansion of hospice and palliative care services worldwide.

Palliative care defined

In the 1960s Balfour Mount, a Canadian physician, coined the term palliative care. The word 'palliative' is derived from the Latin word '*pallium*', meaning a 'cloak'. Twycross (2003) identifies that within palliative care, symptoms such as pain are 'cloaked' with treatments whose primary aim is to promote comfort. The WHO (2002, p. 84) defines palliative care as follows: 'Palliative care is an approach that improves the quality of life of patients and their families facing the problems associated with life threatening illness, through the prevention and relief of suffering by means of early identification and impeccable assessment and treatment of pain and other problems, physical, psychosocial and spiritual.'

The WHO (2002) further elaborates on the definition by stating that palliative care affirms life and regards dying as a normal process. It intends neither to hasten nor postpone death, whilst aiming to provide relief from pain and other distressing symptoms. This is achieved by attending not only to the physical aspects of the person, such as managing pain, but also to the psychological, social and spiritual dimensions of the individual and their family. Therefore, palliative care extends far beyond merely providing symptom relief. Rather than waiting to relieve the suffering, the WHO (2002) advocates the *prevention* of suffering by means of early identification and impeccable assessment. Central to the philosophy of palliative care is the family. It offers a support system to help the family to cope during the person's illness and in their own bereavement.

Palliative care proudly encompasses end-of-life care. However, it is now widely acknowledged that a palliative care approach focuses on the active *living* rather than merely on the anticipated dying. The involvement of palliative care aims to improve the quality of life of people with life-threatening illnesses such as cancer, and may also positively influence the course of the illness. Consequently, palliative care 'is also applicable early in the course of a person's illness, in conjunction with other therapies that are intended to prolong life, such as chemotherapy or radiation therapy, and includes those investigations needed to better understand and manage distressing clinical complications' (WHO, 2002, p. 84).

Palliative care professionals 'share' the care of patients who are receiving therapies to modify the disease, as outlined in Figure 11.1. For this reason, this chapter focuses on alleviating pain not just at the end of life, but throughout the person's illness trajectory.

Finally, the philosophy of palliative care advocates the use of a team approach to address the needs of patients and their families. Interdisciplinary care is provided by a many and varied team each providing a function in reaction to the person's needs.

It is evident from the WHO's (2002) definition that palliative care is not confined to any specific disease or illness. However, approximately 95% of patients referred to specialist palliative care services have cancer. Payne *et al.* (2004) partly proportion this inequity of service to the historical developments in hospice care and key funding from cancer charities. The global report 'Suffering at the End of Life – The State of the World' (Help the Hospices, 2005) claims that there are currently six million cancer deaths and over 10 million new cases of cancer every year, and this is expected to rise to 15 million by 2020. It also reports that 100 million people worldwide need access to hospice and palliative care at the current time. For millions more, access to palliative care *will* become their core essential need.

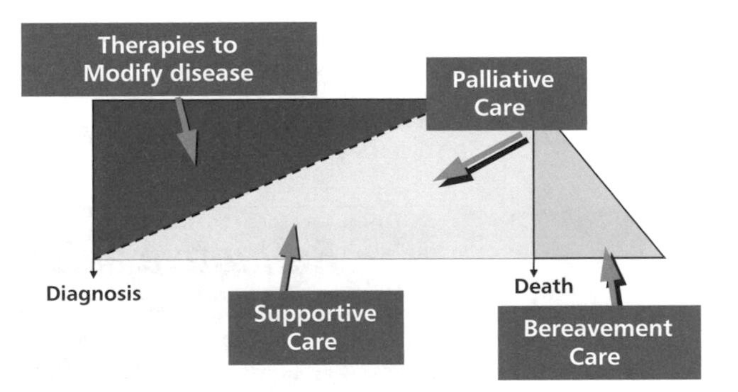

Figure 11.1 'Sharing' the care during the person's illness trajectory

Interdisciplinary approach to pain management in palliative care

The WHO (2002) states that palliative care uses a team approach to address the needs of patients and their families. Thereby, the philosophy of palliative care advocates an interdisciplinary approach to the practice of pain management. The terms 'interdisciplinary teamwork' and 'multidisciplinary teamwork' are used interchangeably. Both terms refer to individual members within a team, each with specific roles and skills. However, interdisciplinary work involves collaboratively intertwined work within a connected team, each member aware of the other's role, and all members sharing and striving towards providing the best care for the individual and their family. Effective continuous communication between the team members is therefore essential.

The National Cancer Institute (NCI) (2005) maintains that assessment involves both the clinician and the patient. However, true interdisciplinary teamwork necessitates the *whole* team conducting pain assessment. All assessment activities should be documented and notes should be accessible to the entire interdisciplinary members. This will enable everyone to understand and appreciate the person's experience (Fink and Gates, 2001) and consequently plan a comprehensive management strategy (Paz and Seymour, 2004).

Key points Top tips

Care within the palliative philosophy involves the following:

- Treating the person, not the disease.
- Individualised, holistic and person focused care by:
  - Attending not only to the physical, but to also the psychological, social and spiritual dimensions of each person
  - Appreciating that palliative care is appropriate throughout the person's illness trajectory
  - Interdisciplinary teamwork
  - Including the family

Reflective activity

Imagine that a person on your ward with an advanced disease is waiting for the palliative care team to visit them for the first time.

- How do you think they might be feeling and thinking?
- What might the family be feeling, or does it matter?

Evidence base

For further information and facts on palliative care, visit the following websites:

- The National Council for Palliative Care (2005) *Palliative Care Explained.* Available online at: www.ncpc.org.uk/palliative_care.html.
- World Health Organization (2004) *Palliative Care: The Solid Facts.* Available online at: www.euro.who.int/document/E82931.pdf.

The concept of 'total pain'

Dame Cicely Saunders transformed pain management for people with life-threatening illnesses. She sought to help dying people to meet the challenges that they face, but without pain. Hence, she identified the concept of *total pain*. She emphasised that effective, individualised pain management is promoted through attention not only to the physical, but to also the psychological, social and spiritual dimensions of distress, which are intertwined in the fabrics of holistic care (Figure 11.2). Saunders opened the door to current advances in pain management by demonstrating the safety and efficacy of opioid drug therapy such as morphine, and its impact on improving the quality of life for patients suffering from moderate to severe pain (Doyle, 2005).

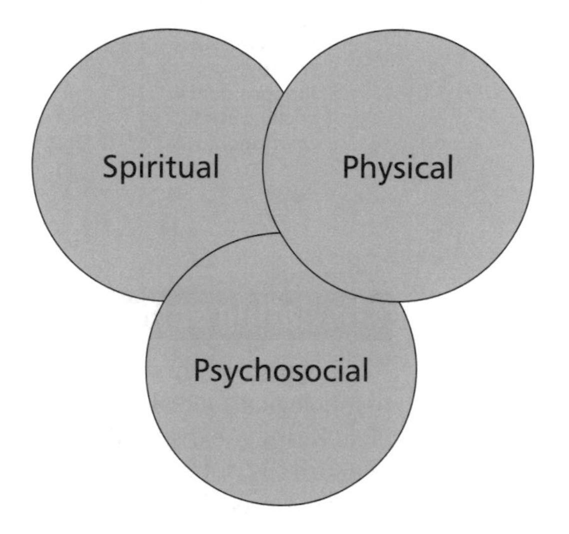

Figure 11.2 Intertwined holistic care as recognised by Dame Cicely Saunders (1918–2005)

The concept of total pain has been a pivotal tool in helping to explain the holistic nature of palliative care. It serves to remind all health care professionals that pain is a deeply personal experience that requires more than merely considering the physiological causes (Paz and Seymour, 2004). Attending to the person holistically requires attentive listening not only to the words but also to the message behind the words. This may enable us to begin to understand the person's experience of total pain (Figure 11.3).

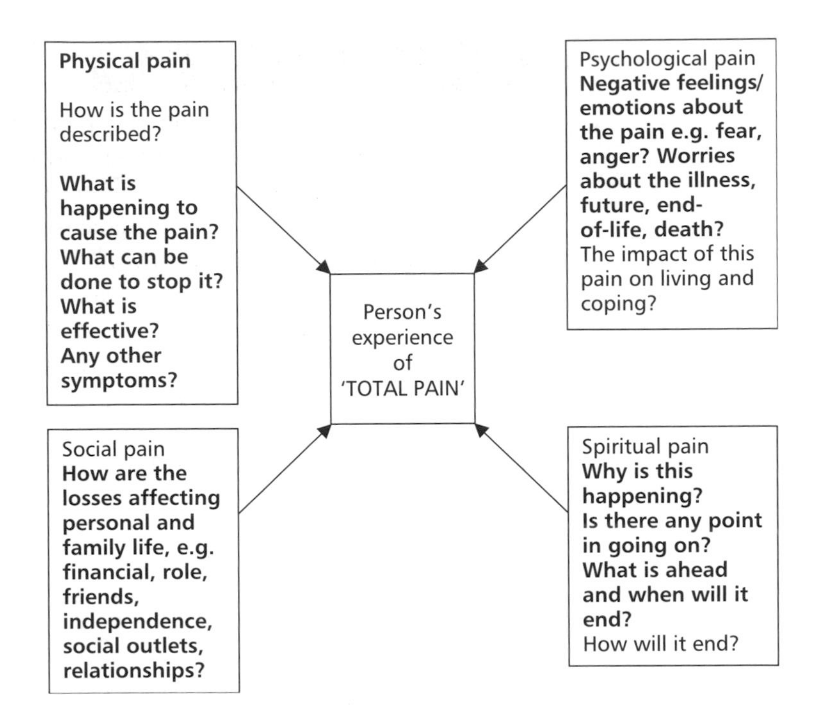

Figure 11.3 Concept of total pain: understanding the person's psychological, social and spiritual issues that impinge on their experience of pain

Alleviating suffering

The anguish of physical pain is often made more complex by the extra psychological, social or spiritual suffering. 'Helping' a person in this type of suffering means finding ways of enabling that person to endure and live with his or her experience. Besides the pharmacological interventions, this involves staying with the person in their suffering. Our willingness to support and empathise with the person may help them find meaning in their suffering. The seminal work of Victor Frankyl (1984) demonstrates that if people can find meaning in their suffering, then it ceases to be meaningless suffering. Although this may appear complex, often it is simply about just 'being there' for the person. Kearney (2000) reassures that rather than having all the answers, helping a person has a lot to do with 'who we are' as carers.

Reflective activity

Imagine that you are asked to explain the concept of 'total pain' to a nursing colleague. Despite your best efforts, the nurse fails to see beyond the physical sensation of pain.

Hints: To assist you in explaining the concept of total pain, consider the image of an iceberg. Imagine that the visible tip of the iceberg represents the physical aspect of pain. In order to care holistically for the person, it is necessary to look beyond the visible and physical, and to assess the deeper psychological, social and spiritual dimensions to the person's experience of pain.

What questions might you consider asking in order to assess what the person's pain experience means to them, hence gaining a holistic understanding of their pain?

Over to you

View the following website to support your answer and to introduce the process of holistic pain assessment.

- Cancer Pain Release. Available at: www.whocancerpain.wisc.edu/contents.html (www.whocancerpain.wisc.edu/eng/14_2/resources.html)

For more information on Dame Cicely Saunders and the concept of total pain, visit the following websites:

- Cicely Saunders Foundation Available at: www.cicelysaundersfoundation.org/ (The Foundation website has an audio interview of Dame Cicely by Professor Irene Higginson.)
- Obituary. Dame Cicely Saunders. 2005. Available at: http://bmj.bmjjournals.com/cgi/content/full/331/7509/DC1

Definition of pain within a palliative care context

Many palliative care professionals espouse the definition 'pain is whatever the patient says hurts' (Twycross and Wilcock, 2002) to capture the subjective, individual experience of pain. However, it is useful to consider a definition that encompasses the pain theories, which have evolved throughout the centuries. The International Association for the Study of Pain (IASP) defines pain as an unpleasant sensory and emotional experience associated with actual or potential tissue damage, or described in terms of such damage (Merskey and Bogduck, 1994).

This definition implies that the pain process is much more than a physiological mechanism that causes the person to experience an unpleasant sensation. It also implies that a person may have pain without a known cause. Therefore, the physiological mechanism of pain is *not* equivalent to pain. Most importantly, the IASP definition recognises that pain is an emotional experience.

This was not always the case. Up until the first half of the twentieth century, pain was viewed solely by the 'specifity theory of pain.' This mechanistic understanding of pain was merely based on the rigid and direct relationship between physical stimulus causing the pain and the sensation felt by the pain. The psychological, affective and social factors influencing pain were ignored. However, Paz and Seymour (2004) cite Leriche's, Beecher's and Bonica's pioneering research on pain in the post-war period, and Melzack and Wall's 'gate-control theory of pain' in the 1960s as greatly contributing to our present day comprehension of pain. Pain is now understood as 'a multidimensional and individual experience with physical, social, psychological, emotional and cultural components' (Paz and Seymour, 2004).

Incidence of pain in palliative care

Portney and Lasage (2001) report that pain is experienced by 70–90% of people with advanced disease. Ross (2004) reports that in the USA, pain ocurs in 25% of all patients at time of diagnosis, and two-thirds of patients receiving anti-cancer therapy. As the stage of the disease advances, the pain also increases, resulting in 75% of patients hospitalised with advanced cancer reporting unrelieved pain (Ross, 2004). Nonetheless, it is imperative to be positive in our interpretation; cancer and pain are not synonymous. Twycross (2003) emphasises the fact that approximately 25% of patients with advanced cancer do not have pain. However, it is worth noting that one-third of those who do suffer pain, one-third experience a single pain, one-third have two pains, and a further one-third have more than two pains (Grond *et al.*, 1996). It is also important to identify the cause, as cancer does not account for all pains. Treatments for the cancer, secondary effects of the cancer, and concurrent disorders must also be considered (Twycross, 2003).

It is also crucial to examine the incidence of non-malignant pain, which Regnard and Kindlen (2004) report is often underestimated. The authors surmise that approximately 60% of patients with cardiac disease, neurological disorders and AIDS suffer with troublesome pain. However, fewer patients with non-malignant conditions are being treated for pain compared to those with malignant pain (Regnard and Kindlen, 2004). Solano *et al.*'s (2005) review of the findings of 64 studies on symptoms experienced in palliative care echo this statement. The smallest and greatest percentage of pain experienced by patients with far advanced cancer, AIDS, heart disease, chronic obstructive pulmonary disease (COPD) and renal disease ranges from 34% to 96% (Solano *et al.*, 2005) as listed in Table 11.1.

Table 11.1 Prevalence of pain in cancer and other diseases (Source: Solano *et al.*, 2005)

Symptom	Cancer	AIDS	Heart disease	Renal disease	COPD
Pain	35–96%	63–80%	41–77%	34–77%	47–50%

Evidence base

For information on the global efforts to tackle pain management in palliative care, visit the following website:

- Help the Hospices (2005) *Suffering at the End of Life – The State of the World.* Available on line at: www.helpthehospices.org.uk/documents/state_of_the_world.pdf.

Classification of pain

Chapter 4 describes pain mechanisms and types of pain. In addition to the definitions described there the following are relevant for palliative care.

Breakthrough pain:
- pain that is more severe than the baseline pain
- common, transitory, exacerbation of pain that is present in patients whose pain is relatively stable and adequately controlled.

Incident pain:
- predictable pain
- occurs in approximately 30% of palliative care patients
- always caused by specific activities (example: someone who is pain free at rest but develops left leg pain on sudden movement).

End of dose failure pain
- occurs towards end of dosing interval of regular analgesic
- present in approximately 16% of palliative care patients
- indication that the regular analgesics require increasing.

Pain assessment in palliative care

The National Cancer Institute (2005) maintains that failure to assess pain is a critical factor leading to under-treatment. Palliative care prevents and relieves suffering by means of *early identification and impeccable assessment* and treatment of pain and other problems, physical, psychosocial and spiritual (WHO, 2002). Pain assessment is therefore pivotal to effective pain management within a palliative care context. In relation to palliative nursing, Fitzsimons and Ahmedzai (2004) concur that the purpose of assessment is to identify *individual* patients' needs or problems in order to set appropriate, achievable and effective goals. Utilising the most appropriate diagnostic and therapeutic considerations, pain assessment will contribute towards defining the cause and directing treatment (Fink and Gates, 2001).

McCaffrey (1983) proclaims that assessment of pain and its relief is no simple matter. More than two decades later, assessment remains problematic and inconsistent (Gordon *et al.*, 2005).

Fitzsimons and Ahmedzai (2004) identify certain fundamental concepts that underpin nursing assessment in palliative care in an attempt to surmount the difficulties with assessment. According to the authors, assessment should be dynamic, individualised, patient and family centred, holistic, therapeutic, sensitive and appropriate to the patient and their family's needs, comprehensive, contextual, and provide reliable and valid information. Furthermore, pain assessment should be evidence-based, and focus upon process and outcomes of care, as evidence of effectiveness of interventions, treatments or services in palliative care is now vital (Fitzsimons and Ahmedzai, 2004).Therefore, the use of an appropriate pain assessment tools is essential in order to objectively measure and capture the sensory and emotional aspects of the individual's pain. Chapter 6 reviews pain assessment tools in detail. As no one scale is suitable for all persons experiencing pain, health care professionals are advised to choose a tool that appropriately meets the needs and capabilities of each person, whilst conscious of the fact that people with advanced disease may tire easily. It is also imperative to facilitate the patient's own story through the use of effective communication skills such as open questions and active listening.

The National Cancer Institute (2005) assert that assessment should occur at regular intervals after initiation of treatment; at each new report of pain; and at a suitable interval after pharmacological or non-pharmacological intervention, e.g., 15–30 minutes after parenteral drug therapy and one hour after oral administration. Impeccable assessment involves assessing the person's description, physical causes and non-physical causes, whilst considering the psychosocial and spiritual aspects of pain. Table 11.2 illustrates the key points to assessing pain within a palliative care context.

Table 11.2 Key points to assessing pain within a palliative care context

Assess Description Quality of the pain: **attentive listening to the individual's descriptive words provides valuable clues to its aetiology/mechanism** Location: **may be more than one pain** Temporal features: **including duration and onset** Severity/intensity: **e.g. numeric/visual analogue scales. Ask about breakthrough and incident pain** Aggravating factors: **ask the individual to identify factors that makes the pain worse** Interventions/Relieving factors: **ask the individual to identify factors that ease the pain, including the effects of analgesia** Associated symptoms: **e.g. dizziness, nausea, sweating, vomiting, shortness of breath**
Assess Physical Cause Disease: **including malignant +/– non-malignant disease** Treatment: **e.g. chemotherapy** Debility: **effects of disease on independence, mobility etc.** Concurrent disorders: **e.g. osteoarthritis**
Assess Mechanism **1. Nocioceptive** **(a) Somatic** **(b) Visceral** **2. Neuropathic**
Assess Psychological, Social and Spiritual Aspects *Impact on the person's life* *Meaning of pain to the person* **Their experience of suffering**
To aid impeccable assessment by. . . A**nticipating pain and early identification** B**elieving the person's experience and rating of pain** C**lassifying pain** D**efining the cause and** D**irecting treatment** E**valuating treatment** F**orwarding (redirecting) treatment**

Student nurse

'It definitely takes time to assess pain properly. I never knew there was so much to it! I used to ask the patient had they got pain, that was it. Rating the pain using the numerical scale is useful, but it's not enough. Now I try to listen to the patient and also to what the family have to say. With the help of the other nurses, I try to work out the type of pain, document how the pain is affecting the person and then we work on what makes it better. But you're never finished. Pain assessment is ongoing, a real challenge, but it's worth it when you see the patients more comfortable.'

Pain management guidelines

According to two WHO publications, *Palliative Care: The Solid Facts* (2004b) and *Better Palliative Care for Older People* (2004c), many Europeans die in unnecessary pain and discomfort because health systems lack the skills and services to provide care towards the end of life. There is strong evidence to suggest that palliative care, through the early identification and treatment of pain and other health problems, can make a dramatic difference in the quality of life of the dying and those who care for them. In 1986, the WHO published guidelines on pain management titled *Cancer Pain Relief*. The guidelines indicate that the appropriate use of simple therapies can effectively relieve pain in the majority of patients.

The guidelines recommend that health care professionals use a three-step analgesic ladder developed for cancer pain relief. The successful implementation of this analgesic ladder depends on the availability of drugs that are safe and effective in relieving chronic severe pain, such as morphine or other opioids (Joranson *et al.*, 2004).

The WHO (1996) advocate that analgesics should be administered in the following way:

- 'by the ladder' (in standard doses at regular intervals in stepwise fashion)
- 'by the mouth' (the oral route is the preferred route for administration when possible)
- 'by the clock' (to maintain freedom from pain, drugs should be given regularly, rather than waiting for the individual to request pain relief)
- 'for the individual' (the analgesic ladder is a framework of principles rather than a rigid protocol)
- 'attention to detail' (attention to each individual's unique experience of pain is required when directing treatment).

The WHO (2005) website states: 'If a pain occurs, there should be a prompt oral administration of drugs in the following order: non-opioids (aspirin or paracetamol); then, as necessary, mild opioids (codeine); or the strong opioids such as morphine, until the patient is free of pain. To maintain freedom from pain, drugs should be given "by the clock", that is every three to six hours, rather than "on demand". This three-step approach of administering the right drug in the right dose at the right time is inexpensive and 80–90% effective'. See Figure 7.5 on page 109 for WHO's Analgesic Ladder.

The analgesic ladder provides a safe, consistent and standard method of pain control (Hanks *et al.*, 2001). However, since its introduction in 1986 there have been major developments in the field of palliative medicine. Thus, some of the recommendations have been modified. The Expert Working Group of the Research Network of the European Association for Palliative Care (EAPC) revised and updated guidelines for the use of opioids (Hanks *et al.*, 2001). The recommendations recognise that pain relief should be provided to all seriously ill and dying patients, not only cancer patients. Additionally, the middle step of the ladder using mild opioids is often skipped in seriously ill and dying patients as their pain is so severe that strong opioids are needed. Furthermore, a range of **adjuvant analgesics** can also be used to treat neuropathic pain and other specific pain conditions at all stages of the ladder (Hanks *et al.*, 2001).

Keywords

Adjuvant analgesics
Drugs that have a primary indication other than pain but are used to enhance analgesia in specific circumstances. They are mainly for neuropathic pain, also used for bony pain

Morphine

The opioid of first choice for moderate to severe cancer pain is morphine, as it remains the most researched and most widely used opioid (Hanks *et al.*, 2001). In the UK, diamorphine is the preferred injectable opioid when patients are unable to take oral morphine. However, there are alternatives for when morphine or diamorphine are unsuitable (e.g. hydromorphone, oxycodone, fentanyl, methadone and buprenorphine). It is important to address the various misconceptions, the cognitive effects and the limitations to using morphine as discussed in Chapter 7.

The safe and appropriate use of morphine

Morphine is a safe drug when used properly. The Europen Association for Palliaitive Care guidelines for the use of opioids state that ideally, two types of morphine formulation are required: normal release (for dose titration) and modified release (for maintenance treatment). The simplest method of dose titration is with a dose of normal release morphine given every four hours and the same dose for breakthrough pain. This rescue dose may be given as often as required (up to hourly) and the total daily dose of morphine should be reviewed daily. One-sixth of the daily 24-hour dose is regarded as an appropriate rescue dose. The regular dose can then be adjusted to take into account the total amount of rescue morphine (Hanks *et al.*, 2001). Alternatively, modified release morphine may be started at 20–30 mg twice daily. Normal release morphine tablets (e.g. sevredol) or solution (e.g. oramorph) are provided for breakthrough use. The dose of the modified release morphine should be increased every two to three days until there is adequate relief throughout each 12-hour period, taking pre re nata (PRN) use into account (i.e. the rescue doses for the breakthrough pain). However, caution must be exercised with the elderly population due to the likelihood of reduced clearance of opioids.

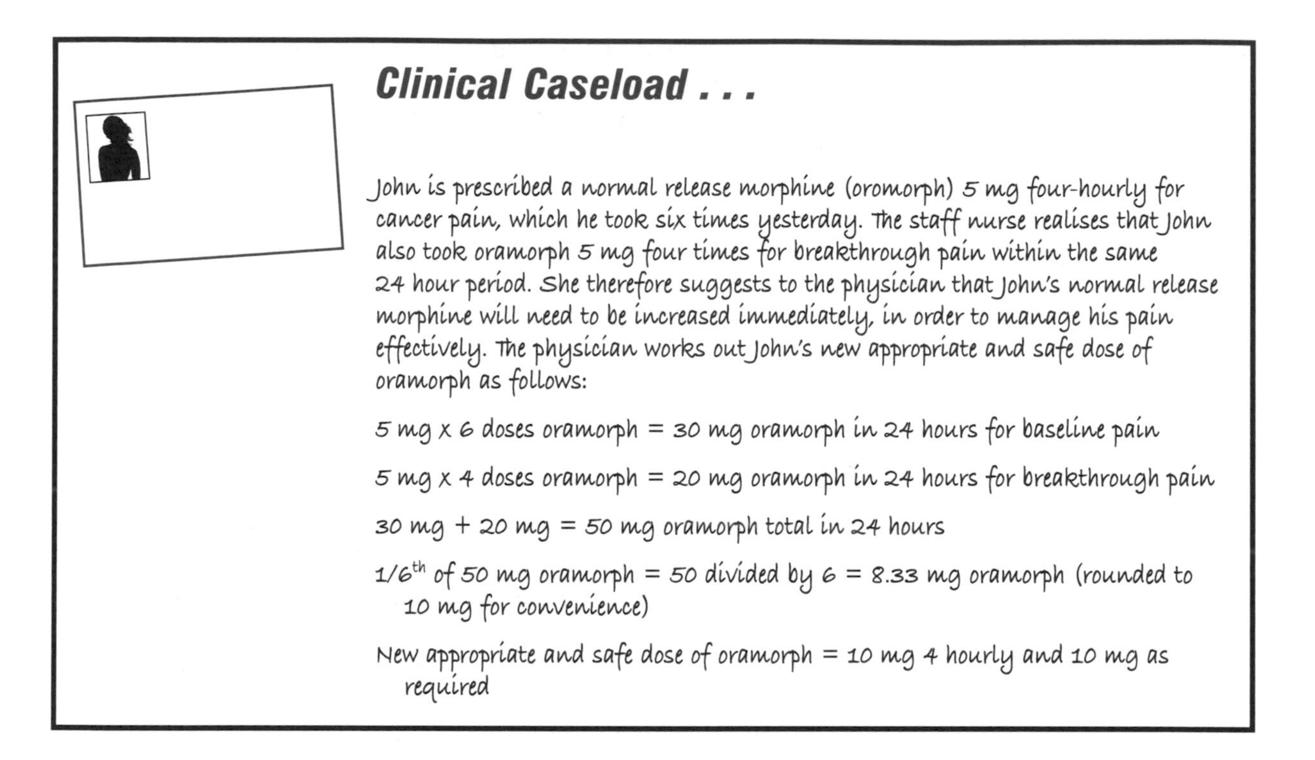

Clinical Caseload . . .

John is prescribed a normal release morphine (oromorph) 5 mg four-hourly for cancer pain, which he took six times yesterday. The staff nurse realises that John also took oramorph 5 mg four times for breakthrough pain within the same 24 hour period. She therefore suggests to the physician that John's normal release morphine will need to be increased immediately, in order to manage his pain effectively. The physician works out John's new appropriate and safe dose of oramorph as follows:

5 mg x 6 doses oramorph = 30 mg oramorph in 24 hours for baseline pain

5 mg x 4 doses oramorph = 20 mg oramorph in 24 hours for breakthrough pain

30 mg + 20 mg = 50 mg oramorph total in 24 hours

$1/6^{th}$ of 50 mg oramorph = 50 divided by 6 = 8.33 mg oramorph (rounded to 10 mg for convenience)

New appropriate and safe dose of oramorph = 10 mg 4 hourly and 10 mg as required

Evidence base

Visit the following WHO website to see the relevant 2004 reports, which aim to ensure that all individuals and families have the opportunity to be appropriately supported towards the end of their lives and not to die in unnecessary pain.

- World Health Organization Website (2004d) *People die in unnecessary pain*. Available at: www.kcl.ac.uk/depsta/palliative/research/who.html.

Continuous subcutaneous infusion (CSCI)

A continuous subcutaneous infusion (CSCI) is an effective method of drug administration that is particularly suited to palliative care when the oral route is inappropriate (Dickman *et al*., 2005). British practitioners commonly refer to a CSCI as a 'syringe driver' (Graham and Clark, 2005). It is a lightweight, portable, battery-operated infusion pump, that delivers drugs over a specified period of time (usually 24 hours).

There are two types of syringe driver manufactured by SIMS Graseby that are widely used in palliative care:

- MS16A with an *hourly* rate (mm/h) syringe driver and a *blue* front panel

- MS26 with a *daily* rate (mm/24h) syringe driver and a *green* front panel.

The majority of centres in the UK use the MS26 because setting the rate is simpler and perhaps safer. However, the MS16A is more flexible, particularly when larger volumes of drugs are required. It is safer if only one or other type is used in a given locality in order to minimise the likelihood of errors (Carlisle *et al.*, 1996). Confusion between the two, in relation to the setting, has led to fatal errors (Dickman *et al.*, 2005). Staff training is therefore essential to highlight the differences (Carlisle *et al.*, 1996).

The use of a CSCI is often incorrectly seen as 'the last resort' and wrongly associated with imminent death. Rather, a CSCI is an effective method of relieving certain symptoms by injection, for example intractable nausea and vomiting (Dickman *et al.*, 2005). Other indications for using a CSCI include dysphagia, intestinal obstruction, severe weakness, unconsciousness, poor absorption of oral drugs (rare), and patient preference (e.g. large amounts of tablets).

As drugs are delivered continuously over the specified period, the CSCI maintains constant plasma levels of the drugs. A great benefit is that more than one medication can be mixed in the CSCI, thereby allowing management of multiple symptoms with a combination of drugs. Most of all, the CSCI avoids the need for repeated injections, thus reducing patient discomfort and requirement for repeated injections. Before setting up a syringe driver, it is important to explain to the patient and the family the rationale for its use, and how it works.

Knowledge of drug compatibility is essential when using a CSCI. The prescription should be checked to determine whether the drug combination, and the solution that the drug is diluted in, is physically compatible (refer to medication information leaflet). As a general rule, drugs with similar pH are more likely to be compatible than those with widely differing pH. As most drug solutions are acidic, alkaline drugs such as ketorolac, diclofenac, phenobarbitol and dexamethasone often cause compatibility problems when used in a combination with other drugs. It is therefore recommended to administer these drugs in a separate infusion. Dexamethasone should always be added last to an already dilute combination of drugs in order to reduce the risk of incompatibility.

Dickman *et al.* (2005) report that a greater number of drugs are now being administered by a CSCI. Usually a maximum of three drugs are mixed in the one syringe, although some centres combine up to four drugs. Table 11.3 lists the drugs that are suitable for delivery by a CSCI. However, it is important to note that although these drugs have product licences, the majority of the drugs are unlicensed for administration via a CSCI (Dickman *et al.*, 2005).

Until recently, the recommended diluent for CSCI was water for injection because it was thought that there was less chance of precipitation. However, Dickman *et al.* (2005) recommend that NaCL should be used for dilution as the majority of drugs commonly used in palliative care have an osmolarity, the same or less than NaCL. There are two exceptions to this general rule. Water for injection must be used as a diluent for solutions containing cyclizine (as cyclizine causes crystal formation in the presence of chloride ions), and for solutions of high concentrations of diamorphine (>40 mg/ml). If in any doubt, the prescribing physician or pharmacist should be contacted.

It is also imperative to note the equal analgesic dose of subcutaneous opioids compared to oral opioids. For example, subcutaneous morphine is twice as potent as oral morphine, therefore morphine sulphate 30 mg orally in 24 hours is equivalent to 15 mg of morphine sulphate subcutaneously over 24 hours.

Table 11.3 Drugs suitable for delivery by a CSCI

Alfentanil	Haloperidol	Octreotide
Clonazepam	Hydromorphone	Ondanestron
Cyclizine	Hyoscine butylbromide	Oxycodone
*Dexamethasone	Hyoscine	*Phenobarbitol
Diamorphine	Hydrobromide	Promethazine
*Diclofenac	Ketamine	Ranitidine
Dihydrocodine	*Ketorolac	Sufentanil
Dimenhydrinate	Levomepromazine	Tramadol
Fentanyl	Methadone	
Glycoprronium	Metoclopramide	
	Midazolam	
	Morphine	

* Indicates drugs to be administered in a separate infusion

Evidence base

For further information on the safe and appropriate use of CSCI, and drug compatibilities, refer to:

- Dickman A., Schneider, J., Varga J. (eds.) (2005) *The Syringe Driver: Continuous subcutaneous infusions in palliative care,* 2nd edn. Oxford University Press, Oxford. Alternatively, visit www.palliative drugs.com, a website that provides essential, comprehensive and independent information for health professionals about the use of drugs in palliative care. It highlights drugs given for unlicensed indications or by unlicensed routes and the administration of multiple drugs by continuous subcutaneous infusion.

Key points Top tips

See Table 11.4 for the key points of the analgesic ladder

Table 11.4 Key points of the analgesic ladder

Step	Class	Drug example	Additional drugs
STEP 1	Non-opioid	Paracetamol Aspirin NSAIDs (nonsteroidal anti-inflammatory drugs)	+/– adjuvant drugs
If pain persists or increases, go to step 2			
STEP 2	Mild opioid for mild to moderate pain (Codeine)	Solpadeine Solpadol Tramadol	+/– adjuvant drugs +/– non-opioid
If pain persists or increases, go to step 3			
STEP 3	Strong opioid for moderate to severe pain	Morphine Hydromorphone Buphenorphine Oxycodone Methadone Fentanyl	+/– adjuvant drugs +/– non-opioid

A palliative care approach to pain management

Jane is a 42-year-old accountant receiving chemotherapy for malignant breast cancer and local bone metastasis. She is graduating with a law degree next week and just wants the pain in her ribcage sorted now so she can plan her graduation party with her fiancé James. She is taking modified 12-hourly release morphine sulphate tablets 30 mg twice daily and wants something for breakthrough pain. However, you note that she is not prescribed normal release morphine. When you report her pain to the physician, he tells her that the palliative care team are coming to visit her this morning. Shocked and horrified, she admits to you that she has no idea why they are visiting or what they plan to offer. She incorrectly presumes that her disease is advancing and that she is going to die very shortly.

You are asked to:

1 Explain the role of the palliative care team to Jane.
2 Discuss how their input may positively influence the course of Jane's illness.
3 What do you do for her pain whilst waiting for the palliative care team to visit?

continued

Hints:

- To assist you in explaining the concept of palliative care, find out what palliative care means to Jane and her family (James). Explore why she is shocked and horrified, including any previous experience of hospice/palliative care involvement.
- Offer Jane accurate information in relation to the benefits of shared supportive/palliative care, in order to empower her.
- Refer to the analgesic ladder and the EAPC recommendations for guiding Jane's pain management, in particular to opioid titration.

Key points Top tips

- People die in unnecessary pain and discomfort because health systems lack the skills and services to provide care towards the end of life
- There is strong evidence to suggest that palliative care, through the early identification and treatment of pain and other health problems, can make a dramatic difference in the quality of life of the dying and those who care for them
- Pain management is fundamental in creating the conditions for a person with advanced disease to address their personal healing
- Failure to identify, assess and treat pain will result in mismanagement of the other elements of a person's total suffering

RRRRRRapid recap

1. Outline the philosophy of a palliative care approach.
2. What is the purpose of pain assessment within a palliative care context?
3. Following the analgesic ladder guidelines, how should analgesics be administered?

References

Archbishop Desmond Tutu (2005) State of the World Report – Launch. BBC World Service Broadcast. Available online at: www.helpthehospices.org.uk/news/index.asp?submenu=5&newsid=171.

Cancer Pain Release. Available at: www.whocancerpain.wisc.edu/contents.html.

Carlisle D., Upton D. and Cousins D. (1996) Infusion Confusion. *Nursing Times*, **92**: 48–49.

Dickman A., Schneider, J. and Varga J. (eds.) (2005) *The Syringe Driver: Continuous subcutaneous infusions in palliative care*. 2nd edn. Oxford University Press, Oxford.

Doyle, D. (2005) Tribute to Dame Cicely Saunders. *IAHPC News On-line* (Hospice & Palliative Care News & Information), **6**(8).

Fink R. and Gate R. (2001) Pain Assessment. In: *Textbook of Palliative Nursing* (eds. Rolling Ferrell, B. and Coyle, N.) Oxford University Press, Oxford.

Fitzsimons D. and Ahmedzai S.H. (2004) Approaches to assessment in palliative care. In: *Palliative Care Nursing: Principles and Evidence for Practice* (eds. Payne S., Seymour J. and Ingleton C.) Open University Press, Berkshire.

Frankyl V. (1984) *Man's Search for Meaning.* New York, Washington Square Press.

Gordon D.B., Dahl J.L., Miaskowski C., McCarbera B., Todd K.H., Paice J.A. *et al.* (2005) American Pain Society recommendations for Improving the Quality of Acute and Cancer Pain Management. *Archives of Internal Medicine*, **65**: 1574–1580.

Graham F. and Clark D. (2005) The syringe driver in palliative care; the inventor, the history and the implications. *Journal of Pain & Symptom Management*, **29**(1): 32–40.

Grond S. *et al.* (1996) Assessment of cancer pain: a prospective evaluation in 2266 cancer patients referred to a pain service. *Pain*, **64**: 107–114.

Hanks G.W., De Conno F., Cherney N., Hanna M., Kalso E., McQuay H.J. *et al.* (2001) Morphine and alternative opioids in cancer pain: the EAPC recommendations. *British Journal of Cancer*, **84**(5): 587–593.

Help the Hospices (2005) 'Suffering at the End of Life – The State of the World'. Available at: www.helpthehospices.org.uk/documents/state_of_the_world.pdf.

Joranson D., Ryan K. and Jorenby R. (2004) Availability of Opioid Analgesics in Romania, Europe, and the World (2001 data). University of Wisconsin Pain and Policy Studies Group/WHO Collaboration Center for Policy and Communication in Cancer Care. Madison, Wisconsin, USA.

Kearney, M. (2000) *A Place of Healing.* Oxford, Oxford University Press.

McCaffrey M. (1983) *Nursing the Patient in Pain*, 2nd edn. Harper and Row, London, p. 276.

Merskey H. and Bogduck N. (1994) Classification of Chronic Pain: Description of Chronic Pain Syndromes and Definition of Pain Terms. Report by the International Association for the Study of Pain Task Force on Taxonomy, 2nd edn. IASP Press, Seattle, WA.

National Cancer Institute (2005) Available at: www.nci.nih.gov/cancertopics/pdq/supportivecare/pain/HealthProfessional/page2#Section_44.

Payne S., Seymour J. and Ingleton C. (eds.) (2004) *Palliative Care Nursing: Principles and Evidence for Practice.* Open University Press, Berkshire.

Paz S.and Seymour J. (2004). Pain: Theories, evaluation and management. In: Payne S., Seymour J. and Ingleton C. (eds.) *Palliative Care Nursing: Principles and Evidence for Practice*. Open University Press, Berkshire.

Portney R.K. and Lesage P. (2001) Management of cancer pain. The Pain Series. Available at: www.thelancet.com/journal/vol357/issss1.

Regnard C. and Kindlen M. (2004) What is pain? In: *Helping the Patient with Advanced Disease* (ed. Regnard C.). Radcliffe Medical Press, Oxford.

Ross E. (2004) *Pain Management*. Hanley & Belfus, Philadelphia.

Solano J.P., Games B. and Higginson I.J. (2005) A comparison of symptom prevalence in far advanced cancer, AIDS, heart disease, chronic obstructive pulmonary disease (COPD) and renal disease. *Journal of Pain and Symptom Management*, **31**(1): 58–69. Also available at: www.worldday.org/documents/state_of_the_world.pdf (page 13).

The National Council for Palliative Care (2005) *Palliative care explained.* Available at: www.ncpc.org.uk/palliative_care.html.

Twycross R. (2003) *Introducing Palliative Care*, 4th edn. Radcliffe Medical Press, Oxford.

Twycross R.G. and Wilcock A. (2002) *Symptom Management in Advanced Cancer*, 3rd edn. Radcliffe Medical Press, Oxford.

Twycross R.G., Wilcock A., Charlesworth S. and Dickman A. (2002) *Palliative care formulary,* 2nd edn. Radcliffe Medical Press, Oxford.

World Health Organization (1986) *Cancer Pain Relief*. WHO, Geneva.

World Health Organization (1996) *Cancer Pain Relief: with a Guide to Opiod Availability*, 2nd edn. WHO, Geneva.

World Health Organization (2002) *National cancer control programmes: policies and managerial guidelines,* 2nd edition. WHO, Geneva. Available at: www.who.int/cancer/en/

World Health Organization (2004a) *Global Day Against Pain*. WHO, Geneva. Available at: www.who.int/mediacentre/news/releases/2004/pr70/en/index.html.

World Health Organization (2004b) *Palliative Care: The Solid Facts*. WHO, Geneva. Available at: www.euro.who.int/document/E82931.pdf.

World Health Organization (2004c) *Better Palliative Care for Older People*. WHO Regional Office for Europe. Available at: www.euro.who.int/document/E82933.pdf.

World Health Organization (2004d) *People die in unnecessary pain*. Available at: www.kcl.ac.uk/depsta/palliative/research/who.html.

World Health Organization (2005) Pain Ladder Website. Available at: www.who.int/cancer/palliative/painladder/en/index.html.

12 Conclusion – Nurses, health care professionals and organisational roles

Learning outcomes

By the end of this chapter you should be able to:

★ Understand the role of nurses and health professionals in pain management

★ Describe the importance of organisational commitment to pain management

★ Describe the benefits of a multidisciplinary approach to pain.

Introduction

This chapter will highlight the role of nurses and health professionals in pain management. It will draw on all other chapters and act as a concluding chapter to the book. It will detail the need for a consistent approach to pain management, the role of professional judgement and the need for education and continuing professional development in pain management for all health professionals. The chapter will support the concept that patients/clients get a better service from a multidisciplinary approach to care. The most important message emerging from this book is the *critical role of evidence-based practice*.

Pain as an important issue for society

Pain has become a world public health issue. Pain is widely prevalent and it is evident that both acute and chronic pain are often poorly managed. The reasons for this vary from cultural to attitude of health professionals and patients, education of staff and lack of multidisciplinary team working. Patients do appear to be generally satisfied with their pain management (Tasso and Behar-Horenstein, 2004). However, as has been seen in the various chapters throughout this book, patients both in hospitals and communities continue to suffer moderate to high levels of pain. Decreased lung volumes, myocardial ischaemia, decreased gastric and bowel motility and anxiety are all harmful side effects associated with under-treated severe pain (Macintyre and Ready, 2002). Pain and in particular chronic and cancer pain affects both quality of life and functional status. There is thus an economic and social burden associated with pain. Pain management has over time become a priority for health care policy makers and providers. Indeed organisations such as the World Health Organization, the International Association for the Study of Pain and the British Pain Society have taken lead roles in promoting best practice in pain management.

Pain as an important issue for individuals

Pain is a personal and unique experience for the individual. While much is known about the physiology of pain, the emotive and contextual aspects of pain continue to give rise to debate. It is interesting that even with increasing knowledge of the physiological theories such as the Gate Control Theory, knowledge of pain physiology could be considered still in its infancy. It is likely that with increased technology such as magnetic resonance imaging (MRI) scanning and increasingly sophisticated research our understanding of pain will increase at a fast pace. Research into genetics and highly developed medications and administration devices will continue to contribute to improving the patient's pain experience.

The cause of an individual's pain will relate highly to how that pain is experienced by any individual, i.e. the extent of pain will relate to the level of nociception involved. Even with greater understandings of the physiology of pain the context of that pain will continue to shape the experience and subsequent management of that pain. Patients' knowledge of pain and analgesia, their expectations of pain, past experiences of pain, fear of addiction, anxiety, culture, age, lack of information, the influence of public and organisational policy and health professionals responsible for care all form part of the pain experience. Worldwide mobility has raised the profile of culture, race and ethnicity in health care in general and in pain management specifically. It emerges from Chapter 5 that in order to enhance communication with patients and improve their pain management, health care professionals should:

- be sensitive to individual needs of patient and their family
- utilise a comprehensive pain history and assessment appropriate to classification of pain
- pay constant attention to communications
- embrace the underlying philosophy of pain management that 'pain is what the patient says it is'.

Reflective activity

Think back to when you looked after your first patient in pain. How did you empathise with the patient? How did you assess their pain? Have you developed more sophisticated methods of assessing pain over time?

Research and evidence-based practice

Pain management, like any aspect of patient care, should be based on the best available evidence. The management of pain is the subject of ongoing research and debate. As such evidence changes rapidly all health professionals should continuously update and review developments. Good evidence is likely to come from good systematic reviews of sound clinical trails. This means that health professionals must be able to access, understand and interpret research in order to provide evidence-based practice.

It is evident throughout this book that ongoing research is necessary in a number of areas. In particular the use of multimodal pain interventions and their evidence base needs continued ongoing research as new information, medications and therapeutic interventions emerge. New technologies continue to develop which will support the development of analgesic management.

Nurses and health care professionals need to be able to lead, initiate and understand research and evidence-based practice. Education and ongoing professional development are key to this as not only does the evidence change but methods used to design and review research studies continuously evolve over time. In the past, one good well-designed randomised controlled trial (RCT) was considered very strong evidence for practice. Now it is evident that even stronger evidence can come from systematic reviews and meta-analysis of research studies. Additionally search tools have become more sophisticated and powerful and there is increasing access to research from all countries regardless of language.

> ***Over to you***
>
> Having read through each of the chapters in the book what areas of pain management do you think should be prioritised for research? What are the gaps in our knowledge with regard to pain?

Barriers to effective pain management

Numerous barriers to effective pain management are detailed throughout this book. It is important that as health professionals we seek solutions to these barriers and become proactive in our outlook and approaches to pain management. Table 12.1 summarises some of these barriers as detailed by Carr (2002).

Table 12.1 Barriers to effective pain management (Source: Carr, 2002)

Prevalence and incidence of pain	• Little has changed over 25 years. • Problem is likely to increase as population grows older
Patients	• Often reluctant to take analgesia (have concerns about its potential harm, constipation and not being in control) • Widespread public education is needed to change lack of public knowledge regarding pain in order to shift pain culture and empower patients to participate in decisions about their pain treatment
Professional knowledge	• Deficits in knowledge and lack of regular assessment often blamed for poor management of pain • Multiple demands compete for nurses' time and pain is not viewed as a priority
Organisational and hospital management priorities	• May impact on nurses' ability to deliver effective pain management.
Perceived harmlessness of pain	• Deleterious effects of pain such as immobility, deep vein thrombosis, depression and anxiety are frequently not seen as consequences of omitting to assess pain and administer appropriate analgesics.

The identification of these barriers is the first step to improving pain management. There is sufficient well-documented pain information available to any service or individual professional who wishes to review their own service area and begin the process of improving approaches to pain management. The literature and evidence available provides as many solutions as barriers.

Reflective activity

Did you complete the 'over to you' exercise in Chapter 2, page 18? If yes consider the results, if not this is a reminder so that you can complete the exercise.

Consider your patient profile. Do you audit your pain prevalence?

If not complete this simple quick task:

- What would you estimate is the prevalence of pain in your clinical area?
- Take a simple numeric rating scale and do a point prevalence for 10 patients (this will take about five minutes)

Are the pain levels what you expected? How do they compare to national and international statistics?

Having completed the exercise are you satisfied that the levels of pain on your clinical area are in line with best practice? What do you think are the barriers that limit effective pain management in your organisation?

Multidisciplinary teams

Responsibility for pain management lies with all of the health care team: nurse, doctor, anaesthetist, physiotherapist, pharmacist, the patient and others as appropriate. Chapter 2 describes the various teams (acute, chronic, palliative care) and the evidence to support their effectiveness. It is evident through the various research studies presented in this book that patients benefit from care from teams.

Over to you

Consider the evidence for multidisciplinary team management for pain in this book. What do you think are the key benefits for patients?

The key desirable characteritsics within these teams are outlined in Figure 12.1.

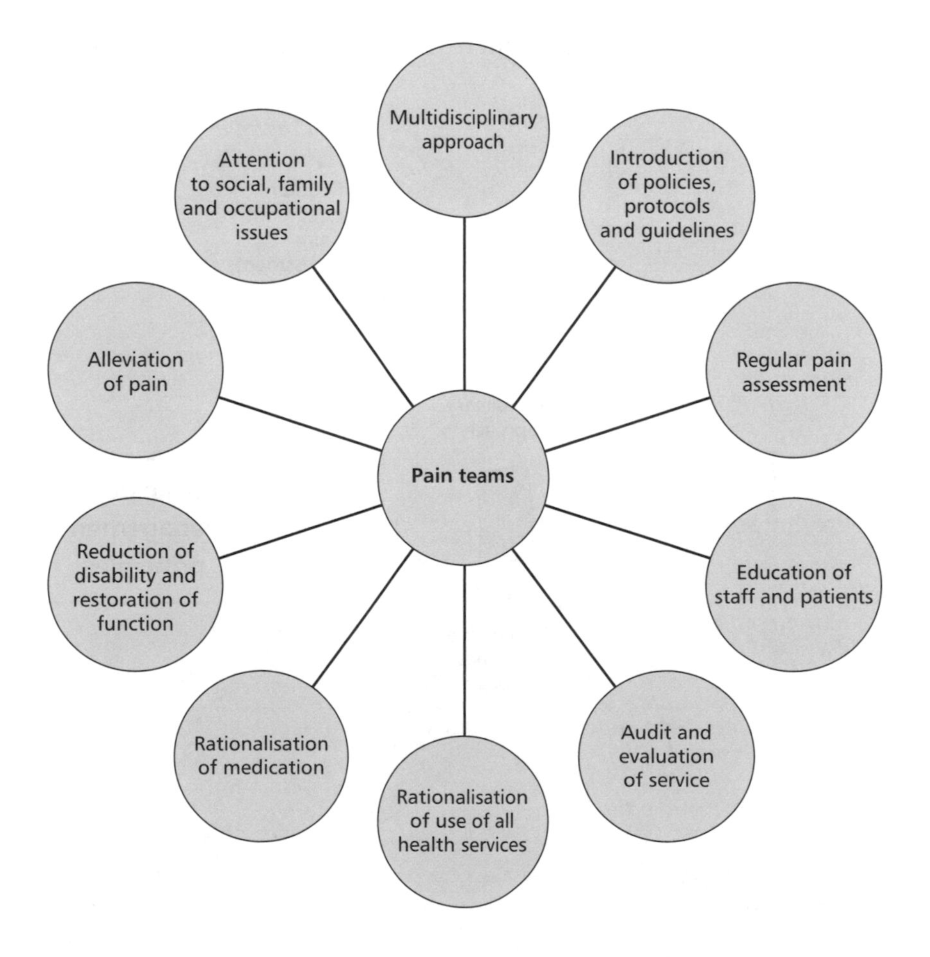

Figure 12.1 Key desirable characteristics of a pain team

In order to manage pain effectively, health professionals should have a firm understanding of the physiology of pain including the concept that each individual perceives his/her pain uniquely. This provides the basis for management and modulation of pain. The options for modulation of pain at each stage of the pain process and pain pathway should be understood. The principles of pain assessment and the need for ongoing monitoring of pain are part of the key knowledge base for health care professionals to be able to manage pain effectively. A basic knowledge base is needed for all health professionals to embrace their role supported by organisational policies and guidelines. Additionally health professionals need to be able to interact with each other effectively and efficiently. This means developing relationships between primary, secondary and tertiary care. It is of little benefit to the patient if a sophisticated pain management plan is developed in the hospital but is not transferred to and supported by the health care professionals in the community. The role of complementary and alternative therapies in a mixed-method approach to holistic pain care management should also be appreciated. This is discussed in Chapter 8.

Drawing together the various experts provides the patient with the most comprehensive pain plan. In addition modern medicine may have brought new understanding and improved technology, but even without

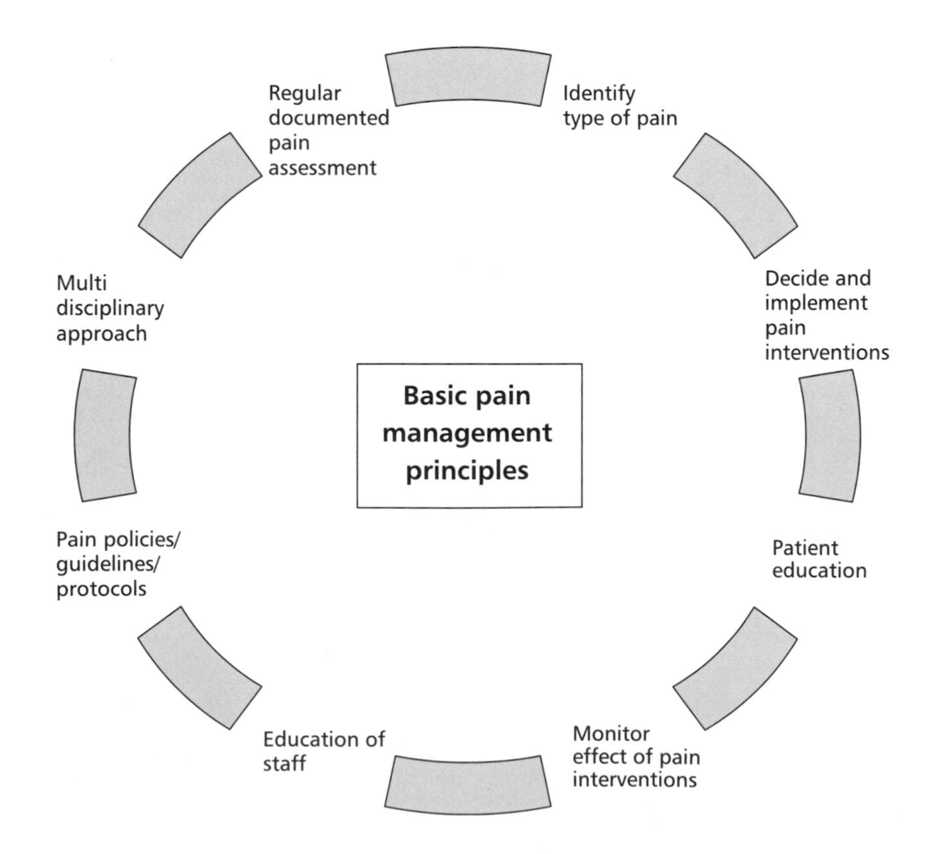

Figure 12.2 Basic pain management principles

new techniques pain management can be improved through an understanding and application of basic pain management principles. Again this evidence is presented throughout the book and is detailed in Figure 12.2.

Nurse's role

The nurse's role as part of the multidisciplinary team is of particular importance in pain management, given that they provide care on a 24 hour basis. Increasingly, nurses have taken on more specialist and advanced practice roles in pain management. They are seen as the lead in some pain teams (Mackintosh and Bowles, 1997). Titles such as pain resource nurses (Holley *et al.*, 2005), nurse specialists and pain nurse consultants have all emerged. Nurses are seen are key members of the multidisciplinary team in managing pain (Clinical Standards Advisory Group (CSAG, 2000).

The nurses' role as detailed in Chapter 6 is very important in pain assessment. It is increasingly evident that appropriate, regular pain assessment is key to pain management. The nurse interprets pain, administers and evaluates pain management procedures, provides information to patients and works as a member of the multidisciplinary team. Certain skills and knowledge are needed in order that the nurse can embrace effectively this role in pain management. Ongoing strategic in-service education is necessary to keep nurses updated. Maximising the nurses' role and utilising basic pain management principles have been shown to decrease patients' pain scores (Bardiau *et al.*, 2003; Mac Lellan, 2004).

Table 12.2 details facets of the nurse's role in pain management. These facets relate to the key areas of functioning necessary for the nurse to support evidence-based care and a holistic approach to pain management. The knowledge and skills required by nurses to manage pain are outlined. Use of the table by matching knowledge and skill to role facet provides guidelines on how nurses can perform an effective role in pain management by applying basic pain management principles.

The Pain Society (2002) produced recommendations for nursing practice in pain management. Nursing competencies for a career pathway in pain management from novice to intermediate leading to higher-level practice are outlined. Seven standards are outlined:

1 Providing effective health care.
2 Improving quality and health outcomes.
3 Evaluation and research.
4 Leading and developing practice.
5 Innovation and changing practice.
6 Developing self and others.
7 Working across professional and organisational boundaries.

Table 12.2 The nurse's role in pain management (Source: Adapted from Mac Lellan, 2000)

Role facet	Knowledge	Skills
Information giving to patients – Methods of pain relief – Assessment of pain – The roles and functions of the health care team in pain management – Role of the patient and family	Pain relief methods Pain assessment methods	● Effective communication skills
Assessment of pain – Verbal assessment – Non-verbal assessment	Pain assessment tools Pain physiology Pain theories	● Ability to use assessment tools reliably ● Ability to interpret non-verbal cues of pain and undertake holistic assessment of patient
Development of therapeutic programmes	Pain management processes Quality of life issues Supports to improve functional status	● Ability to complete assessment of patient and develop holistic pain management plans
Administration and monitoring of effect of analgesia – Opioids – NSAIDs – Anti-convulsants – Paracetamol – Other	Analgesic medications (action, time of effect, side-effects)	● Ability to administer by many routes ● Ability to use technology, e.g. administration pumps ● Ability to integrate theory with practice
Documentation of pain	Best practice in documentation	● Ability to write clearly and concisely
Team work	Team dynamics	● Effective communication skills
Complementary approaches to pain management	Other approaches to pain relief, e.g. relaxation, cold, heat, distraction	● Competency in alternative approaches to pain management
Holistic approach to patient care	Pain management Integrated approach to care Health promotion	● Ability to empower and support patient to improve their health and wellbeing ● Ability to manage patient as a service user and as a member of a wider community
Evidence-based practice	Clinical audit Sources of evidence	● Ability to use results of clinical audit to evaluate pain management service ● Ability to access best evidence and integrate it into practice ● Ability to integrate theory with practice

Reflective activity

Consider your patient caseload. How many suffer moderate to severe levels of pain? What is your role in their pain management? Do you have the knowledge to be effective in your role?

Knowledge of health care staff

A basic knowledge base is needed for all health professionals to embrace their role in pain management. It can be seen by the extent of information provided on pain definitions, pain physiology, pain assessment and pain management techniques that health professionals require ongoing continuing professional development as well as a sound foundation in undergraduate education in pain management. The literature, however, indicates that health care professionals' knowledge and attitudes are not always adequate to ensure a strong evidence-based culture of pain management (Hamilton and Edgar, 1992; Clarke *et al.*, 1996; Furstenberg *et al.*, 1998; McCaffrey, 2002; Jastrzab *et al.*, 2003; Horbury, 2005). Jastrzab *et al.* (2003) found the best knowledge scores for nurses were in the 'nursing assessment and management' section. An interesting small-scale survey (sample 101, response rate 81%) (Coulling, 2005) highlighted misconceptions and inadequate knowledge of both doctors and nurses. What was interesting to note was that nurses were more knowledgeable in assessment (similar to Jastrzab *et al.*, 2003) and analgesic delivery systems and doctors were more knowledgeable in pharmacology. It would seem that traditional roles in pain management continue to exist which may not be supporting holistic patient care.

There is a need to recognise new ways of learning and working to improve delivery of care. It is important to note that education in pain management improves knowledge, as evidenced by both Jones *et al.* (2004) and Innis *et al.* (2004). This is particularly important given the increased blurring of traditional role boundaries when working within a team. One example is the expanded role of nurses in prescribing established in the UK, USA and Australia and just emerging in Ireland.

Service providers, educators and policy makers

It is evident that individual health care professionals and indeed health care teams can improve their knowledge base and subsequently improve practice. However, this enduring health issue requires both service providers and policy makers to take the issue seriously and propose and implement measures to improve the patients' experience. There are staff and resource implications, but economic justification can be found in the reduced morbidity, faster convalescence and improved satisfaction in patients who receive adequate relief from pain.

Organisational managers, to assure that best practice in pain management is being carried out, should support quality assurance procedures. Throughout this book numerous ways of improving pain management have been detailed, from education for staff, information for patients, use of evidence-based guidelines and the utilistation of all the skills available in interprofessional teams.

Education and training of many health care professionals, including nursing staff, may not be providing the skills and competencies necessary to provide a modern health service. Education needs to ensure that enough emphasis on contemporary methods of pain assessment and management is contained within pre- and post-registration curricula.

Pain management cannot be effective if its implementation is based on inadequate knowledge and erroneous beliefs. Brunier (1995) showed that nurses with a university education scored significantly higher in a knowledge and attitude survey. She also demonstrated that nurses who had attended educational sessions on pain management within the preceding year scored significantly higher than those who did not.

Undergraduate courses need to highlight the importance of pain management and treat pain management as a holistic module. Postgraduate courses should include a pain module, while advanced clinical practice courses should include pain management as a specialty. Collaboration between educationalists, organisational managers and clinicians is necessary to ensure such education is available and that practice goals can be met. Such collaboration can and should be supported by health policy at national and international level – see Figure 12.3.

Figure 12.3 Collaboration to support best practice in pain management

Patient benefits

All health systems exist to improve the health of their nation. Each national health system is influenced by international and national governmental policy. Investment in particular areas of health care are often at the cost benefit of other care areas and as such governments must make choices. These choices are ideally based on evidence for best outcomes for patients and the population utilising sophisticated cost-benefit processes. This is complex given the difficulties in health economics and measurement of quality of life. How do you choose one area over the other? The dynamics of this choice are often multifaceted and you should not underestimate the influence of strong evidence based on scientific rigour. You will find these dynamics evident at all levels of society and the health care system. They exist at a macro level (pain as a public health issue), meso level (organisational responses to pain) and a micro level (how the individual experiences his/her pain).

Reflective activity

Consider your role in pain management. Is it as a staff nurse caring for patients, a clinical nurse manager with ward responsibility, or other? Reflect on your level of influence on how a person experiences their pain. Is it at macro, meso or micro level? Are there other ways you could influence pain management?

The evidence in this book and the tools and processes you have encountered to support you to continually update your knowledge on pain should give you a clear and unambiguous 'armoury' for building up a case for resources, time, education etc. to improve pain management for your patients. It is important to manage and treat an individual's pain directly but it is as important to ensure that best practice in pain management is performed in your organisation. You as an individual as well as being part of a team have a responsibility to your patient caseload to utilise your knowledge to improve their experiences. In this way you are an important part of the process for improving the health of the nation. You may ask yourself why should you be proactive? Why is it not someone else's responsibility? As a qualified (or soon to be qualified) professional you have the knowledge and duty to improve your patients' pain experiences in the best way you can.

Put simply, what will an improved pain service mean for the individual patient who could be your mother, your sister, your partner or even you? Figure 12.4 details the benefits to individual patients when pain management is improved.

Figure 12.4 Patient benefits from improving pain management

Key points Top tips

- The economic and social cost of pain is high
- The prevalence of pain is high in both the hospital and the community
- Multidisciplinary pain teams support high-quality responsive patient care
- Clinical audit of pain supports high-quality pain services
- Evidence-based approaches should be used to manage pain
- Further research in pain management is needed

A multidisciplinary appraoch to pain management

Anne has just qualified as a nurse and begins work on a care of the older person unit. She is allocated a senior nurse as a preceptor who will provide her with orientation and be available for any queries she might have. One of the patients allocated to her has limited mobility and an extensive leg ulcer on her right ankle. The leg ulcer is causing quite a bit of pain and Anne thinks that this may be one of the reasons why her mobility is so poor. The lady also has arthritis. The patient is prescribed paracetamol and an NSAID pre re nata (PRN) for pain. Anne speaks to her preceptor who tells her that the lady has had the leg ulcer for six months and is being seen regularly by the tissue viability nurse who has prescribed specific dressings for the ulcer. In fact the tissue viability nurse is now considering referring the patient to a Consultant Physician specialising in vascular problems. Anne is asked to accompany the patient for an afternoon appointment at the pain clinic for review of her arthritis pain. Anne considered that when she qualified she had a lot of new knowledge and was ready to manage a group of patients. It is beginning to dawn on her that many patients have complex histories and require the expertise of a number of professionals to support good-quality care. She also realises that she is going to have to continually develop her own knowledge in order to develop expertise in the area of clinical specialty that she is working. She reviews the hospital in-service propectus.

- How will the patient benefit from the expertise of more than one health professional?
- What areas of continuing professional development should Anne engage in?
- How can Anne's preceptor help her develop the expertise she needs to work in the care of older people?

Rapid recap

1. Why is pain management important?
2. What are the main qualities of a pain service?
3. Name and give examples of three areas of knowledge that nurses should have in order to care for patients in pain.

References

Bardiau F.M., Taviaux N.F., Albert A., Boogaerts J.G. and Stadler M. (2003) An intervention study to enhance postoperative pain management. *Anesthesia and Analgesia*, **96**: 179–185.

Brunier G. (1995) What do nurses know and believe about patients with pain? Results of a hospital survey. *Journal of Pain and Symptom Management*, **10**(6): 436–445.

Carr E.C.J. (2002) Removing barriers to optimize the delivery of pain management. *Journal of Clinical Nursing*, **11**: 703–704.

Clarke E.B., French B., Bilodeau M.L., Capasso V.C., Edwards A. and Empoliti J. (1996) Pain Management Knowledge, Attitudes and Clinical Practice: The Impact of

Nurses' Characteristics and Education. *Journal of Pain and Symptom Management*, **11**(1): 18–31.

Clinical Standards Advisory Group (2000) *Services for Patients with Pain*. Department of Health, London.

Coulling S. (2005) Nurses' and doctors' knowledge of pain after surgery. *Nursing Standard,* **19**(34): 41–49.

Furstenberg C.T., Ahles T.A., Whedon M.B., Pierce K.L., Dolan M., Roberts L. *et al.* (1998) Knowledge and attitudes of health care providers toward cancer pain management. A comparison of physicians, nurses and pharmacists in the state of New Hampshire. *Journal of Pain and Symptom Management*, **15**: 335–349.

Hamilton J. and Edgar L. (1992) A survey examining nurses' knowledge of pain control. *Journal of Pain and Symptom Management*, **7**(1): 18–26.

Holley S., McMillan S.C., Hagan S.J., Palacios P. and Rosenberg D. (2005) Training pain resource nurses: Changes in their knowledge and attitudes. *Oncology Nursing Forum*, **32**(4): 843–8.

Horbury C., Henderson A. and Bromley B. (2005) Influences of patient behaviour on clinical nurses' pain assessment: Implications for continuing education. *The Journal of Continuing Education in Nursing*, **36**(1): 18–24.

Innis J., Bikaunieks N., Petryshen P., Zellermeyer V. and Ciccarelli L. (2004) Patient satisfaction and pain management: an educational approach. *Journal of Nursing Care Quality*, **19**(4): 322–327.

Jastrzab G., Fairbrother G., Kerr S. and McInerney M. (2003) Profiling the 'pain-aware' nurse: acute care nurses' attitudes and knowledge concerning adult pain management. *Australian Journal of Advanced Nursing*, **21**(2): 27–32.

Jones K.R., Fink R., Petter G., Hutt E., Vojir C.P., Scott J. *et al.* (2004) Improving nursing home staff knowledge and attitudes about pain. *The Gerontologist*, **44**(4): 469–478.

McCaffrey M. and Robinson E.S. (2002) Your patient is in pain here's how you respond. *Nursing*, **32**(10): 36–45.

Macintyre P.E. and Ready L.B. (2002) *Acute Pain Management, A Practical Guide*, 2nd edn. Saunders Elsevier Science, Philadelphia.

Mackintosh B.A. and Bowles S. (1997) Evaluation of a nurse-led acute pain service. Can clinical nurse specialists make a difference? *Journal of Advanced Nursing*, **25**: 30–37.

Mac Lellan (2000) An Evaluation of Pain Management Post Surgery. Unpublished PhD thesis. University of Dublin, Trinity College, Ireland.

Mac Lellan (2004) Postoperative pain: strategy for improving patient experiences. *Journal of Advanced Nursing*, **46**(2): 179–185.

Pain Society, The British Chapter of the International Association for the Study of Pain (2002) *Recommendations for Nursing Practice in Pain Management.* The Pain Society, London.

Tasso K. and Behar-Horenstein L.S. (2004) Patients' perceptions of pain management and use of coping strategies. *Hospital Topics*, **82**(4): 10–19.

Appendix: Rapid recap answers

Chapter One

1 What are main dimensions of pain?

1 The main pain dimensions are: sensory, emotional and intensity dimensions.

2 What is the main message for health care professionals coming from the more common definitions of pain?

2 The main message from the more common definitions of pain is that pain is always subjective.

3 Which parts of the nervous system are concerned with the transmission of pain?

3 Both the central and peripheral nervous system are concerned with the transmission of pain.

4 List five desirable characteristics of a pain care service.

4 A pain service should comprise a multidisciplinary team, it should have a policy of continuous quality improvement and education and empowerment of other health care staff, it should provide diagnosis, treatments and interventions as appropriate.

Chapter Two

1 How do you define prevalence of pain?

1 Prevalence of pain is defined as the proportion of individuals in a population who have pain at a specific instant.

2 How would you describe the economic burden of pain?

2 The economic burden of pain can be described in terms of direct costs (such as pain management medications), indirect costs (lost earnings due to a patient being unable to work) and intangible costs (reduced quality of life).

3 What are the sources of research evidence?

3 Sources of research evidence for pain management include evidence-based journals, the Cochrane database and databases such as Medline and CINAHL.

Chapter Three

1 Name two new technologies that have emerged in relation to analgesic administration in the last 30 years.

1 Two new technologies that have emerged in relation to the administration of analgesia in the last 30 years are: patient-controlled analgesia and patient-controlled transdermal analgesia.

2 What are the main elements of a pain service?

2 The main qualities of a pain service include the introduction of policies, protocols and guidelines, education of staff and patients, audit and evaluation, regular pain assessment, introducing and monitoring new analagesic techniques.

3 What new roles in pain management are emerging for nurses?

3 New roles for nurses in pain management include: pain nurse consultants, pain nurse specialists, pain resource nurses.

Chapter Four

1 What are the main ascending pain pathways?

1 The main ascending pain pathways are the neospinothalamic tract (fast ascending fibres) and paleospinothalamic tract (slow ascending fibres).

2 How do the various chemicals at the site of pain work?

2 Chemicals at the site of pain are thought to excite nociceptors and thus increase the pain sensation.

3 What is the Gate Control Theory of Pain?

3 The Gate Control Theory of Pain proposes that a neural mechanism in the dorsal horns of the spinal cord acts like a gate and can increase or decrease the flow of nerve impulses from the peripheral fibres to the central nervous system.

Chapter Five

1 Name three factors which form the context of patients' pain.

1 Three factors that form the context of patients' pain are the meaning of the pain, culture and anxiety.

2 Define what is meant by culture.

2 Culture can be defined as the beliefs, customs and behaviours of a group of individuals due to ethnicity, religion, origin or current residence.

3 Name two organisations who view pain as a public health issue.

3 The World Health Organization and International Association for the Study of Pain are two organisations which view pain as a public health issue.

Chapter Six

1 What qualities should a pain assessment tool have?

1 The qualities that a pain assessment tool should have are: validity, reliability and ease of use.

2 What are the three main pain intensity tools used?

2 The three main pain intensity tools used are the Visual Analogue Scale, Verbal Rating Scale and Numeric Rating Scale.

3 What can a pain audit contribute when deciding which pain assessment tool to use?

3 When deciding which pain assessment tools to use a pain audit will provide patterns and baseline of pain for clinical area.

Chapter Seven

1 Name two side-effects of opioids.

1 Two side-effects of opioids are constipation and respiratory depression.

2 What are the principles of the WHO analgesic ladder?

2 The principles of the WHO analgesic ladder are that if a pain occurs, there should be prompt oral administration of drugs in the following order: non-opioids (aspirin or paracetamol); then, as necessary, mild opioids (codeine); or the strong opioids such as morphine, until the patient is free of pain. To maintain freedom from pain, drugs should be given 'by the clock', that is every 3–6 hours, rather than 'on demand'.

3 What is the medication of choice for chronic neuropathic pain?

3 Anticonvulsants are the class of medication of choice for chronic neuropathic pain.

4 What are the major effects of NSAIDs?

4 NSAIDs have anti-inflammatory, analgesic and antipyretic effects and inhibit thrombocyte aggregation.

5 What are the limiting factors for use of NSAIDs?

5 A major limiting factor is the risk of gastrointestinal toxicity. Additionally there are possible cardiovascular risks and NSAIDs should be used with caution with asthma, impaired renal function and those on anti-platelet regimes including aspirin.

Chapter Eight

1 Outline the main differences between complementary medicine and alternative medicine.

1 Complementary medicine therapies tend to be more widely accepted and are therefore used in an integrated way with conventional medicine. On the other hand, alternative medicine tends to be used as a collective term for treatments that are used in place of conventional medicine.

2 List four underpinning concepts held by CAM practitioners.

2 Illness is a disharmony or deviation from health, affecting the person as part of the larger

environment/universal system; health is a balance of opposing forces, and treatment is about strengthening healing forces. Holistic in nature, strengthening the wellbeing of the whole person. Less high-technology interventions or diagnostics are used. The patient should be active in their cure.

3 What CAM treatments currently have the strongest empirical evidence base for their use in pain care?

3 Acupuncture, some herbal medicines and manual therapies (osteopathy, chiropractic).

4 What is a placebo effect?

4 The placebo effect occurs when a sham treatment without any known biological activity induces biological and/or psychological effects in people, in line with a therapeutic treatment. There are thought to be two main reasons why this can occur either as a conditioned response or due to patient expectations.

5 What would a health care practitioner need to do if they wanted to integrate and deliver complementary therapies within their care setting?

5 If administering complementary or alternative treatments, a health care practitioner must have undertaken an approved course and be deemed competent to offer this treatment. CAM should be discussed with the multidisciplinary team, and the patient must always have granted consent for CAM use. The health care practitioner should ensure that their practice is in line with national guidance from their professional body and their NHS Trusts' local guidance. They may have to have their competency assessed formally using a skills framework.

6 Define a therapeutic practitioner–patient relationship.

6 A therapeutic practitioner–patient relationship requires the practitioner to respect and have genuine interest in the patient and also to show emotional warmth, tolerance and a non-judgemental acceptance. It requires the practitioner to use 'self' in their interactions, whilst maintaining awareness of their limitations and adherence to ethical codes. The patient also plays a part in a therapeutic relationship by trusting and co-operating with the intervention and being motivated to understand the treatment/care.

Chapter Nine

1 Define acute pain.

1 Traditionally, acute pain is described as pain that subsides as healing takes place, i.e. it has a predictable end and it is of brief duration, usually less than six months.

2 Name and give examples of three modalities for reducing acute pain.

2 Table 1 below details three modalities for reducing acute pain.

3 What special considerations are there with the use of morphine in the accident and emergency department?

3 Special considerations with the use of morphine in the emergency department include that it:
a. is usually administered intravenously in the emergency department
b. causes sedation
c. should not be given to shocked patients
d. should not be used in pancreatitis
e. if used in coronary thrombosis it may further slow pulse and lower blood pressure.

4 When assessing chest pain what does the acronym OLDCART stand for?

4 When assessing chest pain the acronym OLDCART stands for:
Onset
Location
Duration
Characteristics
Accompanying symptoms
Radiation
Treatment

Table 1 Modalities for reducing pain

	Examples
Pharmacological preparations	Opioids NSAIDs Paracetamol
Physical modalities	Physiotherapy Manipulation TENs
Complementary modalities	Acupuncture Relaxation

5 What are the main qualities of an acute pain service?

5 Acute pain services consist of a multidisciplinary team including specialist nurses, anaesthetists and pharmacists who support seamless liaison with other health care teams. Their role is to introduce policies, protocols, guidelines and regular pain assessment. It is part of their remit to introduce, monitor and evaluate new pain-relieving techniques as well as ongoing audit of service provided. Additionally, acute pain services provide education of staff and patients.

Chapter Ten

1 Define chronic pain.

1 Chronic pain tends to be considered as pain without apparent biological value that has persisted beyond the normal tissue healing time (usually taken to be three months).

2 What are the main qualities of a chronic-pain service?

2 The main qualities of a chronic-pain service include the introduction of policies, protocols and guidelines, education of staff and patients, audit and evaluation, regular pain assessment, introducing and monitoring new analgesic techniques.

3 Name and give examples of three modalities for reducing chronic pain.

3 Table 2 below details three modalities for reducing chronic pain.

Table 2 Modalities for reducing chronic pain

	Examples
Pharmacological preparations	Opioids NSAIDs Anticonvulsants Tricyclic antidepressants NMDA antagonists Topical preparations
Physical modalities	Ultrasound TENs Diathermy
Comprehensive pain rehabilitation programme	Interdisciplinary pain centres Functional restoration programmes

4 Identify four principles of good practice to guide the prescribing of opioids for chronic non-cancer pain.

4 Four principles of good practice to guide the prescribing of opioids for chronic non-cancer pain include: *evaluation of patient* to include pain history, assessment of impact of pain on the patient, a directed physical examination, a review of previous diagnostic studies, a review of previous interventions, a drug history and an assessment of coexisting conditions. *Treatment plan* which is tailored to both the individual and the presenting problem. Consideration should be given to different treatment modalities. *Periodic review of treatment efficacy* to occur periodically to assess the functional status of the patient, continued analgesia, opioid side-effects, quality of life and indication of medication misuse. Periodic re-examination is warranted to assess the nature of the pain complaint and to ensure that opioid therapy is still indicated. *Documentation* to support the evaluation, the reason for opioid prescribing, the overall pain management treatment plan, any consultation received and periodic review of the status of the patient.

5 How strong is the evidence for the use of TENS with low back pain?

5 In relation to low back pain there is limited and inconsistent evidence to support the use of TENS as an isolated intervention in the management of chronic lower back pain.

Chapter Eleven

1 Outline the philosophy of a palliative care approach.

1 A palliative care approach focuses on an individualised approach to treating the person, not the disease. It includes care of the family as well as the person. It aims to improve the individual's quality of life by preventing and relieving suffering by means of early identification and impeccable assessment and treatment of pain and other symptoms associated with life-threatening illness. In addition, palliative care offers the following:

It provides relief from pain and other distressing symptoms.

It affirms life and regards dying as a normal process.

It intends neither to hasten nor postpone death.

It integrates the psychological, social and spiritual aspects of patient care.

It offers a support system to help patients live as actively as possible until death.

It offers a support system to help the family cope during the patient's illness and in their own bereavement.

It uses a team approach to address the needs of patients and their families, including bereavement counselling, if indicated.

It will enhance quality of life, and may also positively influence the course of illness.

It is applicable early in the course of illness, in conjunction with other therapies that are intended to prolong life, such as chemotherapy or radiation therapy, and includes those investigations needed to better understand and manage distressing clinical complications (WHO, 2002).

2 What is the purpose of pain assessment within a palliative care context?

2 The purpose of assessment within a palliative care context is to identify *individual* patients' needs or problems in order to set appropriate, achievable and effective goals.

Utilising the most appropriate diagnostic and therapeutic considerations, pain assessment will contribute towards defining the cause and directing treatment.

3 Following the analgesic ladder guidelines, how should analgesics be administered?

3 Following the analgesic ladder guidelines, the WHO (1996) advocates that analgesics should be administered in the following way:

'By the ladder' (in standard doses at regular intervals in stepwise fashion.) If a pain occurs, there should be a prompt oral administration of drugs in the following order: non-opioids (aspirin or paracetamol); then, as necessary, mild opioids (codeine); or the strong opioids such as morphine, until the patient is free of pain.

'By the mouth' (the oral route is the preferred route for administration when possible) 'By the clock' (to maintain freedom from pain, drugs should be given regularly, rather than waiting for the individual to request pain relief).

'For the individual' (the analgesic ladder is a framework of principles rather than a rigid protocol).

'Attention to detail' (attention to each individual's unique experience of pain is required when directing treatment).

Chapter Twelve

1 Why is pain management important?

1 Pain management is important because of the adverse effects and social and economic burden of pain.

2 What are the main qualities of a pain service?

2 The main qualities of a pain service include the introduction of policies, protocols and guidelines, education of staff and patients, audit and evaluation, regular pain assessment, introducing and monitoring new analgesic techniques.

3 Name and give examples of three areas of knowledge that nurses should have in order to care for patients in pain.

3 Table 3 details three areas of knowledge that nurses should have in order to care for patients in pain.

Table 3 Areas of knowledge

Role facet	Knowledge
Assessment of pain – Verbal assessment – Non-verbal assessment	Pain assessment tools Pain physiology Pain theories
Administration and monitoring of effect of analgesia – Opioids – NSAIDs – Anti-convulsants – Paracetamol – Other	Analgesic medications (action, time of effect, side-effects)
Evidence-based practice	Clinical audit Sources of evidence

Index

Page reference in italics indicate figures or tables